LONDON MATHEMATICAL SOCIETY LECT

Managing Editor: Professor J.W.S. Cassels, Department o
University of Cambridge, 16 Mill Lane, Cambridge CB2 1

The books in the series listed below are available from boo ... difficulty, from Cambridge University Press.

4 Algebraic topology, J.F. ADAMS
5 Commutative algebra, J.T. KNIGHT
16 Topics in finite groups, T.M. GAGEN
17 Differential germs and catastrophes, Th. BROCKER & L. LANDER
18 A geometric approach to homology theory, S. BUONCRISTIANO, C.P. ROURKE & B.J. SANDERSON
20 Sheaf theory, B.R. TENNISON
21 Automatic continuity of linear operators, A.M. SINCLAIR
23 Parallelisms of complete designs, P.J. CAMERON
24 The topology of Stiefel manifolds, I.M. JAMES
25 Lie groups and compact groups, J.F. PRICE
27 Skew field constructions, P.M. COHN
29 Pontryagin duality and the structure of LCA groups, S.A. MORRIS
30 Interaction models, N.L. BIGGS
31 Continuous crossed products and type III von Neumann algebras, A. VAN DAELE
34 Representation theory of Lie groups, M.F. ATIYAH *et al*
36 Homological group theory, C.T.C. WALL (ed)
38 Surveys in combinatorics, B. BOLLOBAS (ed)
39 Affine sets and affine groups, D.G. NORTHCOTT
40 Introduction to $H_p$ spaces, P.J. KOOSIS
41 Theory and applications of Hopf bifurcation, B.D. HASSARD, N.D. KAZARINOFF & Y-H. WAN
42 Topics in the theory of group presentations, D.L. JOHNSON
43 Graphs, codes and designs, P.J. CAMERON & J.H. VAN LINT
44 Z/2-homotopy theory, M.C. CRABB
45 Recursion theory: its generalisations and applications, F.R. DRAKE & S.S. WAINER (eds)
46 p-adic analysis: a short course on recent work, N. KOBLITZ
48 Low-dimensional topology, R. BROWN & T.L. THICKSTUN (eds)
49 Finite geometries and designs, P. CAMERON, J.W.P. HIRSCHFELD & D.R. HUGHES (eds)
50 Commutator calculus and groups of homotopy classes, H.J. BAUES
51 Synthetic differential geometry, A. KOCK
52 Combinatorics, H.N.V. TEMPERLEY (ed)
54 Markov processes and related problems of analysis, E.B. DYNKIN
55 Ordered permutation groups, A.M.W. GLASS
56 Journees arithmetiques, J.V. ARMITAGE (ed)
57 Techniques of geometric topology, R.A. FENN
58 Singularities of smooth functions and maps, J.A. MARTINET
59 Applicable differential geometry, M. CRAMPIN & F.A.E. PIRANI
60 Integrable systems, S.P. NOVIKOV *et al*
61 The core model, A. DODD
62 Economics for mathematicians, J.W.S. CASSELS
63 Continuous semigroups in Banach algebras, A.M. SINCLAIR
64 Basic concepts of enriched category theory, G.M. KELLY
65 Several complex variables and complex manifolds I, M.J. FIELD
66 Several complex variables and complex manifolds II, M.J. FIELD
67 Classification problems in ergodic theory, W. PARRY & S. TUNCEL
68 Complex algebraic surfaces, A. BEAUVILLE
69 Representation theory, I.M. GELFAND *et al*
70 Stochastic differential equations on manifolds, K.D. ELWORTHY
71 Groups - St Andrews 1981, C.M. CAMPBELL & E.F. ROBERTSON (eds)
72 Commutative algebra: Durham 1981, R.Y. SHARP (ed)

73 Riemann surfaces: a view towards several complex variables, A.T. HUCKLEBERRY
74 Symmetric designs: an algebraic approach, E.S. LANDER
75 New geometric splittings of classical knots, L. SIEBENMANN & F. BONAHON
76 Spectral theory of linear differential operators and comparison algebras, H.O. CORDES
77 Isolated singular points on complete intersections, E.J.N. LOOIJENGA
78 A primer on Riemann surfaces, A.F. BEARDON
79 Probability, statistics and analysis, J.F.C. KINGMAN & G.E.H. REUTER (eds)
80 Introduction to the representation theory of compact and locally compact groups, A. ROBERT
81 Skew fields, P.K. DRAXL
82 Surveys in combinatorics, E.K. LLOYD (ed)
83 Homogeneous structures on Riemannian manifolds, F. TRICERRI & L. VANHECKE
84 Finite group algebras and their modules, P. LANDROCK
85 Solitons, P.G. DRAZIN
86 Topological topics, I.M. JAMES (ed)
87 Surveys in set theory, A.R.D. MATHIAS (ed)
88 FPF ring theory, C. FAITH & S. PAGE
89 An F-space sampler, N.J. KALTON, N.T. PECK & J.W. ROBERTS
90 Polytopes and symmetry, S.A. ROBERTSON
91 Classgroups of group rings, M.J. TAYLOR
92 Representation of rings over skew fields, A.H. SCHOFIELD
93 Aspects of topology, I.M. JAMES & E.H. KRONHEIMER (eds)
94 Representations of general linear groups, G.D. JAMES
95 Low-dimensional topology 1982, R.A. FENN (ed)
96 Diophantine equations over function fields, R.C. MASON
97 Varieties of constructive mathematics, D.S. BRIDGES & F. RICHMAN
98 Localization in Noetherian rings, A.V. JATEGAONKAR
99 Methods of differential geometry in algebraic topology, M. KAROUBI & C. LERUSTE
100 Stopping time techniques for analysts and probabilists, L. EGGHE
101 Groups and geometry, ROGER C. LYNDON
102 Topology of the automorphism group of a free group, S.M. GERSTEN
103 Surveys in combinatorics 1985, I. ANDERSON (ed)
104 Elliptic structures on 3-manifolds, C.B. THOMAS
105 A local spectral theory for closed operators, I. ERDELYI & WANG SHENGWANG
106 Syzygies, E.G. EVANS & P. GRIFFITH
107 Compactification of Siegel moduli schemes, C-L. CHAI
108 Some topics in graph theory, H.P. YAP
109 Diophantine Analysis, J. LOXTON & A. VAN DER POORTEN (eds)
110 An introduction to surreal numbers, H. GONSHOR
111 Analytical and geometric aspects of hyperbolic space, D.B.A.EPSTEIN (ed)
112 Low-dimensional topology and Kleinian groups, D.B.A. EPSTEIN (ed)
113 Lectures on the asymptotic theory of ideals, D. REES
114 Lectures on Bochner-Riesz means, K.M. DAVIS & Y-C. CHANG
115 An introduction to independence for analysts, H.G. DALES & W.H. WOODIN
116 Representations of algebras, P.J. WEBB (ed)
117 Homotopy theory, E. REES & J.D.S. JONES (eds)
118 Skew linear groups, M. SHIRVANI & B. WEHRFRITZ
119 Triangulated categories in the representation theory of finite-dimensional algebras, D. HAPPEL
120 Lectures on Fermat varieties, T. SHIODA
121 Proceedings of *Groups - St Andrews 1985*, E. ROBERTSON & C. CAMPBELL (eds)
122 Non-classical continuum mechanics, R.J. KNOPS & A.A. LACEY (eds)
123 Surveys in combinatorics 1987, C. WHITEHEAD (ed)
124 Lie groupoids and Lie algebroids in differential geometry, K. MACKENZIE
125 Commutator theory for congruence modular varieties, R. FREESE & R. MCKENZIE
126 Van der Corput's method for exponential sums, S.W. GRAHAM & G. KOLESNIK
127 New directions in dynamical systems, T.J. BEDFORD & J.W. SWIFT (eds)
128 Descriptive set theory and the structure of sets of uniqueness, A.S. KECHRIS & A. LOUVEAU

London Mathematical Society Lecture Note Series. 128

# Descriptive Set Theory and the Structure of Sets of Uniqueness

Alexander S. Kechris
*Department of Mathematics, California Institute of Technology*

and

Alain Louveau
*Centre National de la Recherche Scientifique,*
*Université Pierre et Marie Curie, Paris*

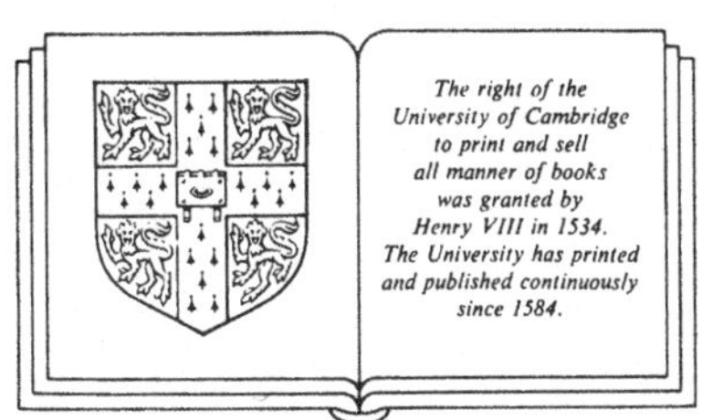


CAMBRIDGE UNIVERSITY PRESS
Cambridge
New York New Rochelle Melbourne Sydney

CAMBRIDGE UNIVERSITY PRESS
Cambridge, New York, Melbourne, Madrid, Cape Town, Singapore, São Paulo

Cambridge University Press
The Edinburgh Building, Cambridge CB2 2RU, UK

Published in the United States of America by Cambridge University Press, New York

www.cambridge.org
Information on this title: www.cambridge.org/9780521358118



First published 1987
Reprinted 1989

*A catalogue record for this publication is available from the British Library*

ISBN-13 978-0-521-35811-8 paperback
ISBN-10 0-521-35811-6 paperback

Transferred to digital printing 2005

To Olympia and Lise

# Preface

This book grew out of a set of notes prepared during the course of a joint Caltech-UCLA Seminar in Descriptive Set Theory and Harmonic Analysis, organized by the authors during the academic year 1985–86. We appreciate very much the help as well as the patience of the participants in this seminar.

We are grateful to G. Debs, R. Dougherty, S. Jackson, R. Kaufman, R. Lyons, and J. Saint Raymond for many valuable comments and suggestions. The first author is indebted to S. Pichorides for introducing him to the subject of uniqueness for trigonometric series. We would like also to thank N. O'Connor for her efficiency, care and patience in typing the manuscript.

The work of A. S. Kechris has been partially supported by NSF Grant DMS84-16349. A. Louveau has been supported by CNRS, France and by UCLA during his visit in the academic year 1985–86. He takes this opportunity to thank the Mathematics Department for its hospitality.

Alexander S. Kechris — Pasadena

Alain Louveau — Paris

June 1987

# Contents

# Introduction

This book is devoted to the study of sets of uniqueness for trigonometric series and a few related concepts from classical harmonic analysis. However, as the title indicates, our approach to the subject is rather unusual, as it is chiefly guided by our main goal, which is to bring forward the relationship between this subject and descriptive set theory. And it is this use of descriptive set theory, both as a tool and as a source of inspiration, that gives this book its specific flavor.

These connections have been discovered rather recently, during the past 3–4 years, and this monograph is primarily an exposition of these new results. On the other hand, our aim of a reasonably self-contained presentation gave us the opportunity to cover many topics of the classical as well as the modern theory of sets of uniqueness. Needless to say we did not try to be exhaustive, and we have left out many important results that were not central to our goal.

******

i). The concept of a set of uniqueness arose from the <u>problem of uniqueness</u>, originating in the work of Riemann and Heine in the mid-19th century: If a function f admits a trigonometric expansion $f(x) = \sum_n c_n e^{inx}$, is this expansion unique? An affirmative solution to this problem was obtained by Cantor in 1870. He then went on to show essentially that uniqueness follows even if one relaxes the hypothesis by allowing a countable closed set of exceptional points x. (Actually Cantor formally established and published only the special case of countable closed sets of finite Cantor-Bendixson rank). As is well known, it is through this work on trigonometric series that Cantor was led to the creation of set theory (see for example Dauben [1]).

As the convergence of a trigonometric series at some point x depends only on x (mod $2\pi$), let us work in the unit circle $\mathbb{T} = \mathbb{R}/2\pi\mathbb{Z}$. A set $P \subseteq \mathbb{T}$ is a set of uniqueness if every trigonometric series which converges to 0 outside P is identically 0. Otherwise P is a set of multiplicity. We will denote by $\mathcal{U}$ and $\mathcal{M}$ the corresponding classes of sets.

Countable closed sets are in $\mathcal{U}$, by Cantor's theorem. This was later extended by W. H. Young (1909) to arbitrary countable sets, and by N. K. Bary who showed in 1923 that the union of countably many closed sets of uniqueness is also a set of uniqueness, one of the most important structural results about $\mathcal{U}$.

Clearly subsets of sets of uniqueness are also of uniqueness, and it is easy to prove that $\mathcal{U}$-sets are small: measurable $\mathcal{U}$-sets are of measure 0. However, and this came as a big surprise at the time, there are also closed $\mathcal{M}$-sets of measure 0, as proved by D. E. Menshov in 1916. (So a trigonometric series can converge to 0 a.e. without being identically 0). There are also perfect sets in $\mathcal{U}$, as proved independently by A. Rajchman and Bary in 1921–23. Rajchman's results imply for example that the Cantor (1/3 —) set is in $\mathcal{U}$.

The closed $\mathcal{M}$-set E of measure 0 in Menshov's construction is of a special type: There exists a probability measure $\mu$ concentrated on E such that its Fourier(-Stieltjes) coefficients $\hat{\mu}(n) = \int e^{-inx} d\mu(x)$ converge to 0 as $|n| \to \infty$, so that

E is in $\mathcal{M}$ with witness the series $\Sigma\hat{\mu}(n)e^{inx}$. Sets which carry such measures are called sets of <u>strict multiplicity</u> ($\mathcal{M}_0$-sets) and those that don't are called sets of <u>extended uniqueness</u> ($\mathcal{U}_0$-sets). (Actually the definition of $\mathcal{U}_0$, $\mathcal{M}_0$ that we adopt officially in Chapter II is slightly different, but coincides with the above for universally measurable sets, like Borel, analytic, etc., which include all the sets we are interested in anyway).

So we have the inclusions (for measurable sets)

$$\text{countable} \subsetneqq \mathcal{U} \subseteq \mathcal{U}_0 \subsetneqq \text{measure } 0.$$

These facts are studied in Chapter I. Contrary to prevailing expectations again, it turned out that the inclusion $\mathcal{U} \subseteq \mathcal{U}_0$ is also strict, even for closed sets, as proved by I. I. Piatetski-Shapiro in 1954 (we will give a proof of this in Chapter VII).

By the 1920's, the results obtained so far indicated that the concept of a set of uniqueness was hard to delineate, and led to some basic open questions, discussed for example in Bary's memoir [1]. These problems will play an important role in this book.

A) <u>The Union Problem</u>. By Bary's theorem, the union of countably many $F_\sigma$ sets in $\mathcal{U}$ is in $\mathcal{U}$. What about more general Borel sets?

B) <u>The Interior Problem</u>. It is easy to check that a set of multiplicity always contains a $G_{\delta\sigma}$ set of multiplicity. Does it always contain a closed one?

C) The Category Problem. Every measurable $\mathcal{U}$-set is of measure 0. Is the same true for category, i.e. is every $\mathcal{U}$-set with the Baire property of the first category?

D) The Characterization Problem. "Trouver la condition nécessaire et suffisante pour qu'un ensemble parfait donné soit un ensemble (U)" (Bary [1]). This is unfortunately a vague problem. Somehow the intended meaning seems to have been that of asking for geometric, analytic or (as we will see shortly) number theoretic structural properties of a perfect set E, expressed explicitly in terms of some standard specification of E, like e.g. its contiguous intervals, that will determine whether it is a $\mathcal{U}$-set or a $\mathcal{M}$-set.

Since the twenties, there has been very little progress on problems A and B, with the exception of an extension of Bary's theorem due to Kholshchevnikova in 1981, presented in Chapter I, and a result due independently to Kechris-Louveau and Debs-Saint Raymond, presented in Chapter VII, which links the two problems. (See also Section VIII.4 for some interesting connections with problems of so-called synthesis). The Union Problem is open even for the case of two $G_\delta$ sets, and the Interior Problem is again open even for $G_\delta$ sets.

As we shall see soon, problem C has been solved positively by Debs and Saint Raymond in 1986, using the kind of descriptive set theoretic methods this book is about.

Our own initial interest in the theory of sets of uniqueness came through problem D, i.e. the Characterization Problem, which has the kind of heuristic and metamathematical flavor logicians like most. Hence this problem will be in various guises one of the central themes of this book.

ii). We will concentrate in the sequel on the structure of closed sets of uniqueness, where most of the theory lies. However we will see important consequences for the structure of sets in more general classes, like Borel, analytic, etc. The modern theory of closed sets of uniqueness is based on a reformulation of this notion in terms of concepts of functional analysis, which goes back to work of Piatetski-Shapiro in the mid-1950's.

Let us denote by A the Banach algebra of continuous functions on **T** with absolutely convergent Fourier Series, and norm $\|f\|_A = \sum_n |\hat{f}(n)|$. Thus as a Banach space, A is the same as $\ell^1 \equiv \ell^1(\mathbf{Z})$. For E a closed subset of **T**, let J(E) be the ideal of those functions in A which vanish in an open nbhd of E. Piatetski-Shapiro proved in 1954 that for closed E

$E \in \mathcal{U} \Leftrightarrow$ J(E) is dense in A, for the weak*-topology (coming from the duality $A \cong \ell^1 = c_0^*$, where $c_0 \equiv c_0(\mathbf{Z})$ is the Banach space of sequences converging to zero at infinity).

From the point of view of the Characterization Problem, this result is not of the expected kind. But its fundamental importance is to bring in the tools from functional analysis, especially Banach space theory.

In particular, it leads to a break down of the class U of <u>closed</u> sets of uniqueness into a transfinite hierarchy of at most $\omega_1$ (= the first uncountable ordinal) levels. This is done by associating to each $E \in U$ a countable ordinal, its Piatetski-Shapiro rank $[E]_{PS}$, which measures the number of steps it takes to get the space A, starting from J(E) and closing at each step under weak*-limits of sequences. (Since A with the weak*-topology is not metrizable, one cannot obtain the weak*-closure of J(E) by taking weak*-limits of sequences from J(E). One can do however the next best thing, namely iterate this process into the transfinite, following ideas of Banach). Indeed, by a result of O. C. McGehee in 1968, $[E]_{PS}$ is unbounded in $\omega_1$, so there are exactly $\omega_1$ levels.

The simplest sets in this hierarchy are those of rank 1, for which a sequence of functions in J(E) weak*-converges to the function 1. For example all countable closed sets have rank 1, and so does the Cantor set. In fact most explicit examples of closed $\mathcal{U}$-sets have rank 1.

The rank $[E]_{PS}$ was originally used by Piatetski-Shapiro to prove by transfinite induction a decomposition theorem of the form:

If E is a closed $\mathcal{U}$-set, E can be written as $E = \bigcup_n E_n$, where each $E_n$ is a closed $\mathcal{U}$-set of a particular type (that we do not spell out in detail here—we discuss this result in Chapter VI).

This kind of result seems to us clearly relevant to the Characterization Problem, especially if the sets in the decomposition do admit a simpler description. For example, one can ask whether they could be sets of uniqueness of rank 1. This question and other similar ones will be discussed later in this introduction.

A major advance in the Characterization Problem was achieved also in the 1950's. Following up on earlier work of Bary, Salem, and Piatetski-Shapiro, R. Salem and A. Zygmund characterized in 1955 when a perfect symmetric set of constant ratio of dissection is a $\mathcal{U}$-or a $\mathcal{M}$-set. We denote by $E_\xi$ the symmetric perfect set of ratio of dissection $\xi$, where $0 < \xi < \frac{1}{2}$. This is constructed (on the interval $[0, 2\pi]$) like the Cantor set, except that at each dissection stage the closed intervals that are kept have ratio $\xi$ instead of $\frac{1}{3}$. Salem and Zygmund proved that $E_\xi$ is a $\mathcal{U}$-set iff $\theta = \frac{1}{\xi}$ is a Pisot number (i.e. an algebraic integer $> 1$ all of whose conjugates have absolute value $< 1$). We discuss this result in Chapter III. So it appeared that number theoretic structural properties of closed sets ought to be coming into the picture for a solution of the Characterization Problem.

Except for certain interesting generalizations of the Salem-Zygmund Theorem, not much progress has been achieved since then towards a positive solution of the Characterization Problem. In particular, the case of symmetric perfect sets with varying ratios of dissection $E_{\xi_1,\xi_2,\ldots}$ is still open.

In 1971–73 important work of Körner and Kaufman amplified the result of Piatetski-Shapiro on the difference between the concepts of uniqueness and extended uniqueness. First Körner proved the existence of closed sets of multiplicity which are very particular types of extended uniqueness sets (called Helson sets), and Kaufman simplified and strengthened Körner's work to show that such sets exist within any closed set of multiplicity.

iii). We come now to the recent results relating descriptive set theory and the theory of sets of uniqueness. Descriptive set theory is the study of definable sets in Polish (i.e. complete metrizable, separable) spaces. Sets in such spaces are classified in hierarchies according to the complexity of their definitions and the structure theory of the sets in each level of these hierarchies is systematically developed. The simplest sets in this classification are the Borel sets with the corresponding Borel hierarchy within them. Beyond that by alternating the operations of taking continuous images and complementations we have the much more complicated projective sets and the corresponding projective hierarchy of Lusin. The first level of this hierarchy consists of the two classes of analytic or $\underset{\sim}{\Sigma}_1^1$ sets, i.e. the continuous images of Borel sets, and coanalytic or $\underset{\sim}{\Pi}_1^1$ sets, i.e. the complements of $\underset{\sim}{\Sigma}_1^1$ sets. We will be primarily concerned about Borel, $\underset{\sim}{\Sigma}_1^1$ and $\underset{\sim}{\Pi}_1^1$ sets in this book.

By classifying exactly a given set within a level of these hierarchies one obtains twofold information: By being one of the sets in a particular level our set shares many important properties revealed by the structure theory of this level. On the other hand by not being in any lower level, potential definitions or "characterizations of membership in this set", which are of a "too simple" kind are ruled out. There are several examples of such classifications. We mention an old one due to Mazurkiewicz: the set of differentiable functions is $\underset{\sim}{\Pi}_1^1$ but not Borel, in the space C([0, 1]) of continuous functions on [0, 1].

We can apply these ideas in the context of closed sets of uniqueness: The space K(T) of closed subsets of $\mathbf{T}$ has a natural Polish (in fact compact, metrizable) topology, and it is not hard to compute that the class of closed sets of uniqueness, $U = \mathcal{U} \cap K(\mathbf{T})$ is a $\underset{\sim}{\Pi}_1^1$ subset of K(T). In 1982, motivated by a discussion with S. Pichorides on the Characterization Problem, A. S. Kechris proposed to find the exact complexity of U, and conjectured that U is not Borel (or equivalently by a theorem of Suslin is not $\underset{\sim}{\Sigma}_1^1$). This conjecture was proved by R. M. Solovay at the end of 1983, both for U and for $U_0 = \mathcal{U}_0 \cap K(E)$, the class of closed sets of extended uniqueness, using among other things one of the standard extensions of the Salem-Zygmund Theorem. Independently of all this, R. Kaufman who was studying the problem of representing bounded $\underset{\sim}{\Sigma}_1^1$ subsets of $\mathbb{C}$ as point spectra of operators on Banach spaces, arrived at the same result in early 1984. Kaufman's methods have been very useful in subsequent work.

The Solovay-Kaufman Theorem, which we present in Chapter IV, rules out the possibility of any characterization of when an arbitrary closed (or perfect) set is in $\mathcal{U}$ or in $\mathcal{M}$, in terms of geometric, analytic or number theoretic properties of the closed set, which are of "explicit enough" form so that they can be expressed by countable operations given any reasonable specification of the closed set, like e.g. in terms of its contiguous intervals. This is because any such characterization would yield a Borel definition of U. We note that all positive results obtained so far in special cases, like the Salem-Zygmund Theorem and its extensions, are of this type. Of course nothing can rule out less explicit characterizations or the possibility of establishing non-trivial reformulations of the concept of uniqueness in terms of other notions.

We turn now to the other aspect of classification as we discussed it earlier. The $\underset{\sim}{\Sigma}_1^1$ and $\underset{\sim}{\Pi}_1^1$ sets have been extensively studied since their introduction by Suslin and Lusin in the early 1920's, and results from this theory can be applied to U and $U_0$.

One of the main properties of $\underset{\sim}{\Pi}_1^1$ sets P, isolated in its current form by Y. N. Moschovakis, is that they carry a $\underset{\sim}{\Pi}_1^1$-rank, i.e. a function $\varphi : P \to \omega_1$ assigning to each element x of P a countable ordinal $\varphi(x)$, with the crucial definability property that the initial segments $\{x \in P : \varphi(x) \leq \alpha\} = P_\alpha$, $\alpha < \omega_1$, are Borel "uniformly on the ordinal $\alpha$". Such $\underset{\sim}{\Pi}_1^1$-ranks allow one to study membership in P, prove results by transfinite induction, or use more subtle definability arguments (boundedness arguments, overspill arguments, etc., see Chapter IV) to get existence results—e.g. in our case to prove the existence of an $M_0$-set of measure 0 (Menshov's theorem) by only constructing U-sets. (We denote here by M the class $\mathcal{M} \cap K(\mathbb{T})$ and similarly for $M_0$).

In the specific context of U, one already has a "natural" rank, the Piatetski-Shapiro rank $[E]_{PS}$, and Kechris conjectured that this rank should have the right definability properties to be a $\underset{\sim}{\Pi}_1^1$-rank on U. (This would also give a different proof, using McGehee's result that the rank is unbounded, that U cannot be Borel, by a basic fact about $\underset{\sim}{\Pi}_1^1$-ranks called the boundedness theorem). This conjecture was again proved by Solovay in early 1984.

The sets U and $U_0$ are not only $\underset{\sim}{\Pi}_1^1$ subsets of $K(\mathbb{T})$, they are also $\sigma$-ideals, i.e. closed under subsets and countable unions (which are closed). In 1984, motivated by the results of Solovay and Kaufman, A. S. Kechris, A. Louveau, and W. H. Woodin started a systematic descriptive set theoretic study of $\underset{\sim}{\Pi}_1^1$ $\sigma$-ideals of closed sets in compact, metrizable spaces. In general, such $\sigma$-ideals occur very often in analysis as notions of "thin" or "exceptional" sets. Some examples are the nowhere dense sets, the measure 0 sets, the countable sets and of course U and $U_0$. As an example of the kind of results obtained in the Kechris, Louveau and Woodin paper, we mention the Dichotomy Theorem, which shows that the structural property of being a $\sigma$-ideal imposes severe definability constraints: A $\underset{\sim}{\Pi}_1^1$ or $\underset{\sim}{\Sigma}_1^1$ $\sigma$-ideal of closed sets in a compact, metrizable space is either very complicated, i.e. $\underset{\sim}{\Pi}_1^1$ but not Borel, or else very simple, i.e. $G_\delta$.

One of the main themes in this joint paper was the following "weak characterization" problem: Let I be a $\sigma$-ideal of closed sets. A basis B for I is a subfamily $B \subseteq I$ closed under subsets, and such that every element of I is a countable union of sets in B—i.e. it is a subfamily for which a decomposition result à la Piatetski-Shapiro holds. Even if the $\sigma$-ideal I is itself complicated ($\underset{\sim}{\Pi}_1^1$ but not Borel) it could be that it admits a simply describable (at worst Borel) basis. This led to the following weak form of the Characterization Problem:

Do U and $U_0$ admit Borel bases?

By subsequent work of Kechris and Louveau, the Piatetski-Shapiro basis for U is not Borel. Of course a particularly natural potential candidate for a basis for U would be the family of U-sets of rank 1.

In their paper, aiming at a negative answer to the above question, Kechris, Louveau and Woodin gave a criterion on a $\sigma$-ideal I of closed sets which implies that it has no Borel basis. It is based on a method of constructing perfect sets, as unions of $G_\delta$ and $F_\sigma$ sets, quite different from the classical Cantor-type construction. This criterion had several applications, but it was not clear at the time how it could be applied directly to U, $U_0$.

During the academic year 1985–86, Kechris and Louveau ran a joint Caltech-UCLA seminar in Descriptive Set Theory and Harmonic Analysis, an outgrowth of whose notes is this book. First, they studied the Piatetski-Shapiro rank on U, and established its equivalence with two other natural ranks, one corresponding to the predual $c_0$ of $A(\cong \ell^1)$ and the other to the dual $\ell^\infty$ of A. This analysis is discussed, together with the Piatetski-Shapiro rank itself, in Chapter V. One of these reformulations of the rank on U was then used to obtain a natural rank on $U_0$, denoted by $[E]_0$, which was shown to be a $\underset{\sim}{\Pi}^1_1$-rank as well. The sets in $U_0$ of rank 1 form a natural class, previously considered in a different context by R. Lyons. It then turned out surprisingly (to the authors at least) that this fairly canonical class is a Borel basis for $U_0$—so that the "weak characterization" problem is solved positively for $U_0$.

This result had some immediate consequences, e.g. that the class of perfect symmetric sets $E_{\xi_1,\xi_2,\ldots}$ in $U_0$ is Borel ($G_{\delta\sigma}$)—suggesting the possibility of some explicit characterization for them. And in a totally different direction it could be applied, combined with the Borel basis criterion mentioned earlier, to give a new proof and extension of results of O. S. Ivashev-Musatov and Kaufman on the relation between $U_0$ and Hausdorff measures.

But the most striking applications of these ideas and methods were to come soon after, in March and April 1986, in work of G. Debs and J. Saint Raymond. Using a variant of the Kechris-Louveau-Woodin Borel basis criterion (which is indeed a special case of it), they showed that under certain assumptions on a $\sigma$-ideal of closed sets I, which were known for the $\sigma$-ideal $U_0$ from work of Kaufman, the existence of a Borel basis for I implies that $G_\delta$ (in fact $\underset{\sim}{\Sigma}^1_1$) sets all of whose closed subsets are in I, can actually be covered by countably many sets in I. Using the Kechris-Louveau basis theorem for $U_0$, they concluded therefore that a $G_\delta$ (in fact $\underset{\sim}{\Sigma}^1_1$) set in $\mathcal{U}_0$ can be covered by countably many (closed) sets in $U_0$, from which a positive solution to the Category Problem follows immediately. This had also consequences in seemingly unrelated problems. It provided for example a new proof of a recent result of R. Lyons, who resolved an old conjecture of J.-P. Kahane and R. Salem, by showing that the set of non-normal in base 2 numbers is in $\mathcal{M}_0$. These results on $U_0$ and $\mathcal{U}_0$ are discussed in Chapter VIII. In this Chapter we also present another, very elementary and direct, approach (due to Kechris and Louveau) to the covering theorem for $\underset{\sim}{\Sigma}^1_1$ $\mathcal{U}_0$-sets, which leads to very simple proofs of this result as well as its consequences, including the solution to the Category Problem and the above mentioned results of Ivashev-Musatov, Kaufman and Lyons.

The technical assumptions in the Debs-Saint Raymond Theorem were also known for U, by independent work of Kaufman, Kechris-Louveau and Debs-Saint Raymond. Finally, using the Körner Theorem on the existence of Helson sets of multiplicity, Debs and Saint Raymond constructed a $G_\delta$ set all of whose closed subsets are in U, but which cannot be covered by countably many (closed) U-sets. It follows that U itself <u>does</u> <u>not</u> admit a Borel basis, i.e. it is not "weakly characterizable".

In particular the U-sets of rank 1 are not a basis for U. Since every "known" until now U-set was indeed a countable union of such sets, one has many new examples of closed sets of uniqueness. The above results of Kechris-Louveau for $U_0$ and Debs-Saint Raymond for U indicate a strong structural difference between the $\sigma$-ideals U and $U_0$. In some sense $U_0$ is well-behaved, but U is not.

The Körner-Kaufman construction of Helson sets of multiplicity and the Debs-Saint Raymond non-basis result are given in Chapter VII.

iv). In Chapter IX, we study a topic related to $U_0$ sets, the so-called Rajchman measures. A (positive) Rajchman measure is a positive measure on $\mathbb{T}$ whose Fourier coefficients tend to 0 at $\infty$, so that the Borel $\mathcal{U}_0$-sets are those Borel sets which are null for all (positive) Rajchman measures. The question is whether conversely the class of Borel $\mathcal{U}_0$-sets characterizes the Rajchman measures, in the sense that $\mu$ is Rajchman iff every Borel $\mathcal{U}_0$-set is null for $\mu$. This was shown to be true in 1983 by R. Lyons. During the course of the Caltech-UCLA seminar, Louveau realized that Lyons' result follows from a theorem of G. Mokobodzki in measure theory, which (among other things) describes completely the $\Sigma_1^1$ sets of measures which can be characterized in the above sense by their common null sets. We present Mokobodzki's result in Chapter IX, together with Lyons original proof, which gives more information about Rajchman measures, by establishing that an even smaller subclass of $\mathcal{U}_0$, the so-called Weyl sets, suffices to characterize the Rajchman measures.

The last Chapter X is more tentative. We study there two notions closely related to the theory of sets of uniqueness, those of synthesis and resolution. A closed set $E \subseteq \mathbb{T}$ is a set of synthesis if every function in A vanishing on E can be approximated, in the norm of A, by functions vanishing in a nbhd of E. A set of resolution is a closed set all of whose closed subsets are of synthesis. It follows from P. Malliavin's work (1962) on the existence of sets which are not of synthesis, that sets of resolution are uniqueness sets. Definability questions related to the classes RE and S of sets of resolution and synthesis are discussed in Chapter X.

However, most of the basic structural questions about these notions are open, so that it is not clear yet how the methods introduced for U and $U_0$ might be useful in this context, and Chapter X can be only considered as a preliminary introduction to this area.

v). Traditionally the classical theory of real functions has had a metric component (trigonometric series, integration, differentiation, etc.) and a descriptive component (descriptive set theory; theory of analytic and projective sets). Although often in the first few decades of this century these two aspects have been studied by the same groups of mathematicians, the interaction between them has been rather limited. For example, it is interesting to read the reminiscenses of A. N. Kolmogorov in a volume devoted to N. N. Lusin (Russian Math. Surveys 40:3(1985), 5–7), according to which Lusin divided his students into those who were to study the metric theory and those who were to study the descriptive theory. (Kolmogorov by the way violated this rule by doing both). Later on, say from the 1940's, descriptive set theory found many interesting applications in several areas of modern analysis, like potential theory (capacitability arguments), probability and measure theory and functional analysis. On the other hand, the main development of the subject was primarily carried out by logicians, who incorporated the classical theory into the extensive framework of definability theory. They discovered also important connections between the classical concepts and those of computability theory on the one hand and on the other with the deep foundational issues of modern set theory, related to metamathematical problems of unsolvability from the standard ZFC axioms and the use of strong axioms of set theory, like determinacy and large cardinals, which transcend the limits of classical mathematics.

Thus eventually the two subjects of the metric and the descriptive theory of functions drifted further apart as each was being practiced by mathematicians working even in widely different areas of mathematics. It is hoped that the ideas developed in this book will remedy this situation by bringing forward what we think are quite interesting relationships between descriptive set theory and the study of sets of uniqueness of trigonometric series, a traditional subject in the metric theory. Such work is potentially interesting for the set theorist as it provides a new context in which the ideas and methods of descriptive set theory are relevant or directly applicable, and whose problems could lead to the discovery of new results or techniques needed to handle them. It also opens up the possibility for new foundational discoveries related to these areas of analysis. For the analyst it brings the benefit of a new point of view and the availability of novel tools. This could be useful in attacking old problems as well as bringing to light new ones.

There are indeed many interesting open questions that the kind of work explained here suggests, and we discuss them at the relevant parts of this book. We hope therefore that the fruitful interplay between ideas of definability theory and harmonic analysis which is described here will continue to contribute to further progress in these areas.

# About this Book

Since we view this book as addressed to both logicians or set theorists and analysts, we tried to keep the prerequisites from each field to a minimum, so we believe it is essentially self-contained modulo some basic material. As far as the analysis is concerned we assume only a familiarity with the fundamental results of real and complex analysis (complex methods are actually used rarely), and the elements of functional analysis, i.e. with topics covered in the first or second year of graduate studies. Typical references for them are Stromberg [1], Hewitt and Stromberg [1], Rudin [1] and Rudin [2, Chapters 1–4]. Also for becoming a little more familiar with the basics of harmonic analysis the reader can consult the first chapter of Katznelson [1]. From set theory the reader should be acquainted with its elements, including transfinite induction and ordinals. Some references here are Halmos [1], Enderton [1] or the first 50 pages or so of Jech [1]. Beyond that one needs some classical results in descriptive set theory concerning basic properties of Borel, analytic and coanalytic sets. These are all summarized in Chapter IV.1 and can be found with detailed proofs in any standard text, like Kuratowski [1, Vol. I, §33–39] or Moschovakis [1, Ch. 1, 2].

Except for this background, as a general rule we develop in detail all the material from analysis or descriptive set theory that we need. This necessitates the inclusion here of standard results from books or other references for which we claim no originality in either the context or the presentation. In particular there is no implication that results for which no specific credit or reference is given, usually because of lack of relevant information, are due to the authors. Moreover having in mind readers with diverse backgrounds, we have given many more detailed arguments than one would find necessary to include in a book for experts in each area. This has been also responsible for our preference for more elementary arguments when sleeker or shorter, but perhaps more advanced, ones were possible. We have also avoided discussing many interesting extensions, strengthenings and generalizations of various results, unless we felt they were necessary to our main concerns, referring the reader instead to the appropriate sources for more information on them.

There are however some exceptions to our rule of trying to be as self-contained as possible. We quote for example a couple of results from number theory and another one or two from descriptive set theory without proof. In these particular instances we feel that nothing would be lost by not including the arguments, which are not really relevant to our main subject, and moreover adequate references are given. More seriously, in Chapter X we make use of some crucial results from the literature without again giving the proofs. This is because the development of the subject treated in this chapter is from our point of view still preliminary and we think of this only as an introduction to an area in which one awaits a fuller development before a more complete expository account can be written. Finally, we make throughout the book a lot of comments or side remarks concerning related results by giving references to the relevant literature, but not complete accounts.

There is of course an extensive literature dealing with the subject of uniqueness for trigonometric series and related topics. The reader can consult the following books and the references contained therein for a further study of this subject: N. K. Bary [2], J. J. Benedetto [1, 2], C. C. Graham and O. C. McGehee [1], J.-P. Kahane [1], J.-P. Kahane and R. Salem [1], Y. Katznelson [1], L.-Å. Lindahl and F. Poulsen [1], L. Loomis [2], Y. Meyer [1], R. Salem [1], A. Zygmund [1]. As far as descriptive set theory and related areas are concerned one can consult F. Hausdorff [1], J. Hoffman-Jorgensen [1], T. Jech [1], K. Kuratowski [1], N. N. Lusin [1], Y. N. Moschovakis [1], C. A. Rogers et al. [1].

Concerning notation and terminology, we have tried to follow the traditions of people working in each of the fields which are discussed (harmonic analysis, Banach space theory and descriptive set theory) and at the same time tried to avoid the use of very close terminology or notation for totally unrelated notions. At the few places where this was not possible—e.g. the notions of norms in Banach spaces, and norms for $\underset{\sim}{\Pi}_1^1$ sets—we changed one of the two terminologies. We apologize to those who created them. There is a comprehensive index of notation and terminology at the end of the book.

Concerning the numbering of results, we use the minimality method: The book is divided in chapters and sections, and items are numbered by one number in each section. Reference to "Lemma 6" refers to the same section, "Theorem 3.4" to Theorem 4 in Section 3 of the same chapter, and "Definition II.1.7" to Definition 7 in Section 1 of Chapter II.

We finally give at the end a list of references—which we did not try to make complete and a list of the various open problems discussed in the book, together with the place where they are discussed.

# Chapter I. Trigonometric Series and Sets of Uniqueness

In this chapter we will introduce the basic concepts concerning sets of uniqueness for trigonometric series and we will prove some results of the classical theory due to Riemann, Cantor, Young, Rajchman and Bary. We will also discuss some of the old problems in the subject.

## §1. Trigonometric and Fourier Series

A trigonometric series $S$ is a formal expression

$$S \sim \sum_{n=-\infty}^{+\infty} c_n e^{inx}$$

where $x$ varies over $\mathbb{R}$ and $c_n \in \mathbb{C}$, for each $n \in \mathbb{Z}$. We can write this also as

$$S \sim \frac{a_0}{2} + \sum_{n=1}^{\infty} a_n \cos(nx) + b_n \sin(nx)$$

where $a_n = c_n + c_{-n}$ and $b_n = i(c_n - c_{-n})$. We call $S$ real if the $a_n, b_n$ are real, i.e. if $c_n = \overline{c}_{-n}$.

We say that the series $S$ as above converges at a point $x \in \mathbb{R}$ if the (symmetric) partial sums

$$S_N(x) = \sum_{n=-N}^{+N} c_n e^{inx}, \qquad N = 0, 1, 2, \ldots$$

converge. If their limit is s we write

$$s = \sum_{n=-\infty}^{+\infty} c_n e^{inx}.$$

Equivalently one has

$$S_N(x) = \frac{a_0}{2} + \sum_{n=1}^{N} a_n \cos(nx) + b_n \sin(nx)$$

and

$$s = \frac{a_0}{2} + \sum_{n=1}^{\infty} a_n \cos(nx) + b_n \sin(nx).$$

Clearly all that matters here is only x(mod $2\pi$), i.e. x modulo $2\pi\mathbb{Z}$. Let $\mathbb{T}$ be the unit circle in $\mathbb{R}^2$. We identify $\mathbb{T}$ with $\mathbb{R}/2\pi\mathbb{Z}$ via the map $x \mapsto e^{ix}$. Many times it will be convenient to think of $\mathbb{T}$ as $[0, 2\pi)$ or $[0, 2\pi]$ with 0 and $2\pi$ identified. Similarly functions on $\mathbb{T}$ can be viewed as functions on $[0, 2\pi)$, as functions f on $[0, 2\pi]$ with $f(0) = f(2\pi)$ or as $2\pi$-periodic functions on $\mathbb{R}$.

We will denote by $\lambda$ the Lebesgue measure on $\mathbb{T}$ (i.e. that induced by its above identification with $[0, 2\pi)$) normalized so that $\lambda(\mathbb{T}) = 1$. As usual $L^1(\mathbb{T})$ will be the Banach space of (complex Lebesgue) integrable functions on $\mathbb{T}$ with norm

$$\|f\|_{L^1} = \|f\|_{L^1(\mathbb{T})} = \int_{\mathbb{T}} |f| d\lambda$$

$$= \frac{1}{2\pi}\int_0^{2\pi} |f(x)|dx, \text{ if f is viewed as a function on } [0,2\pi).$$

The Fourier series of a function $f \in L^1(\mathbb{T})$ is the trigonometric series

$$S(f) \sim \sum_{n=-\infty}^{+\infty} \hat{f}(n)e^{inx}$$

where $\hat{f}(n)$, the Fourier coefficients of f, are given by

$$\hat{f}(n) = \frac{1}{2\pi}\int_0^{2\pi} f(x)e^{-inx}dx.$$

We denote by

$$S_N(f,x) = \sum_{n=-N}^{N} \hat{f}(n)e^{inx}$$

the partial sums of $S(f)$.

## §2. The problem of uniqueness

The problem of uniqueness for trigonometric series originating in the work of Riemann and Heine in the mid-nineteenth century is the following:

Let $S \sim \Sigma\, c_n\, e^{inx}$, $T \sim \Sigma\, d_n\, e^{inx}$ be two trigonometric series which converge everywhere to the same value, i.e.

$$f(x) = \sum c_n\, e^{inx} = \sum d_n\, e^{inx}, \qquad \forall\, x \in \mathbb{R}.$$

Is it true then that $c_n = d_n$, $\forall n \in \mathbb{Z}$?

Equivalently, if $S \sim \Sigma\, c_n\, e^{inx}$ is a trigonometric series and $\Sigma\, c_n\, e^{inx} = 0$, for all $x \in \mathbb{R}$, does it follow that $c_n = 0$, for all $n \in \mathbb{Z}$, i.e. S is identically 0, in symbols $S = 0$?

This problem was solved affirmatively by Cantor in 1870. He then went on to show essentially that the conclusion $S = 0$ followed even if $\Sigma\, c_n\, e^{inx} = 0$ holds except perhaps at a countable closed set. (Actually Cantor formally established and published only the special case of countable closed sets of finite Cantor-Bendixson rank). As is well-known, it was through this work on trigonometric series that Cantor was led to the creation of set theory.

It is high time now to introduce the basic concept of a set of uniqueness.

Definition 1. A set $P \subseteq \mathbb{T}$ is called a set of uniqueness if every trigonometric series which converges to 0 outside P is identically 0. Otherwise P is called a set of multiplicity.

We denote by $\mathcal{U}$ the class of sets of uniqueness and by $\mathcal{M}$ the class of sets of multiplicity. (Thus $\mathcal{M}$ = power($\mathbb{T}$) $-$ $\mathcal{U}$). We will also say sometimes $\mathcal{U}$-set or $\mathcal{M}$-set for "set in $\mathcal{U}$" or "set in $\mathcal{M}$".

In this terminology Cantor's theorem asserts that every countable closed set is in $\mathcal{U}$. This was generalized by W. H. Young in 1909 who showed that every countable set is in $\mathcal{U}$. Finally in 1923 N. Bary showed that the union of countably many closed sets in $\mathcal{U}$ is also in $\mathcal{U}$. We will prove these results later in this chapter.

It is obvious that if $P \in \mathcal{U}$ and $Q \subseteq P$ then $Q \in \mathcal{U}$ as well, i.e. $\mathcal{U}$ is hereditary. Somehow sets in $\mathcal{U}$ ought to be "small" and indeed it is easy to see (and we will prove it shortly) that every measurable set of uniqueness has (Lebesgue) measure 0. For a while it was thought that the converse was also true. It thus came as a big surprise when Menshov in 1916 constructed closed sets of multiplicity that had measure 0. Later on in 1921–23 Rajchman and (independently) Bary discovered many examples of perfect sets of uniqueness. (Rajchman's results implied for example that the Cantor ($\frac{1}{3}$ —) set is a set of uniqueness).

## §3. The Riemann Theory and the Cantor Uniqueness Theorem

First let us dispose of the easy fact that every measurable $\mathcal{U}$-set has measure 0.

Proposition 1. If $P \subseteq \mathbf{T}$ is a measurable $\mathcal{U}$-set, then P has measure 0.

Proof. Assume $\lambda(P) > 0$ towards a contradiction. Then there is a closed set $F \subseteq P$ with $\lambda(F) > 0$. Let $\chi_F$ be its characteristic function, and consider the Fourier series $S(\chi_F) = S$. Then S converges to 0 off F by the localization principle for Fourier series, which asserts that the Fourier series of a function f converges to 0 in any open interval in which f vanishes. Since F is also a set of uniqueness this implies that S is identically 0, i.e. $\hat{\chi}_F(n) = 0$ for all n. But this is absurd, since

$$\hat{\chi}_F(0) = \frac{1}{2\pi}\int_0^{2\pi} \chi_F(x)dx = \lambda(F) > 0. \qquad \square$$

Our main goal in this section will be to show that every countable closed set is in $\mathfrak{U}$. For that we will need some basic results of the Riemann theory of trigonometric series.

Definition 2. Let $S \sim \Sigma\, c_n e^{inx}$ be a trigonometric series with bounded coefficients (i.e. $\sup |c_n| < \infty$). The Riemann function of S is defined by

$$F_S(x) = \frac{c_0 x^2}{2} - \sum_{n=-\infty}^{+\infty}{}' \frac{1}{n^2} c_n e^{inx}, \quad x \in \mathbb{R}$$

where the $'$ on the summation sign indicates that the sum excludes 0. Observe that $F_S$ is obtained by formally integrating $\Sigma\, c_n e^{inx}$ twice. Notice also that, since $\{c_n\}$ is bounded, $F_S$ is a continuous function on $\mathbb{R}$.

Given a function F let

$$\Delta^2 F(x,h) = F(x+h) + F(x-h) - 2F(x)$$

and define the second Schwarz derivative of F at x by

$$D^2F(x) = \lim_{h \to 0} \frac{\Delta^2 F(x,h)}{h^2}$$

if this limit exists. It is not hard to check that if $F''(x)$ exists so does $D^2F(x)$ and they are equal, but the converse fails.

We have now the following important relationship between $F_S$ and $\Sigma\, c_n e^{inx}$.

Lemma 3 (Riemann's First Lemma). Suppose $S \sim \Sigma\, c_n e^{inx}$ is a trigonometric series with bounded coefficients and $\Sigma\, c_n e^{inx} = s$. Then $D^2F_S(x)$ exists and $D^2F_S(x) = s$.

Proof. By direct calculation

$$\frac{\Delta^2 F_S(x,2h)}{4h^2} = \sum \frac{\sin^2(nh)}{(nh)^2}\, c_n\, e^{inx}.$$

It will thus suffice to show the following

Lemma 4. For any sequence $a_n \in \mathbb{C}$,

$$\text{if } \sum_{n=0}^{\infty} a_n = a, \text{ then } \quad \lim_{h\to 0}\left[\sum_{n=0}^{\infty} \frac{\sin^2(nh)}{(nh)^2}\, a_n\right] = a.$$

Proof. Let $A_N = \sum\limits_{n=0}^{N} a_n$. Then

$$\sum_{n=0}^{\infty} a_n \left(\frac{\sin nh}{nh}\right)^2 = \sum_{n=0}^{\infty} A_n \left\{\left(\frac{\sin nh}{nh}\right)^2 - \left(\frac{\sin(n+1)h}{(n+1)h}\right)^2\right\}.$$

Let $h_k \to 0$, $h_k$ positive and put

$$s_{kn} = \left(\frac{\sin nh_k}{nh_k}\right)^2 - \left(\frac{\sin(n+1)h_k}{(n+1)h_k}\right)^2.$$

Then we have to show that

$$(*) \quad \sum_{n=0}^{\infty} A_n\, s_{kn} \xrightarrow{k} a = \lim A_n.$$

What we are dealing with here is a summability method given by the matrix $(s_{kn})$. So to prove (*) it is enough to verify that this matrix is <u>regular</u>, i.e. satisfies the usual <u>Toeplitz</u> <u>conditions</u>:

$$\text{(i)} \quad \lim_k s_{kn} = 0, \qquad n = 0, 1, 2, \ldots$$

$$\text{(ii)} \quad \sum_{n=0}^{\infty} |s_{kn}| \le C < \infty, \qquad k = 0, 1, 2, \ldots$$

$$\text{(iii)} \quad \sum_{n=0}^{\infty} s_{kn} \xrightarrow{k} 1.$$

Now (i), (iii) are obvious. To verify (ii) let

$$u(x) = \left(\frac{\sin x}{x}\right)^2$$

and notice that $\int_0^\infty |u'(x)|dx < \infty$, so that

$$\begin{aligned}
\sum_{n=0}^{\infty} |s_{kn}| &= \sum_{n=0}^{\infty} \left| \left(\frac{\sin nh_k}{nh_k}\right)^2 - \left(\frac{\sin(n+1)h_k}{(n+1)h_k}\right)^2 \right| \\
&= \sum_{n=0}^{\infty} \left| \int_{nh_k}^{(n+1)h_k} u'(x)dx \right| \\
&\le \int_0^\infty |u'(x)|dx = C < \infty
\end{aligned}$$

and we are done. □

The second Schwarz derivative shares the following basic property with the usual second derivative.

Lemma 5 (Schwarz). Suppose F is real valued continuous on (a,b) and $D^2F(x) \geq 0$ for all $x \in (a,b)$. Then F is convex on (a,b). In particular, if now F is complex-valued and continuous on (a,b) and $D^2F(x) = 0$ for all $x \in (a,b)$, F is linear on (a,b).

Proof. We can assume $D^2F(x) > 0$ for all $x \in (a,b)$ (since if necessary we can replace F by $F(x) + \epsilon x^2$, $\epsilon > 0$ and let $\epsilon \to 0$). If F is not convex, towards a contradiction, then there is a linear function $\mu x + \nu$ and $a < c < d < b$, such that if $G(x) = F(x) - (\mu x + \nu)$, then $G(c) = G(d) = 0$ and $G(x) > 0$ for some x in (c,d). Let $x_0$ be a point where G achieves its maximum in [c,d]. Then $x_0 \in (c,d)$ and for small enough h, $\Delta^2F(x_0,h) \leq 0$, so that $D^2F(x_0) \leq 0$, a contradiction. □

We only need one more fact before we can prove that the empty set is a set of uniqueness.

Lemma 6 (The Cantor-Lebesgue Lemma). If $a_n$, $b_n$ ($n \in \mathbb{N}$) are such that $a_n \cos(nx) + b_n \sin(nx) \to 0$ for x in a set of positive measure, then $a_n, b_n \to 0$. In particular, if $\Sigma c_n e^{inx} = 0$ on a set of positive measure, then $c_n \to 0$.

Proof. We can assume clearly that $a_n, b_n \in \mathbb{R}$. Let $\rho_n = \sqrt{a_n^2 + b_n^2}$ so that we can write

$$a_n \cos nx + b_n \sin nx = \rho_n \cos (nx + \varphi_n)$$

for some $\varphi_n$. Thus $\rho_n \cos(nx + \varphi_n) \to 0$ on a set $E \subseteq [0, 2\pi)$ of positive measure $\alpha$. If $\rho_n \not\to 0$, towards a contradiction, choose $\epsilon > 0$ and $n_0 < n_1 < n_2 < \ldots$ so that $\rho_{n_k} \geq \epsilon$. Then $\cos(n_k x + \varphi_{n_k}) \to 0$, so $2\cos^2(n_k x + \varphi_{n_k}) \to 0$ and therefore $1 + \cos 2(n_k x + \varphi_{n_k}) \to 0$ for $x \in E$. In particular $\int_E (1 + \cos 2(n_k x + \varphi_{n_k}))\, dx \to 0$, i.e.

$$\alpha + \int_0^{2\pi} \chi_E(x) \cos 2(n_k x + \varphi_{n_k})\, dx \to 0.$$

But by the Riemann-Lebesgue Lemma (which states that if $f \in L^1(\mathbb{T})$, then $\hat{f}(n) \to 0$ as $|n| \to \infty$) the above integral tends to 0 as $k \to \infty$, so that $\alpha = 0$, a contradiction. □

We can prove now

**Theorem 7** (Cantor). If $\Sigma c_n e^{inx} = 0$ for all $x \in \mathbb{R}$, then $c_n = 0$ for all n. In other words $\emptyset$ is a $\mathcal{U}$-set.

*Proof*. By the Cantor-Lebesgue Lemma, $c_n \to 0$ as $|n| \to \infty$, so in particular $c_n$ is bounded. By Riemann's First Lemma $D^2F_S(x) = 0$, $\forall x \in \mathbb{R}$ and so by Schwarz's Lemma $F_S$ is linear, i.e.

$$c_0 \frac{x^2}{2} - \sum{}' \frac{1}{n^2} c_n e^{inx} = ax + b$$

for all $x \in \mathbb{R}$. Put $x = \pi$ and $x = -\pi$ and subtract to get $a = 0$. Then put $x = 0$ and $x = 2\pi$ and subtract to get $c_0 = 0$. Thus $\Sigma' \frac{1}{n^2} c_n e^{inx} = b$ for all x, and because of the uniform convergence of this series, term-by-term integration shows immediately that $c_n = 0$ for all $n \neq 0$ as well, and we are done. □

We will prove next the general Cantor theorem on countable closed sets. We need first another important property of the Riemann function.

<u>Lemma 8</u> (Riemann's Second Lemma). Let $S \sim \Sigma\, c_n e^{inx}$ be a trigonometric series with $c_n \to 0$. Then

$$\frac{\Delta^2 F_S(x,h)}{h} = \frac{F_S(x+h) + F_S(x-h) - 2F_S(x)}{h} \longrightarrow 0$$

as $h \to 0$, uniformly on x.

If for a function F we have $\frac{\Delta^2 F(x,h)}{h} \to 0$ as $h \to 0$ we say that F is <u>smooth</u> at x. Note then that if the left and right derivatives of F exist at x they must be equal, i.e. (the graph of) F can have no corner at x.

<u>Proof</u>. We have

$$\frac{\Delta^2 F_S(x,2h)}{4h} = \sum \frac{\sin^2(nh)}{n^2 h}\, c_n e^{inx}.$$

Again let $0 < h_k \leq 1$, $h_k \to 0$, and let

$$t_{kn} = \frac{\sin^2(nh_k)}{n^2 h_k}.$$

We have to show that $\sum_n (c_n e^{inx})\, t_{kn} \to 0$, as $k \to \infty$ (uniformly on x). Since $c_n e^{inx} \to 0$ (uniformly on x) it is enough to verify that the matrix $(t_{kn})$ satisfies the first two Toeplitz conditions (i), (ii) (one need not verify (iii) since the sequence $c_n e^{inx}$ to which the summability method $(t_{kn})$ is applied tends to 0). Again (i) (i.e. $\lim_k t_{kn} = 0$) is obvious, so we verify (ii): Fix k and choose $N > 1$ so that $N - 1 \leq h_k^{-1} < N$. Then

$$\sum_{n=1}^{\infty} |t_{kn}| = \sum_{n=1}^{N-1} \frac{\sin^2(nh_k)}{n^2 h_k} + \sum_{n=N}^{\infty} \frac{\sin^2(nh_k)}{n^2 h_k}$$

$$\leq (N-1) \cdot h_k + \frac{1}{h_k} \sum_{n=N}^{\infty} \frac{1}{n^2}$$

$$\leq 1 + \frac{1}{h_k} \sum_{n=N}^{\infty} \frac{1}{n(n-1)}$$

$$= 1 + \frac{1}{h_k} \cdot \frac{1}{N-1} \leq 3$$

and we are done. □

We are now ready to prove

Theorem 9 (Cantor). Every countable closed set is in $\mathcal{U}$.

Proof. Let $E \subseteq \mathbb{T}$ be countable and closed. By translating if necessary we can assume that $0 \notin E$. So $E \subseteq (0, 2\pi)$. Denote by $E^{(\alpha)}$ the $\alpha^{\text{th}}$ Cantor-Bendixson derivative of E defined as usual by

$$E^{(0)} = E,$$
$$E^{(\alpha+1)} = (E^{(\alpha)})' = \{x \in E^{(\alpha)}: \text{ x is a limit point of } E^{(\alpha)}\},$$
$$E^{(\lambda)} = \cap_{\alpha<\lambda} E^{(\alpha)}, \lambda \text{ limit.}$$

Let also $V^{(\alpha)} = (0, 2\pi) - E^{(\alpha)}$.

Suppose now $S \sim \Sigma c_n e^{inx}$ is such that $\Sigma c_n e^{inx} = 0$ off E. We will show, by induction on $\alpha$, that $F_S$ is linear in each component of $V^{(\alpha)}$. Since $E^{(\alpha_0)} = \emptyset$

for some $\alpha_0$ (E being countable) and thus $V^{(\alpha_0)} = (0, 2\pi)$, it follows that $F_S$ is linear on $(0, 2\pi)$. Of course the same holds in any interval of $\mathbb{R}$ of length $2\pi$, therefore $F_S$ is linear on $\mathbb{R}$ and so, as in the proof of Theorem 7, S is identically 0.

Now clearly $F_S$ is linear on each component of $V^{(0)}$ by Riemann's First Lemma and Schwarz's Lemma. The induction step goes through limit ordinals easily by a compactness argument. Finally, assuming $F_S$ is linear in each component of $V^{(\alpha)}$, consider a component (a,b) of $V^{(\alpha+1)}$. Then in each closed interval $[c,d] \subseteq (a,b)$ there are only finitely many points of $E^{(\alpha)}$, say $c \leq x_1 < x_2 < \ldots < x_n \leq d$. Thus $F_S$ is linear on $(c,x_1), (x_1,x_2), \ldots, (x_n, d)$. But notice that $F_S$ is smooth everywhere by Riemann's Second Lemma, since $c_n \to 0$ (in view of the Cantor-Lebesgue Lemma and the fact that $\Sigma\, c_n e^{inx} = 0$ except on a countable set). Thus $F_S$ can have no corners at $x_1, \ldots, x_n$, i.e. $F_S$ is linear on $[c, d]$ and thus on $(a, b)$. □

## §4. The Rajchman Multiplication Theory. Examples of Perfect Sets of Uniqueness

We will discuss here an elegant and powerful technique of Rajchman dealing with the formal multiplication of trigonometric series by (appropriately nice) functions.

Let $S \sim \Sigma\, c_n e^{inx}$ be a trigonometric series with bounded $\{c_n\}$. Let f be a continuous function on $\mathbb{T}$ with absolutely convergent Fourier series, i.e. such that $\Sigma\, |\hat{f}(n)| < \infty$. Clearly then $f(x) = \Sigma\, \hat{f}(n) e^{inx}$ uniformly. We define the <u>formal product</u> $S(f) \cdot S$ of (the Fourier series of) f by S as follows:

$$S(f) \cdot S \sim \sum C_n e^{inx}$$

where

$$C_n = \sum_k c_k \hat{f}(n - k).$$

Clearly $C_n$ exists and indeed $\{C_n\}$ is bounded (by $\sup_k |c_k| \cdot \sum_n |\hat{f}(n)|$). Moreover we have

<u>Lemma 1</u>. If $c_n \to 0$, then $C_n \to 0$.

<u>Proof</u>. Note that

$$|C_n| = \left|\sum_k c_k \hat{f}(n - k)\right|$$

$$\leq \sum_{|k| \leq \frac{1}{2}|n|} |c_k| \; |\hat{f}(n - k)| + \sum_{|k| > \frac{1}{2}|n|} |c_k| \cdot |\hat{f}(n - k)|$$

$$\leq (\sup_k |c_k|) \sum_{|m| \geq \frac{1}{2}|n|} |\hat{f}(m)| + (\sup_{|k| > \frac{1}{2}|n|} |c_k|) \cdot \sum_m |\hat{f}(m)| \to 0, \text{ as } n \to \infty$$

and we are done. □

If now a function $\varphi$ has rapidly converging to 0 Fourier coefficients then one can establish a nice relationship between the formal product $S(\varphi) \cdot S = \Sigma\, C_n e^{inx}$ and $\varphi(x) \cdot \Sigma\, c_n e^{inx}$.

<u>Lemma 2</u>. Assume $\varphi(x)$ is such that $\sum_{k=0}^{\infty} (\sum_{|n| \geq k} |\hat{\varphi}(n)|) < \infty$ or equivalently $\sum_{n=-\infty}^{\infty} |n\hat{\varphi}(n)| < \infty$ (for example $\hat{\varphi}(n) = O(|n|^{-3})$ is enough). Let $S \sim \Sigma\, c_n e^{inx}$ be a trigonometric series with coefficients $c_n \to 0$. Then if $S(\varphi) \cdot S \sim \Sigma\, C_n e^{inx}$ is the formal product of $\varphi$ by S we have $\sum_{n=-N}^{N} C_n e^{inx} - \varphi(x) \cdot \sum_{n=-N}^{N} c_n e^{inx} \longrightarrow 0$, as $N \to \infty$

uniformly on x, i.e. $\sum_{n=-\infty}^{\infty} (C_n - \varphi(x)c_n)e^{inx} = 0$, uniformly on x.

Proof. We will start with the following sublemma, which is a special case of what we want to prove:

Lemma 3. With the notation of Lemma 2, if $\varphi(x) = 0$ for all $x \in P \subseteq \mathbb{T}$ then $\Sigma\, C_n e^{inx} = 0$ uniformly on $x \in P$.

Proof. For any $x \in P$ and $N = 0, 1, 2, \ldots$ we have

$$\begin{aligned}
\sum_{n=-N}^{N} C_n e^{inx} &= \sum_{n=-N}^{N} \Big(\sum_{k=-\infty}^{+\infty} c_k e^{ikx} \hat{\varphi}(n-k) e^{i(n-k)x}\Big) \\
&= \sum_{k=-\infty}^{+\infty} c_k e^{ikx} \Big(\sum_{n=-N}^{N} \hat{\varphi}(n-k) e^{i(n-k)x}\Big) \\
&= \sum_{k=-\infty}^{+\infty} c_k e^{ikx} \Big(\sum_{m=-N-k}^{N-k} \hat{\varphi}(m) e^{imx}\Big) \\
&= \sum_{|k| \le \frac{1}{2}N} \cdots + \sum_{|k| > \frac{1}{2}N} \cdots \\
&= I_1 + I_2.
\end{aligned}$$

Now $\sum_{m=-\infty}^{+\infty} \hat{\varphi}(m)e^{imx} = \varphi(x) = 0$, so $\sum_{m=-N-k}^{N-k} \hat{\varphi}(m)e^{imx} = -\Big(\sum_{m=-\infty}^{-N-k-1} \hat{\varphi}(m)e^{imx} + \sum_{m=N-k+1}^{+\infty} \hat{\varphi}(m)e^{imx}\Big)$, thus

$$|I_1| \le (\sup_k |c_k|) \cdot 2 \Big(\sum_{k \ge \frac{N}{2}} \sum_{|m| \ge k} |\hat{\varphi}(m)|\Big) \to 0$$

as $N \to \infty$ uniformly on x, since $\sum_{k=0}^{\infty} \sum_{|m| \ge k} |\hat{\varphi}(m)| < \infty$. Also

$$|I_2| \leq \left( \sup_{|k|>\frac{N}{2}} |c_k| \right) \cdot 3 \left( \sum_{k=0}^{\infty} \sum_{|m|\geq k} |\hat{\varphi}(m)| \right) \to 0$$

as $N \to \infty$, uniformly on x. □

We complete now the proof of Lemma 2. For that note that in the preceding proof of Lemma 3 we could actually have $\hat{\varphi}(n)$ depending on x, i.e. have a series of the form $\Sigma\varphi^*(n,x)e^{inx}$ provided that $\sum_{|n|\geq k} |\varphi^*(n,x)| \leq M_k$, $k = 0, 1, 2, \ldots$ for all x and some $M_k$ with $\sum_{k=0}^{\infty} M_k < \infty$. (In this case the hypothesis $\varphi(x) = 0$ is replaced by $\Sigma\varphi^*(n,x)e^{inx} = 0$).

Observe now that

$$\sum_{n=-N}^{N} C_n e^{inx} - \varphi(x) \sum_{n=-N}^{N} c_n e^{inx} = \sum_{n=-N}^{N} [C_n - \varphi(x)\, c_n] e^{inx} = \sum_{n=-N}^{N} D_n e^{inx}$$

where
$$D_n = C_n - \varphi(x)\, c_n$$

$$= \sum_{k=-\infty}^{\infty} c_k\, \hat{\varphi}(n-k) - \varphi(x)\, c_n$$

$$= \sum_{k=-\infty}^{\infty} c_k \varphi^*(n-k, x)$$

with

$$\varphi^*(m,x) = \begin{cases} \hat{\varphi}(m), \text{ if } m \neq 0 \\ \hat{\varphi}(0) - \varphi(x), \text{ if } m = 0. \end{cases}$$

Then $\Sigma\varphi^*(n,x)e^{inx} = \Sigma\hat{\varphi}(n)e^{inx} - \varphi(x) = 0$ for all x and $\sum_{|n|\geq k} |\varphi^*(n,x)| = \sum_{|n|\geq k} |\hat{\varphi}(n)|$, if $k \neq 0$, while $\sum_n |\varphi^*(n,x)| \leq \Sigma\, |\hat{\varphi}(n)| + C$ for some constant C, so we are done. □

The Rajchman theory is very useful in dealing with localization problems. As an illustration let us see how to use it to prove the classical principle of localization of Riemann.

Theorem 4 (Riemann Localization Principle). Let $S \sim \Sigma c_n e^{inx}$ be a trigonometric series with $c_n \to 0$. If $F_S$ is linear in some open interval (a,b) then $\Sigma c_n e^{inx} = 0$ on (a, b) and uniformly on closed subintervals of it.

Proof. Let $[c, d] \subseteq (a, b)$. Take $\varphi \in C^{\infty}(\mathbb{T})$ (so that certainly $\hat{\varphi}(n) = O(|n|^{-3})$ to be 1 on [c, d] and 0 off (a, b). Let $S(\varphi) \cdot S = T \sim \Sigma C_n e^{inx}$. Then

$$\sum_{n=-\infty}^{\infty} (C_n e^{inx} - \varphi(x) c_n e^{inx}) = 0$$

uniformly on x, by Lemma 2. So by Lemma 3.4

$$\lim_{h \to 0} \sum_{n=-\infty}^{\infty} (C_n e^{inx} - \varphi(x) c_n e^{inx}) \left(\frac{\sin nh}{nh}\right)^2 = 0.$$

But since $F_S$ is linear on (a, b), for x in (a, b) and small enough h,

$$\sum_{n=-\infty}^{\infty} c_n e^{inx} \left(\frac{\sin nh}{nh}\right)^2 = \frac{\Delta^2 F_S(x, 2h)}{4h^2} = 0$$

thus for such x

$$(*) \qquad \lim_{h \to 0} \sum_{n=-\infty}^{\infty} \varphi(x)\ c_n e^{inx} \left(\frac{\sin nh}{nh}\right)^2 = 0.$$

On the other hand if $x \notin (a, b)$, $\varphi(x) = 0$ so we actually have (*) for all x. It follows that

$$D^2F_T(x) = \lim_{h\to 0} \sum_{n=-\infty}^{\infty} C_n e^{inx} \left(\frac{\sin nh}{nh}\right)^2 = 0$$

for all x, i.e. $F_T$ is linear and thus $C_n = 0$ for all n (as in the proof of 3.7). It follows by Lemma 2 again that $\varphi(x)\ (\Sigma\ c_n e^{inx}) = 0$ uniformly and so $\Sigma c_n e^{inx} = 0$ uniformly on [c, d]. □

We will now use multiplication theory to exhibit a whole class of perfect $\mathcal{U}$-sets.

Definition 5 (Rajchman). A set $E \subseteq \mathbf{T}$ is an H-set if there is a nonempty interval $I \subseteq \mathbf{T}$ and an infinite sequence $n_0 < n_1 < n_2 < \ldots$ in $\mathbb{N}$ such that for all k, $(n_k E) \cap I = \emptyset$.

As usual if $E \subseteq \mathbf{T}$ and $m \in \mathbb{Z}$ we define, viewing E as a subset of $[0, 2\pi]$,

$$mE = \{mx \pmod{2\pi} : x \in E\}.$$

Also by an interval in $\mathbf{T}$ we mean an (open) arc in $\mathbf{T}$, viewed as the unit circle, or equivalently the corresponding subset of $[0, 2\pi)$.

It is trivial to check that every finite set is an H-set. If E is the Cantor (1/3-)set in $\mathbf{T}$, i.e. the set of all points of the form $2\pi(\Sigma\ \epsilon_n 3^{-n})$, where $\epsilon_n = 0$ or 2, then E is an H-set, since $3^n E$ always avoids the open middle-third interval of $[0, 2\pi]$. One can easily construct also a great variety of other perfect H-sets.

(Note however that not all countable closed sets are H-sets. In fact there are closed sets of the form $E = \{x_1, x_2, \ldots\} \cup \{0\}$ where $x_i \to 0$, which are not H-sets. To see this, enumerate all the rational relative to $2\pi$ intervals of $[0, 2\pi]$ in a sequence $I_1, I_2, \ldots$. (A rational relative to $2\pi$ is a rational multiple of $2\pi$). Look at $\frac{1}{n} I_1, \ldots, \frac{1}{n} I_n$, which are included in $[0, \frac{2\pi}{n}]$, and choose a point $r_i^n \in \frac{1}{n} I_i$ $(i = 1, \ldots, n)$. Let finally $x_1, x_2, \ldots$ enumerate the $r_i^n$'s, so that $x_i \to 0$. Since $m \geq k \Rightarrow m x_k^m \in I_k$, $E = \{x_1, x_2, \ldots\} \cup \{0\}$ is not an H-set. By a result of Salinger [1] though one has that every countable closed set of finite Cantor-Bendixson rank n can be written as the union of $2^{n-1}$ closed H-sets).

We have now the following result which among other things establishes the existence of perfect $\mathcal{U}$-sets.

<u>Theorem</u> <u>6</u> (Rajchman). Any H-set is a set of uniqueness. (In particular the Cantor set is a set of uniqueness).

<u>Proof</u>. Since the closure of an H-set is an H-set, let E be a closed H-set and let $I \neq \emptyset$ be an open interval and $n_0 < n_1 < \ldots$ be positive integers such that $(n_k E) \cap I = \emptyset$, for all k. Let $S \sim \Sigma c_n e^{inx}$ be a trigonometric series with $\Sigma c_n e^{inx} = 0$ off E. Then by the Cantor-Lebesgue Lemma (3.6) $c_n \to 0$. We will show that $c_n = 0$ for all n.

Let $\varphi \in C^\infty(T)$ be such that

$$\mathrm{supp}(\varphi) = \overline{\{x : \varphi(x) \neq 0\}}$$

is contained in I and $\hat{\varphi}(0) = 1$. Let $\varphi_k(x) = \varphi(n_k x)$. Then clearly each $\varphi_k$ is 0 on

E. Let now $\Sigma\, C_n^k\, e^{inx}$ be the formal product $S(\varphi_k) \cdot S$. We claim that $C_n^k \to c_n$ as $k \to \infty$. Since

$$\sum_{n=-N}^{N} C_n^k\, e^{inx} - \varphi_k(x) \sum_{n=-N}^{N} c_n e^{inx} \to 0$$

for all x (Lemma 2), we have (as $\varphi_k(x) = 0$ on E and $\Sigma\, c_n e^{inx} = 0$ off E) $\sum_{n=-\infty}^{\infty} C_n^k e^{inx} = 0$ for all x. Thus by Cantor's Theorem $C_n^k = 0$ for all k, n and assuming our claim $c_n = 0$ for all n.

We prove now the claim. Notice that the Fourier series of $\varphi_k$ is $\Sigma\hat{\varphi}(n)e^{in \cdot n_k x}$, so that

(i) $\sum |\hat{\varphi}_k(n)| \leq C < \infty$, for all k,

(ii) $\hat{\varphi}_k(0) = \hat{\varphi}(0) = 1$,

(iii) $\lim_{k \to \infty} \hat{\varphi}_k(n) = 0$, if $n \neq 0$.

Now

$$C_n^k = \sum_m c_{n-m}\, \hat{\varphi}_k(m)$$

$$= \sum_{|m| \leq N} \cdots + \sum_{|m| > N} \cdots, \text{ where } N > |n|.$$

The first sum converges to $c_n$ as $k \to \infty$ (by (ii), (iii)), while the second is bounded by $(\sup_{|k| \geq N-|n|} |c_k|) \cdot C$ which goes to 0 as $N \to \infty$ and we are done. □

## §5. Countable Unions of Closed Sets of Uniqueness

We prove now the following important closure property of $\mathcal{U}$.

Theorem 1 (Bary). The union of countably many closed sets of uniqueness is also a set of uniqueness.

For the proof we will use a special case of the following extension of Cantor's Theorem, due to de la Vallée-Poussin.

Theorem 2 (de la Vallée-Poussin). Let $S \sim \Sigma\, c_n e^{inx}$ be a trigonometric series with $c_n \to 0$. Let $S^*(x) = \overline{\lim}\, S_n(x)$, $S_*(x) = \underline{\lim}\, S_n(x)$. Suppose that for each x, $|S^*(x)|$, $|S_*(x)|$ are finite and $S^*$, $S_*$ are integrable. Then $S = S(f)$ where

$$f(x) = \underline{D}^2 F_S(x) = \underline{\lim}_{h\to 0} \frac{F_S(x+h) + F_S(x-h) - 2F_S(x)}{h^2}.$$

In particular, if $\Sigma\, c_n e^{inx} = g(x)$ everywhere, where g is integrable, $S = S(g)$.

We will not prove this result as we will not need it in its full generality. A proof can be found in Bary [2] or Zygmund [1]. The special case we use is the following.

Theorem 3 (de la Vallée-Poussin). Suppose $S \sim \Sigma\, c_n e^{inx}$ is a trigonometric series with $c_n \to 0$, $S_n(x)$ bounded at each point x, and $\Sigma\, c_n e^{inx} = 0$ a.e. Then $c_n = 0$ for all n.

Proof. We need first the following

Lemma 4. Let G be a $G_\delta$ set of measure 0 in $[0, 2\pi]$. Then there is a positive nondecreasing continuous function g on $[0, 2\pi]$ such that $g'(x) = +\infty$ for $x \in G$.

Proof of Lemma. Let $G = \bigcap_n G_n$ with $G_n$ open and $\lambda(G_n) < \frac{1}{2^n}$. Let $g_n(x) = \frac{1}{2\pi}\int_0^x \chi_{G_n}(t)dt$. Then $0 \leq g_n \leq 2^{-n}$. Put $g = \Sigma\, g_n$. Clearly g is positive, continuous and nondecreasing. Let now $x_0 \in G$ and $K > 0$ be given. Choose $n_0 > K$ and $\epsilon > 0$ such that $(x_0 - \epsilon,\ x_0 + \epsilon) \subseteq G_0 \cap \ldots \cap G_{n_0}$. Then for all $0 < |h| < \epsilon$

$$\frac{g(x_0+h)-g(x_0)}{h} \geq \frac{n_0 \cdot h}{h} = n_0 > K$$

so that $g'(x_0) = +\infty$. □

We complete now the proof of Theorem 3. Let $P = [0, 2\pi] - \{x : \Sigma\, c_n e^{inx} = 0\}$. Thus $\lambda(P) = 0$. Choose a $G_\delta$ set G of measure 0 with $P \subseteq G$. Let g be given by the preceding lemma and put $f(x) = \int_0^x g(t)dt + C$, where $C < 0$ is such that $f(2\pi) = 0$. Then f(x) is convex and $f'(x) = g(x)$. Let now F(x) on $[0, 2\pi]$ be such that $F(0) = F(2\pi) = 0$ and

$$F(x) = F_S(x) + ax + b.$$

We will show that F is identically 0 and thus $F_S$ is linear on $[0, 2\pi]$, so by Theorem 4.4 and Cantor's Thecrem $c_n = 0$ for all n.

To show $F = 0$ we will show that $F \geq 0$ and $F \leq 0$:

$F \leq 0$: For $\epsilon > 0$, $x \in [0, 2\pi]$ define $F_\epsilon(x) = F(x) - \epsilon x(2\pi - x) + \epsilon f(x)$. Then $F_\epsilon(0) < 0$, $F_\epsilon(2\pi) = 0$, so if $F \leq 0$ fails, towards a contradiction, there is $\epsilon > 0$ and $x_0 \in (0, 2\pi)$ at which $F_\epsilon$ achieves a maximum (which is positive). Then (for small enough h)

$$(*)\ 0 \geq \frac{\Delta^2 F_\epsilon(x_0, h)}{h^2} = \frac{\Delta^2 F(x_0, h)}{h^2} + 2\epsilon + \frac{\epsilon\Delta^2 f(x_0, h)}{h^2}.$$

We have two cases now

<u>Case</u> <u>(i)</u>, $x_0 \notin G$: Then $\Sigma\, c_n e^{inx_0} = 0$, so $D^2F_S(x_0) = D^2F(x_0) = 0$ and thus $\frac{\Delta^2 F(x_0, h)}{h^2} \to 0$ as $h \to 0$. But $\Delta^2 f(x_0, h) \geq 0$ (f being convex), so (*) leads immediately to a contradiction.

<u>Case</u> <u>(ii)</u>, $x_0 \in G$: Since $S_n(x_0)$ is bounded we see as in the proof of 3.4. that

$$\left| \frac{\Delta^2 F(x_0, h)}{h^2} \right| < K < \infty$$

for all small enough h. But $D^2 f(x_0) = g'(x_0) = +\infty$, so for small enough h (*) leads again to a contradiction.

$F \geq 0$: Replace above $\epsilon$ by $-\epsilon$ and repeat the argument. □

We are now ready to prove Theorem 1. In fact we will prove a stronger version due to Carlet and Debs, which subsumes both Theorem 1 and recent results of Kholshchevnikova [1].

Theorem 5 (Carlet-Debs [1]). Let $E_n$ be a sequence of $\mathcal{U}$-sets, which are relatively closed in their union $E = \bigcup_n E_n$. Then E is also a $\mathcal{U}$-set.

Proof. Assume, towards a contradiction, that $S \sim \Sigma\, c_n e^{inx}$ is a trigonometric series which converges to 0 off E but is not identically 0. By the Cantor-Lebesgue Lemma (2.6) and 3.1, $c_n \to 0$ as $|n| \to \infty$. Let

$$G = \{x : S_n(x) \text{ is unbounded}\}.$$

Then $G \subseteq E$, G is a $G_\delta$ set and $G \neq \emptyset$ by Theorem 3. Let $G_i = G \cap E_i$. Then each $G_i$ is closed in G and $G = \bigcup_i G_i$. Since G (being a $G_\delta$ subset of a Polish space) is a Polish space (see Kuratowski [1]), it follows by the Baire Category Theorem that for some open interval I and some i, $G \cap I = G_i \cap I \neq \emptyset$.

We will derive a contradiction by showing that S converges to 0 on I, hence $G \cap I = \emptyset$. To see this, let $\varphi \in C^\infty(\mathbb{T})$ be positive on I and 0 off I. By Lemma 4.2 it is enough to prove that $T = S(\varphi) \cdot S$ converges everywhere to 0. Since $E_i$ is a $\mathcal{U}$-set it is actually enough to prove that T converges to 0 off $E_i$. If $x \notin I$, T converges to 0 at x by the choice of $\varphi$ and if $x \notin E$ by the choice of S (we use again Lemma 4.2 here). So the only remaining case is when $x \in E \cap I$, but $x \notin E_i$. Since $E_i$ is closed in E, we have that $x \notin \overline{E_i}$, so there is a subinterval J of I with $x \in J$ and $\overline{J} \cap \overline{E}_i = \emptyset$. Choose a function $\psi \in C^\infty(\mathbb{T})$ with $\psi(x) = 1$ and supp($\psi$) $\subseteq \overline{J}$. By Lemma 4.2 again, the partial sums of $S(\psi) \cdot T$ are bounded off $\overline{J} \cap G =$

$\bar{J} \cap G_i = \emptyset$, i.e. everywhere. Since $S(\psi) \cdot T$ converges to 0 a.e. (because S does), it follows from Theorem 3 that $S(\psi) \cdot T = 0$, so T converges to 0 at x and we are done. □

Here is an immediate corollary of Bary's Theorem (proved earlier by Young using a different method).

Theorem 6 (W. H. Young). Let $C \subseteq \mathbb{T}$ contain no perfect set. Then C is a set of uniqueness.

Proof. If not, there is a trigonometric series $S \sim \Sigma\, c_n e^{inx}$ with $\Sigma\, c_n e^{inx} = 0$ off C but $S \neq 0$. Let $B = \mathbb{T} - \{x : \Sigma\, c_n e^{inx} = 0\} \subseteq C$. Since B is Borel and contains no perfect set if must be countable (since by a standard result in descriptive set theory (see Kuratowski [1]) all uncountable Borel sets contain perfect subsets). By Theorem 1 $B \in \mathcal{U}$ so $S = 0$, a contradiction. □

It follows from Bary's Theorem that the union of countably many $F_\sigma$ sets of uniqueness is also a set of uniqueness. A partial extension of this to $G_\delta$ sets has been proved relatively recently by Kholshchevnikova [1].

Theorem 7 (Kholshchevnikova [1]). (i) If G, H are two disjoint $G_\delta$ sets of uniqueness, then $G \cup H$ is a set of uniqueness.

(ii) If G, H are sets of uniqueness and H is both a $G_\delta$ and an $F_\sigma$, then $G \cup H$ is a set of uniqueness.

Proof. In both cases, G and H are relatively $F_\sigma$ in their union, so Theorem 5 applies. □

## §6. Four classical problems

We will discuss here some of the classical problems on sets of uniqueness. Some references for them are Bary [1], Bary [2] and Zygmund [1].

A) The Union Problem. It is easy to see that the union of two sets of uniqueness might not be necessarily a set of uniqueness. For example, split $\mathbb{T}$ into two disjoint sets P, Q neither of which contains a perfect subset (a standard result of Bernstein, using the Axiom of Choice). Then by 5.6. P, Q $\in \mathcal{U}$ but $\mathbb{T} = P \cup Q$ $\notin \mathcal{U}$.

Clearly this is not a very illuminating example, dealing as it does with the usual pathological sets. A more relevant question is whether definable sets of uniqueness are closed under unions (of two or countably many sets). Bary's Theorem of §5 handles the case of closed sets, and in Theorem 5.7 an affirmative answer was obtained for a special situation in the case of $G_\delta$ sets, but the following problem remains open.

The Union Problem. Is the union of two (resp. countably many) Borel sets of uniqueness also of uniqueness?

This is still unknown even for two $G_\delta$ sets. It seems also unknown for the union of a $G_\delta$ and a countable set.

B) The Interior Problem. Let us introduce first the following

Definition 1. A set $P \subseteq \mathbb{T}$ is called of interior uniqueness (in symbols $U^{int}$) if every closed subset of P is a set of uniqueness. (We use $U^{int}$ instead of $\mathcal{U}^{int}$ to conform with a general notation to be introduced later on).

The question is whether every set in $U^{int}$ is actually in $\mathcal{U}$ (or equivalently whether every $\mathcal{M}$-set contains a closed $\mathcal{M}$-set). Notice that it is enough to consider only sets which are $G_{\delta\sigma}$. This is because if P is in $U^{int}$ but in $\mathcal{M}$ and S is a non-0 trigonometric series with $\Sigma c_n e^{inx} = 0$ off P, then $Q = [0, 2\pi] - \{x : \Sigma c_n e^{inx} = 0\}$ in a $G_{\delta\sigma}$ subset of P which is also in $U^{int}$ and in $\mathcal{M}$. So we can state the problem as follows.

The Interior Problem. Is every $G_{\delta\sigma}$ set of interior uniqueness a set of uniqueness?

This problem is also still open even for $G_\delta$ sets. (For $F_\sigma$ sets Bary's Theorem implies clearly a positive solution). There is however a connection between this and the Union Problem. By a recent result of Kechris-Louveau and (independently) Debs-Saint Raymond (see Chapter VI. 1.4) the union of countably many $G_{\delta\sigma}$ sets of interior uniqueness is also of interior uniqueness. It follows that an affirmative answer to the Interior Problem implies a similar answer to the Union Problem for $G_{\delta\sigma}$ sets. Or, what seems more likely, a counterexample for the Union Problem for such sets provides also one for the Interior Problem.

In Chapter VIII.4 we discuss also some interesting connections between the

Union and Interior Problems and problems of so-called synthesis.

C) The Category Problem. We have seen that every measurable set of uniqueness has measure 0. Does a corresponding result hold for category? Again it is trivial to see that if we decompose $\mathbb{T}$ into two disjoint sets P, Q none of which contains a perfect set, these two sets will be sets of uniqueness which are not of the first category. So to avoid pathologies we should at least require that the set in question has the property of Baire. Since every such set is either of the first category or else contains a $G_\delta$ subset which is dense in some open interval, we can state the problem as follows.

The Category Problem. Is every $G_\delta$ set of uniqueness of the first category?

This problem has very recently been solved affirmatively by Debs and Saint Raymond [1], using methods developed in Kechris-Louveau-Woodin [1] as well as results of Piatetski-Shapiro [2] and Kechris-Louveau (given in Chapter VIII). This and actually quite a bit stronger results are discussed in Chapter VIII.

D) The Characterization Problem. This is an important but unfortunately vague problem. It asks to characterize in terms of the "structural properties" of a given set P whether it is a $\mathcal{U}$- or an $\mathcal{M}$-set. We will restrict ourselves to the most specific case, namely that of closed sets. Thus we can (vaguely again) phrase the problem as follows:

The Characterization Problem. Characterize in "structural terms" whether a given closed set E is a set of uniqueness or multiplicity.

Progress in this direction has been achieved mainly for the case of the symmetric sets of constant ratio of dissection $E_\xi$ (constructed like the Cantor set except that the two intervals that are kept have length $2\pi \cdot \xi$ instead of $2\pi \cdot \frac{1}{3}$, where $0 < \xi < \frac{1}{2}$), and some generalizations. The famous result of Salem-Zygmund (see Chapter III) asserts that $E_\xi$ is a set of uniqueness iff $\theta = \frac{1}{\xi}$ is a Pisot number (i.e. an algebraic integer of absolute value $> 1$ all of whose conjugates have absolute value $< 1$). This type of result strongly suggested that the solution of the Characterization Problem should involve not only analytic or geometric properties of a closed set, but arithmetic ones as well.

Except for extensions of the Salem-Zygmund Theorem to homogeneous perfect sets (also to be discussed in Chapter III), not much more is known in the positive direction on the Characterization Problem. No characterization has been found even for symmetric perfect sets $E_{\xi_1,\xi_2,\ \ldots}$ of varying ratios of dissection $(0 < \xi_i < \frac{1}{2})$.

Despite the imprecise nature of the Characterization Problem it is safe to assume that if one could achieve a positive solution (i.e. could find such a characterization), this would be recognized as such immediately (as in the case of the Salem-Zygmund Theorem). However recent results rule out any characterizations which are of "explicit enough" form, so that they can be expressed by countable operations involving any standard specification of a closed set like e.g. in terms of its contiguous intervals. Certainly every known characterization in special cases, like the Salem-Zygmund Theorem has indeed this form. This negative conclusion follows from the theorem of Solovay [1] and (independently) Kaufman [5], which asserts that the collection of closed sets of uniqueness is descriptively very complicated, i.e. (coanalytic but) not Borel (in the space of closed subsets of $\mathbb{T}$); see Chapter IV. Because any reasonably "explicit" characterization as above, would easily imply that the collection of closed $\mathcal{U}$-sets is Borel. This is further reinforced by the even more recent result (see Chapter VII) of Debs-Saint Raymond [1], which shows that the class of closed $\mathcal{U}$-sets does not even have a Borel basis, i.e. there is no Borel subclass of closed $\mathcal{U}$-sets which generates all of them by taking countable unions. Thus there is no way to even generate all the closed $\mathcal{U}$-sets by reasonably simply describable types of closed $\mathcal{U}$-sets, like for example the H-sets, the $H^{(n)}$-sets (see Chapter III), the sets satisfying the Piatetski-Shapiro criterion, also called rank 1 closed $\mathcal{U}$-sets (again see Chapter III), etc.

Remark. Sometimes in the Characterization Problem attention is further focused on perfect (as opposed to arbitrary closed) sets (see e.g. Bary [1]). However, all the preceding and subsequent comments and results concerning the Characterization Problem apply readily to this context as well.

# Chapter II. The Algebra A of Functions with Absolutely Convergent Fourier Series, Pseudofunctions and Pseudomeasures.

In the next few chapters we will study primarily the structure of closed sets of uniqueness, where most of the theory lives. Nevertheless we will derive through the study of closed sets important consequences for the structure of sets in more general classes like Borel and analytic. Our goal in this chapter is to introduce the "modern theory" of closed sets of uniqueness, which is based on the reformulation of this notion in terms of concepts of functional analysis. Fundamental in this work are the three spaces of pseudofunctions (PF), functions with absolutely convergent Fourier series (A), and pseudomeasures (PM).

## §1. The Spaces PF, A and PM

Let $c_0 \equiv c_0(\mathbb{Z})$ be the Banach space of (complex $\mathbb{Z}$—) sequences $\{c_n\}$ with $c_n \to 0$ as $|n| \to \infty$, equipped with the sup norm $\|\{c_n\}\|_\infty = \sup_{n\in\mathbb{Z}} |c_n|$. The dual space of $c_0$ is $\ell^1 \equiv \ell^1(\mathbb{Z})$ which consists of all sequences $\{c_n\}$ with $\sum_{n\in\mathbb{Z}} |c_n| < \infty$ and whose norm is given by $\|\{c_n\}\|_1 = \sum_{n\in\mathbb{Z}} |c_n|$. Finally the dual of $\ell^1$ is $\ell^\infty \equiv \ell^\infty(\mathbb{Z})$, the space of sequences $\{c_n\}$ with $\sup_{n\in\mathbb{Z}} |c_n| < \infty$ and the sup norm $\|\{c_n\}\|_\infty = \sup_{n\in\mathbb{Z}} |c_n|$ again. Clearly $c_0$ is a closed subspace of $\ell^\infty$. Also $c_0$, $\ell^1$ are separable (with the eventually 0 sequences from $\mathbb{Q} + i\,\mathbb{Q}$ dense), but $\ell^\infty$ is not. If $\{c_n\} \in \ell^1$, $\{d_n\} \in \ell^\infty$ then $\{d_n\}$ operates on $\{c_n\}$ as follows

$$\langle \{c_n\}, \{d_n\} \rangle = \sum c_n\, d_{-n}.$$

The same formula of course defines the operation of $\{c_n\}$ on $\{d_n\} \in c_0$.

We have also that $\ell^1$ is a Banach algebra with multiplication the operation of <u>convolution</u>

$$\{c_n\} * \{d_n\} = \{e_n\}$$

where $e_n = \sum_{m \in \mathbb{Z}} c_m\, d_{n-m}$. Notice that $1 = \{c_n\}$, where $c_0 = 1$, $c_n = 0$ if $n \neq 0$, is the unit in this algebra.

This is the point of view of standard Banach space theory. For our purposes every $\{c_n\} \in \ell^1$ can be identified with the function

$$f(x) = \sum_{n=-\infty}^{\infty} c_n e^{inx}.$$

Then $\hat{f}(n) = c_n$ are the Fourier coefficients of f, i.e. f is a continuous function with absolutely convergent Fourier series. Moreover if f, g are two such functions with $\hat{f}(n) = c_n$, $\hat{g}(n) = d_n$, then it is easy to check that if $h = fg$ (i.e. $h(x) = f(x) \cdot g(x)$) then $\hat{h}(n) = \Sigma\, c_m\, d_{n-m}$, i.e. $\widehat{fg} = \hat{f} * \hat{g}$. We denote by $A \equiv A(\mathbb{T})$ the Banach algebra of continuous functions on $\mathbb{T}$ with absolutely convergent Fourier series with norm

$$\|f\|_A \equiv \|f\|_{A(\mathbb{T})} = \sum_{n \in \mathbb{Z}} |\hat{f}(n)|.$$

The unit in this algebra is the constant function 1. Note that the <u>trigonometric polynomials</u>, i.e. the functions of the form

$$f(x) = \sum_{n=-N}^{N} c_n e^{inx}$$

are dense in A (so in particular the trigonometric polynomials with coefficients in $\mathbb{Q} + i\mathbb{Q}$ form a countable dense set in A).

Since we have identified A with $\ell^1$ we can view any $S = \{c_n\} \in \ell^\infty$ as operating on A by

$$\langle f, S \rangle = \sum \hat{f}(n)\, S(-n)$$

where $S(n) = c_n$. We define the Fourier coefficients of $S \in \ell^\infty$ by

$$\hat{S}(n) = \langle e^{-inx}, S \rangle.$$

Since for $f(x) = e^{-inx}$, $\hat{f}(m) = 1$ if $m = -n$ and $\hat{f}(m) = 0$ if $m \neq -n$, it is easy to check that $\hat{S}(n) = c_n = S(n)$, so we will just write $S(n)$ from now on.

Viewed this way (i.e. as functionals on A) the elements of $\ell^\infty$ are called in harmonic analysis <u>pseudomeasures</u> (PM). We will thus write from now on PM for $\ell^\infty$ and call this the space of pseudomeasures. We will also write

$$\|S\|_{PM} = \sup_{n \in \mathbb{Z}} |S(n)|$$

instead of $\|S\|_\infty$ for the sup norm. Furthermore the elements of $c_0(\subseteq \ell^\infty)$ are called pseudofunctions (PF) and so we will write PF for $c_0$ and call this the space of pseudofunctions. So the (Banach space) dual of PF is A and the dual of A is PM, in symbols

$$PF^* = A, \; A^* = PM.$$

Let us see now the reason for the choice of this terminology. If $f \in L^1(\mathbb{T})$ then by the Uniqueness Theorem for Fourier series the map $f \mapsto \hat{f} = \{\hat{f}(n)\}$ is one-to-one, thus f is uniquely determined by its Fourier series. Moreover by the Riemann-Lebesgue Lemma $\hat{f}(n) \to 0$ as $|n| \to \infty$, i.e. $\hat{f} \in PF$. Thus we can view any function in $L^1(\mathbb{T})$ as a pseudofunction. Note that $\|\hat{f}\|_{PM} \leq \|f\|_{L^1} = \int |f| \, d\lambda$.

Let now $C(\mathbb{T})$ be the Banach space of (complex) continuous functions on $\mathbb{T}$ with the sup norm

$$\|f\|_C \equiv \|f\|_{C(\mathbb{T})} = \sup_{x \in \mathbb{T}} |f(x)|.$$

(Note that for $f \in A$, $\|f\|_C \leq \|f\|_A$). Let also $M(\mathbb{T})$ be the dual of $C(\mathbb{T})$, i.e. the space of (complex Borel) measures on $\mathbb{T}$ with norm

$$\|\mu\|_M \equiv \|\mu\|_{M(\mathbb{T})}.$$

Recall also that $\|\mu\|_M = |\mu|(\mathbb{T})$, where $|\mu|$ is the total variation of $\mu$. If $\mu \in M(\mathbb{T})$, then $\mu$ operates on $f \in C(\mathbb{T})$ by

$$\langle f, \mu\rangle = \int f d\mu.$$

We define the Fourier coefficients of $\mu$, called also the Fourier-Stieltjes coefficients, by

$$\hat{\mu}(n) = \langle e^{-inx}, \mu\rangle = \int e^{-inx}\, d\mu.$$

Then the map $\mu \mapsto \hat{\mu}$ is again one-to-one (since the trigonometric polynomials are dense in $C(\mathbb{T})$), and as

$$|\hat{\mu}(n)| \leq \|\mu\|_M$$

it follows that $\hat{\mu} \in PM$ and $\|\hat{\mu}\|_{PM} \leq \|\mu\|_M$. Thus every measure on $\mathbf{T}$ can be viewed as a pseudomeasure. Moreover for $\mu \in M(\mathbf{T})$ and $f \in A$

$$\begin{aligned}\langle f, \hat{\mu}\rangle &= \sum \hat{f}(n)\, \hat{\mu}(-n) \\ &= \lim_{N\to\infty} \sum_{n=-N}^{N} \hat{f}(n)\, \hat{\mu}(-n) \\ &= \lim_{N\to\infty} \int \sum_{n=-N}^{N} \hat{f}(n) e^{inx} d\mu \\ &= \int f d\mu = \langle f, \mu\rangle\end{aligned}$$

so that this agrees with the way $\mu$ operates on f. Also if $f \in L^1(\mathbf{T})$ and $d\mu = f d\lambda$ is the corresponding measure (i.e. $\int g d\mu = \int g f d\lambda$), then $\hat{\mu}(n) = \int e^{-inx} d\mu = \int e^{-inx} f d\lambda = \hat{f}(n)$ and for $\varphi \in A$,

$$\langle \varphi, \hat{f}\rangle = \langle \varphi, \hat{\mu}\rangle = \int \varphi f d\lambda.$$

Let us conclude with a small (but very useful) remark. If

$$M^+(\mathbb{T}) \ (\subseteq M(\mathbb{T}))$$

denotes the class of positive measures $\mu \geq 0$ on $\mathbb{T}$ then for $\mu \in M^+(\mathbb{T})$ we have

$$\|\hat{\mu}\|_{PM} = |\hat{\mu}(0)| = \|\mu\|_M \left( = \int d\mu = \mu(\mathbb{T}) \right).$$

§2. Some basic facts about A

We will often need some sufficient condition for a function f to be in A and some estimate of $\|f\|_A$. The following is the simplest and perhaps most useful of them all (and in any case it is all we will need later).

Proposition 1. Let $f \in C(\mathbb{T})$ be absolutely continuous with $f' \in L^2(\mathbb{T})$. Then $f \in A$ and

$$\|f\|_A \leq |\hat{f}(0)| + \|f'\|_{L^2} \cdot C$$

where $C = \sqrt{2 \cdot \sum_{n=1}^{\infty} \frac{1}{n^2}}$.

Proof. This is immediate using integration by parts, the Cauchy-Schwarz inequality, and the fact that $\|g\|^2_{L^2} = \Sigma|\hat{g}(n)|^2$ for $g \in L^2(\mathbb{T})$. □

In particular any $f \in C^1(\mathbb{T})$ (the class of continuously differentiable functions) is in A.

Some particularly useful kinds of functions in A are the triangular functions $\psi_{a,h}$ and trapezoidal functions $\tau_{a,h}$.

For $a \in \mathbb{R}$ and $0 < h < \pi$, let $\psi_{a,h}$ be the $2\pi$-periodic function defined in the period $[a - \pi, a + \pi]$ as follows: It is equal to $\frac{2\pi}{h}$ at a, 0 on $[a + h, a + \pi]$, $[a - \pi, a - h]$ and linear on $[a - h, a]$ , $[a, a + h]$. Its graph then in the period $[a - \pi, a + \pi]$ is

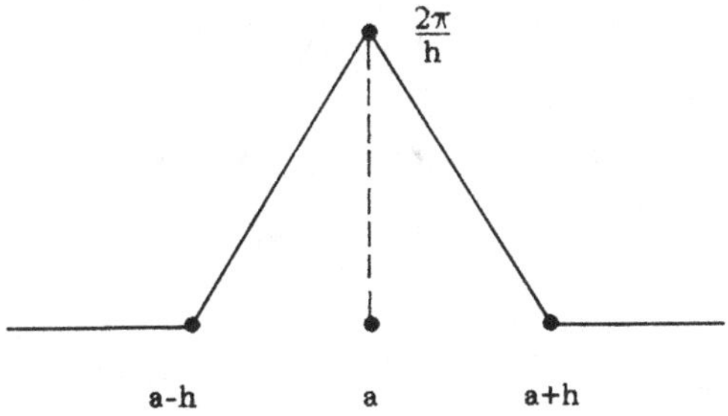


A direct calculation shows that $\psi_{a,h} \in A$ and that the Fourier series of $\psi_{a,h}$ is

$$\psi_{a,h}(x) = \sum_{n=-\infty}^{\infty} e^{-ina} \left(\frac{\sin(nh/2)}{nh/2}\right)^2 e^{inx}$$

so that

$$\|\psi_{a,h}\|_A = \|\psi_{0,h}\|_A = \psi_{0,h}(0) = \frac{2\pi}{h}.$$

(Notice here that this is the type of sum coming up in the Riemann theory. More about that soon).

Assuming also that $h < \frac{\pi}{2}$ define the function $\tau_{a,h}$ by

$$\tau_{a,h} = (4\,h\,\psi_{a,2h} - h\,\psi_{a,h}) \cdot \frac{1}{2\pi}.$$

Then the graph of $\tau_{a,h}$ is

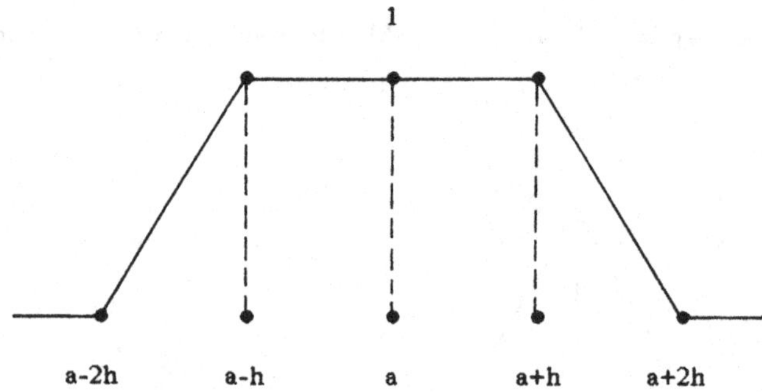


and $\|\tau_{a,h}\|_A \leq 3$.

As an illustration of the use of these functions let us prove the following lemma which we will need often later on.

Lemma 2. Suppose $f \in A$ and assume $f(a) = 0$. Then $\|\tau_{a,h} \cdot f\|_A \to 0$ as $h \to 0$.

Proof. By translation we can assume without loss of generality that $a = 0$. Put $\tau_h = \tau_{0,h}$. We have to show that $\|\tau_h f\|_A \to 0$ as $h \to 0$.

Given $\epsilon > 0$ choose N so that $\sum_{|n|>N} |\hat{f}(n)| < \epsilon$. Let $\epsilon_1 = \sum_{|n|>N} \hat{f}(n)$ and consider the trigonometric polynomial

$$Q(x) = \sum_{|n| \le N} \hat{f}(n)e^{inx} + \epsilon_1.$$

Then $\|f - Q\|_A \le |\epsilon_1| + \epsilon < 2\epsilon$, and $Q(0) = f(0) = 0$. So we can assume that f is a trigonometric polynomial to start with (recall here that $\|\tau_h\|_A \le 3$).

In this case $\tau_h f$ is absolutely continuous and (piecewise) $(\tau_h f)' = \tau_h f' + \tau_h' f$ is uniformly bounded, independently of h. So $\|(\tau_h f)'\|_{L^2} = O(\sqrt{h})$ and $\widehat{(\tau_h f)}(0) = \frac{1}{2\pi}\int_{-2h}^{2h} (\tau_h f)dx = O(h)$. This finishes the proof by Proposition 1. □

Letting in the preceding lemma $g_{a,h} = (1 - \tau_{a,h})f$ we conclude that $\|g_{a,h} - f\|_A \to 0$ as $h \to 0$. Since $g_{a,h}$ is 0 in a nbhd of a it follows that every function in A which is 0 at some point can be approximated in the norm of A by functions which are 0 in a nbhd of this point. (In later terminology this asserts that every singleton is a set of synthesis).

The fundamental fact about the algebra A is the following classical result of Wiener (see for example Rudin [1]).

Theorem 3 (Wiener). If $f \in A$ and $f(x) \neq 0$ for all $x \in \mathbb{T}$, then $1/f \in A$.

Another basic fact is that the property of belonging to A is local.

Theorem 4 (Wiener). Assume f is a function on $\mathbb{T}$ such that for each point $x \in \mathbb{T}$ there is an (open) nbhd $V_x$ and a function $f_x \in A$ with $f = f_x$ in $V_x$. Then $f \in A$.

Proof. By compactness we can cover $\mathbb{T}$ by finitely many open sets $V_1, \ldots, V_n$ such that there are functions $f_1,\ldots,f_n \in A$ with $f = f_i$ on $V_i$. By the Partition of Unity Lemma (which we prove in a moment) there are functions $h_1,\ldots,h_n \in C^\infty(\mathbb{T}) \subseteq A$ with $\mathrm{supp}(h_i) \subseteq V_i$ and $\sum_{i=1}^n h_i(x) = 1$ for all $x \in \mathbb{T}$. Then $f = \sum_{i=1}^n fh_i = \sum_{i=1}^n f_ih_i \in A$, and we are done.

<u>Lemma 5</u> (Partition of Unity). Let $K \subseteq \mathbb{T}$ be closed and $K \subseteq V_1 \cup\ldots\cup V_n$ with $V_i$ open. Then there are $h_i \in C^\infty(\mathbb{T})$ with $\mathrm{supp}(h_i) \subseteq V_i$ and $\sum_{i=1}^n h_i(x) = 1$ for all $x \in K$.

<u>Proof</u>. For each $x \in K$ let $I_x$ be an open interval around $x$ with $\overline{I_x} \subseteq V_i$ (some $i$). By compactness let $x_1, \ldots, x_m$ be such that $K \subseteq I_{x_1} \cup\ldots\cup I_{x_m}$. Let $H_i = \cup\{\overline{I}_{x_j} : \overline{I}_{x_j} \subseteq V_i\}$. Then $H_i$ is a closed subset of $V_i$, so we can find $g_i \in C^\infty(\mathbb{T})$ with $g_i(x) = 1$ on $H_i$ and $\mathrm{supp}(g_i) \subseteq V_i$. Let $h_i = (1 - g_1) \cdot (1 - g_2) \cdot \ldots \cdot (1 - g_{i-1}) \cdot g_i$. Then $\mathrm{supp}(h_i) \subseteq V_i$ and $\sum_{i=1}^n h_i = 1 - (1 - g_1)\cdots(1 - g_n)$. Since $x \in K$ implies $x \in \cup_{i=1}^n H_i$, we have $\sum_{i=1}^n h_i(x) = 1$ for $x \in K$. □

Finally we introduce the important concept of multiplication of a pseudomeasure by a function in A.

<u>Definition 6</u>. For $S \in PM$ and $f \in A$ we define $f \cdot S \in PM$ by

$$\langle g, f \cdot S\rangle = \langle gf, S\rangle.$$

Note that

$$|\langle g, f \cdot S\rangle| \leq \|g\|_A \cdot \|f\|_A \cdot \|S\|_{PM}$$

so we have

$$\|f \cdot S\|_{PM} \leq \|f\|_A \cdot \|S\|_{PM}.$$

We reconcile first this concept with that of Rajchman formal multiplication. *If* $S \sim \Sigma\, c_n e^{inx}$ *is a trigonometric series with bounded coefficients we can identify it with the pseudomeasure* $S(n) = c_n$. If now $f \in A$ then we claim that the trigonometric series $S(f) \cdot S$ is exactly the same as the pseudomeasure $f \cdot S$, i.e.

$$S(f) \cdot S = f \cdot S$$

Indeed if $T = f \cdot S$ we have

$$T(n) = \langle e^{-inx}, f \cdot S \rangle$$

$$= \langle e^{-inx} f, S \rangle$$

$$= \langle \sum_k f(k) e^{i(k-n)x}, S \rangle$$

$$= \langle \sum_m \hat{f}(n+m) e^{imx}, S \rangle$$

$$= \sum_m \hat{f}(n+m)\, S(-m),$$

$$= \sum_m \hat{f}(n-m)\, S(m).$$

It follows in particular, from Lemma I.4.1, that

$$S \in PF,\ f \in A \Rightarrow f \cdot S \in PF$$

i.e. PF <u>is closed under multiplication by functions in A</u>.

Finally observe that $1 \cdot S = S$, $f_1 \cdot (f_2 \cdot S) = (f_1 f_2) \cdot S$, and $\langle f, S\rangle = \langle 1, f \cdot S\rangle = (f \cdot S)(0)$.

§3. <u>Supports of pseudomeasures</u>

Recall that if f is a function on $\mathbf{T}$ its support is the closed set

$$\mathrm{supp}(f) = \overline{\{x \in \mathbf{T} : f(x) \neq 0\}}.$$

<u>Definition 1</u>. Let $S \in PM$ and $V \subseteq \mathbf{T}$ be open. We say that S <u>vanishes on</u> V if $\langle\varphi, S\rangle = 0$ for all $\varphi \in C^{\infty}(\mathbf{T})$ with $\mathrm{supp}(\varphi) \subseteq V$.

We have the following equivalent versions of this basic concept. This result comes from Kahane-Salem [1].

<u>Theorem 2</u>. Let $S \in PM$ and let $V \subseteq \mathbf{T}$ be open. Then the following are equivalent:

(i) S vanishes on V,

(ii) For all $f \in A$ with $\mathrm{supp}(f) \subseteq V$, $\langle f, S\rangle = 0$,

(iii) For all $f \in A$ with $\mathrm{supp}(f) \subseteq V$, $f \cdot S = 0$,

(iv) If $[a - h, a + h] \subseteq V$, then $\langle\psi_{a,h}, S\rangle = 0$ (where $\psi_{a,h}$ is the triangular function of §2),

(v) The Riemann function $F_S$ is linear in each component of V.

Proof. (i) $\Rightarrow$ (ii). Using partition of unity we can assume that $\mathrm{supp}(f) \subseteq I$, where I is an open interval contained in V. Fix $\epsilon > 0$ and let $\varphi \in C^{\infty}(\mathbf{T})$ be such that $\varphi = 1$ on supp(f) and $\mathrm{supp}(\varphi) \subseteq I$. Let then P be a trigonometric polynomial such that $\|f - P\|_A < \epsilon/\|\varphi\|_A$. Then

$$\|f - P\varphi\|_A = \|f\varphi - P\varphi\|_A \text{ (since } \varphi = 1 \text{ on supp(f))}$$

$$= \|(f - P)\varphi\|_A < \epsilon.$$

But supp $(P\varphi) \subseteq I$ and $P\varphi \in C^{\infty}(\mathbf{T})$, so $\langle P\varphi, S\rangle = 0$, therefore

$$|\langle f, S\rangle| = |\langle P\varphi, S\rangle + \langle f - P\varphi, S\rangle|$$

$$= |\langle f - P\varphi, S\rangle|$$

$$\leq \|f - P\varphi\|_A \cdot \|S\|_{PM}$$

$$< \epsilon \cdot \|S\|_{PM}.$$

Since $\epsilon$ was arbitrary $\langle f, S\rangle = 0$.

(ii) $\Rightarrow$ (iii). We have $\langle g, f \cdot S\rangle = \langle gf, S\rangle = 0$, since $\mathrm{supp}(gf) \subseteq V$. So $f \cdot S = 0$.

(iii) $\Rightarrow$ (iv). Clearly $\mathrm{supp}(\psi_{\alpha,h}) \subseteq V$, so $\psi_{\alpha,h} \cdot S = 0$ therefore $\langle\psi_{a,h}, S\rangle = \langle 1, \psi_{a,h} \cdot S\rangle = 0$, and we are done.

(iv) ⇒ (v). Recall that

$$\psi_{a,h}(x) = \sum_{n=-\infty}^{\infty} e^{-ina} \left(\frac{\sin(nh/2)}{nh/2}\right)^2 e^{inx}$$

so that (see the proof of I.3.3)

$$\frac{\Delta^2 F_S(a,h)}{h^2} = \sum \left(\frac{\sin(nh/2)}{nh/2}\right)^2 S(n)e^{ina}$$

$$= \langle \psi_{a,h}, S\rangle = 0.$$

Thus $F_S$ is linear in each component of V.

(v) ⇒ (i). Let $\varphi \in C^{\infty}(\mathbb{T})$ have support contained in V. By partition of unity we can assume that $\text{supp}(\varphi) \subseteq I$, where I is an open interval contained in V. By hypothesis $F_S(x) = ax + b$ on I.

Consider now the following functions of a complex variable

$$F^+(w) = S(0)\frac{w^2}{2} - \sum_{n=1}^{\infty} \frac{S(n)}{n^2} e^{inw}, \text{ defined for } \operatorname{Im}(w) > 0$$

$$F^-(w) = \sum_{-\infty}^{n=-1} \frac{S(n)}{n^2} e^{inw}, \text{ defined for } \operatorname{Im}(w) < 0.$$

They are clearly analytic in the upper and lower half plane respectively. Consider then the function F(w) which is equal to $F^+(w)$ on $\operatorname{Im}(w) > 0$ and equal to $F^-(w) + aw + b$ on $\operatorname{Im}(w) < 0$ or on $w \in I$. Then F is analytic on $I \cup \{w : \operatorname{Im}(w) \neq 0\}$, since F is continuous in that region and analytic on $\{w : \operatorname{Im}(w) \neq 0\}$.

Notice now that the second derivative $F''(w)$ is given by

$$G(w) = F''(w) = \sum_{n=0}^{\infty} S(n)e^{inw}, \quad \text{if } \mathrm{Im}(w) > 0$$

$$= -\sum_{-\infty}^{n=-1} S(n)e^{inw}, \quad \text{if } \mathrm{Im}(w) < 0.$$

For each $\epsilon > 0$ consider the function

$$G_\epsilon(x) = F''(x + i\epsilon) - F''(x - i\epsilon), \qquad x \in \mathbb{R}.$$

Then $G_\epsilon \in C^\infty(\mathbb{T})$ and $\hat{G}_\epsilon(n) = S(n) \cdot e^{-\epsilon|n|}$. So we have

$$<\varphi, S> = \lim_{\epsilon \to 0} <\varphi, \hat{G}_\epsilon>.$$

Now (see §1)

$$<\varphi, \hat{G}_\epsilon> = \int \varphi G_\epsilon \, d\lambda$$

so

$$<\varphi, S> = \lim_{\epsilon \to 0} \int \varphi G_\epsilon \, d\lambda$$

and letting $E = \mathrm{supp}(\varphi)$,

$$<\varphi, S> = \lim_{\epsilon \to 0} \int_E \varphi G_\epsilon \, d\lambda.$$

But $G_\epsilon(x) \to F''(x + i0) - F''(x - i0) = 0$ as $\epsilon \to 0$ and uniformly on compact subsets of $I$, in particular on $E$, so $<\varphi, S> = 0$ and we are done. □

Observe that the proof of (v) ⇒ (i) is straightforward in the case $S \in PF$, using the Riemann Localization Principle (I.4.4) and I.4.2.

Let us note now the following

Proposition 3. Let $S \in PM$. Then there is a largest open set $V \subseteq \mathbf{T}$ on which S vanishes.

Proof. First notice that if S vanishes on $V_1,\ldots,V_n$ it vanishes on $V_1 \cup\ldots\cup V_n$. This follows immediately from partition of unity. So let V be the union of all open sets on which S vanishes. Then by compactness and the previous remark S vanishes on V. □

We can give now the following

Definition 4. Let $S \in PM$. We define the support of S, in symbols supp(S), as $\mathbf{T} - V$, where V is the largest open set on which S vanishes.

Thus supp(S) is the smallest closed E with the property that $\langle\varphi, S\rangle = 0$ for every $\varphi \in C^\infty(\mathbf{T})$ (or in A) with $\text{supp}(\varphi) \cap E = \emptyset$ (i.e. which vanishes in a nbhd of E).

Note here that if $f \in C(\mathbf{T})$, then $\text{supp}(f) = \text{supp}(\hat{f})$. Because if $\varphi \in C^\infty(\mathbf{T})$ has support disjoint from supp(f), then $\langle\varphi, \hat{f}\rangle = \int \varphi f d\lambda = 0$, so $\text{supp}(\hat{f}) \subseteq \text{supp}(f)$. Let now I be an open interval which is disjoint from $\text{supp}(\hat{f})$. Then $\langle g, \hat{f}\rangle = \int gfd\lambda = 0$ for all $g \in A$ with $\text{supp}(g) \subseteq I$, thus for all $g \in C(\mathbf{T})$ with $\text{supp}(g) \subseteq I$. If f(x)

$\neq 0$ for some $x \in I$, then $|f(x)| > \epsilon > 0$ in some open interval J with $\bar{J} \subseteq I$. Let then $\varphi \in C^{\infty}(\mathbf{T})$ be such that $0 \leq \varphi \leq 1$, $\varphi = 1$ on $\bar{J}$, and $\text{supp}(\varphi) \subseteq I$. Then if $\varphi\bar{f} = g$ we have $<g, \hat{f}> = \int g f d\lambda = \int \varphi |f|^2 d\lambda > 0$, a contradiction. So $f = 0$ on I, i.e. $\text{supp}(f) \subseteq \text{supp}(\hat{f})$.

If $\mu \in M(\mathbf{T})$ then

$$\text{supp}(\mu) =^{\text{def}} \text{supp}(\hat{\mu})$$

is also the smallest closed set E such that $<f, \mu> = 0$ for all $f \in A$, or even $f \in C(\mathbf{T})$, which vanish on E itself (as opposed to vanishing in a nbhd of E). This is true since functions in $C(\mathbf{T})$ which vanish on E can be approximated in the norm of $C(\mathbf{T})$ by functions in $C^{\infty}(\mathbf{T})$ which vanish in a nbhd of E. This strong property of measures does not hold however for arbitrary pseudomeasures (see our remarks following the next theorem).

If moreover $\mu \in M^{+}(\mathbf{T})$ is a positive measure, then $\text{supp}(\mu)$ is the smallest <u>closed</u> E with $\mu(\mathbf{T} - E) = 0$, i.e $\mu(E) = \mu(\mathbf{T})$. It is easy to see finally that for any $\mu \in M(\mathbf{T})$, $\text{supp}(\mu) = \text{supp}(|\mu|)$ (since for any closed interval $\bar{I}$ disjoint from $\text{supp}(\mu)$, $\int_{\bar{I}} f d\mu = 0$ for all $f \in C(\mathbf{T})$, thus $|\mu|(\bar{I}) = 0$).

Notice also that if $f \in A$ and $S \in PM$ then

$$\text{supp}(f \cdot S) \subseteq \text{supp}(f) \cap \text{supp}(S).$$

Our final remark concerns a closure property of the set

$$PM(E)$$

which by definition is the set of pseudomeasures whose support is contained in a given closed set E. (If $S \in PM(E)$ we say that S is <u>supported</u> by E). Since PM is the Banach space dual of A we have on PM the <u>weak*</u> ($w^*$)–<u>topology</u> induced by A, i.e. the smallest topology for which the maps $S \mapsto \langle f, S\rangle$ are continuous for each $f \in A$. Then it is obvious that for each closed set $E \subseteq \mathbb{T}$, PM(E) is $w^*$-closed (i.e. closed in the weak*-topology). This is because $\text{supp}(S) \not\subseteq E$ iff there is $f \in A$, with $\text{supp}(f) \cap E = \emptyset$ and $\langle f, S\rangle \neq 0$. In particular if $S_n \xrightarrow{w^*} S$ (i.e. $S_n$ $w^*$ – converges to S) and $\text{supp}(S_n) \subseteq E$, then $\text{supp}(S) \subseteq E$ as well. [Recall that by standard functional analysis (since $PM = \ell^\infty$) $S_n \xrightarrow{w^*} S$ iff $\forall f \in A\ [\langle f, S_n\rangle \to \langle f, S\rangle]$ iff $(\sup \|S_n\|_{PM} < \infty$ and $\forall k \in \mathbb{Z}(S_n(k) \to S(k)))$]. In fact we have something a bit stronger: If $E_0 \supseteq E_1 \supseteq \ldots$ are closed sets, $E = \bigcap_n E_n$, $\text{supp}(S_n) \subseteq E_n$, and $S_n \xrightarrow{w^*} S$, then $\text{supp}(S) \subseteq E$.

Let us relate now the support of a pseudofunction with the convergence of the corresponding trigonometric series. The next result will be the key to the reformulation of the concept of uniqueness in the present framework.

<u>Theorem 5</u>. Let $S \in PF$ and $E \subseteq \mathbb{T}$ be closed. Then the following are equivalent:

(i) $\text{supp}(S) \subseteq E$,

(ii) $\Sigma\, S(n)e^{inx} = 0,\ \forall x \notin E$.

<u>Proof</u>. (i) $\Rightarrow$ (ii). By Theorem 2, the Riemann function $F_S$ is linear in each

component of $\mathbb{T} - E$, so by the Riemann Localization Principle (I.4.4) $\Sigma\, S(n)e^{inx} = 0$ off E.

(ii) ⇒ (i). By the Riemann First Lemma and Schwarz's Lemma (I.3.3 and I.3.5), $F_S$ is linear on each component of $\mathbb{T} - E$, so by Theorem 2 again S vanishes on $\mathbb{T} - E$, i.e. $\text{supp}(S) \subseteq E$. □

Remark. We have that if $S \in PM$, $f \in A$ and f vanishes in a nbhd of supp(S), then $f \cdot S = \langle f, S\rangle = 0$. One cannot replace this by just assuming that $f = 0$ on supp(S) even when $S \in PF$. This is because as we will see in Chapter VII.3 there are examples of closed sets E, which support pseudofunctions S with the property that for some f with $f = 0$ on E, $\langle f, S\rangle \neq 0$. (Then also $f \cdot S \neq 0$, since $\langle f, S\rangle = \langle 1, f \cdot S\rangle$). Notice that this shows that Lemma I.4.2 of the Rajchman multiplication theory cannot hold for any $\varphi \in A$. (Because if it did, then for any $\varphi \in A$, $S \in PF$ with $\varphi = 0$ on supp(S), the trigonometric series corresponding to $\varphi \cdot S$ (i.e. $S(\varphi) \cdot S$) would converge to 0 on supp(S) and also to 0 off supp(S) by Theorem 5, so that $\varphi \cdot S = 0$).

## §4. Description of closed $\mathcal{U}$-sets in terms of pseudofunctions

We can give now the modern formulation of the concept of closed $\mathcal{U}$-set using supports of pseudofunctions which goes back to the mid-1950's. It is more convenient to state it for $\mathcal{M}$-sets.

Theorem 1 (Piatetski-Shapiro [2], Kahane-Salem [1]). Let $E \subseteq \mathbb{T}$ be closed. Then E is a set of multiplicity iff it supports a non-0 pseudofunction.

Thus for closed sets $E \subseteq \mathbb{T}$,

$$E \in \mathcal{M} \Leftrightarrow \exists S \in PF(S \neq 0 \text{ and } \mathrm{supp}(S) \subseteq E).$$

Proof. If $E \in \mathcal{M}$ then there is $S \sim \Sigma c_n e^{inx}$, $S \neq 0$ such that $\Sigma c_n e^{inx} = 0$ off E. If $E = \mathbb{T}$ let $T = \hat{\lambda}$. Then $T \in PF$, $T \neq 0$ and clearly $\mathrm{supp}(T) \subseteq E$. Otherwise, $\lambda(E) < 1$ so by the Cantor-Lebesgue Lemma (I.3.6) $\hat{S}(n) = c_n \to 0$, i.e. $S \in PF$. By Theorem 3.5. then $\mathrm{supp}(S) \subseteq E$ and we have proved one direction of the theorem. Conversely, if E supports a non-0 pseudofunction S, then by Theorem 3.5 again $\Sigma c_n e^{inx} = 0$ off E, where $c_n = \hat{S}(n)$ and we are done. □

Remark. Actually in Piatetski-Shapiro [2] the author gives the dual of the above formulation, see V.4.1, which does not use the notion of support. The above version comes from Kahane-Salem [1].

This way of looking at closed $\mathcal{U}$- and $\mathcal{M}$-sets will be the basis for most of the further theory to be developed in the next chapters. At the moment let us illustrate its use with a couple of applications.

First let us give a quick proof of the following special case of Bary's Theorem: If $E_n$ are closed $\mathcal{U}$-sets and $E = \bigcup_n E_n$ is also closed, then E is a $\mathcal{U}$-set. (This particular case will be quite important later on). For this assume $E \in \mathcal{M}$ towards a contradiction and let $S \in PF$, $S \neq 0$ be such that $F = \mathrm{supp}(S) \subseteq E$. Let $F_n = F \cap E_n$ so that $F_n \in \mathcal{U}$ and $\bigcup_n F_n = F$. By the Baire Category Theorem there is an open interval I with $I \cap F_n = I \cap F \neq \emptyset$, for some n. Then there is $f \in A$ with $\mathrm{supp}(f) \subseteq I$ and $f \cdot S \neq 0$ (by 3.2). So $T = f \cdot S \in PF$, $T \neq$

0 and $\mathrm{supp}(T) \subseteq \mathrm{supp}(f) \cap \mathrm{supp}(S) \subseteq I \cap F = I \cap F_n \subseteq F_n$, so that $F_n \in \mathcal{M}$, a contradiction.

It is clear that $\mathcal{U}$-sets are closed under translation. A result of Marcinkiewicz and Zygmund states that $\mathcal{U}$-sets are closed under dilations and contractions. As our second application we will prove this here for closed sets. This proof comes from Kahane-Salem [1].

Theorem 2 (Marcinkiewicz-Zygmund). Let E, F be closed subsets of $[0, 2\pi]$, $t > 0$ and $F = tE$ ($= \{tx : x \in E\}$). Then $E \in \mathcal{U} \Leftrightarrow F \in \mathcal{U}$.

We will need some preliminary concepts and lemmas.

For each pseudomeasure S with $\mathrm{supp}(S) \subseteq (0, 2\pi)$ (i.e. avoiding 0) define for each real y,

$$\hat{S}(y) = \langle \psi, S \rangle$$

where $\psi \in A$ is equal to $g_y(x) = e^{-iyx}$ in a nbhd of supp(S). Note that $\langle \psi, S \rangle$ is independent of $\psi$. Also such $\psi$'s exist since if $\varphi$ is in $C^\infty(\mathbb{T})$ and is equal to 1 in a nbhd of supp(S) and 0 in a nbhd of 0, then $\psi = \varphi g_y$ is in $C^\infty(\mathbb{T})$ and so in A. If $f \in C(\mathbb{T})$ with $\mathrm{supp}(f) \subseteq (0, 2\pi)$, then $\hat{f}(y) = \int e^{-iyx} f(x) d\lambda(x) = \frac{1}{2\pi} \int_0^{2\pi} e^{-iyx} f(x) dx$. Moreover clearly $\hat{S}(y)$ agrees with our standard definition if $y = n \in \mathbb{Z}$.

Lemma 3. Let $S \in PM$, $\mathrm{supp}(S) \subseteq (0, 2\pi)$. If for $t > 0$ we define

$$S_t(n) = S(tn)$$

then $S_t \in PM$. If $S \in PF$, then $S_t \in PF$.

Proof. If $S_t \notin PM$ let, without loss of generality, $0 < n_0 < n_1 < \ldots$ be such that $|S(tn_j)| \to \infty$. Say $tn_j = m_j + \epsilon_j$ where $m_j \in \mathbb{N}$, $m_j \to \infty$ and $0 \leq \epsilon_j < 1$. We can assume again that $\epsilon_j \to \epsilon$, $0 \leq \epsilon \leq 1$. Then taking $\varphi \in C^\infty(\mathbb{T})$ with $\varphi = 1$ in a nbhd of supp(S) and 0 near 0 we have that

$$\|e^{-im_jx} \cdot e^{-i\epsilon_jx}\varphi - e^{-im_jy} \cdot e^{-i\epsilon x}\varphi\|_A$$

$$\leq \|e^{-i\epsilon_j \cdot x}\varphi - e^{-i\epsilon x}\varphi\|_A \to 0$$

(We use here the fact that $e^{-iyx}\varphi(x)$ is in $C^\infty(\mathbb{T})$ and 2.1). So $S(tn_j) = <e^{-im_jx} e^{-i\epsilon_jx}\varphi, S> = <e^{-im_jx} \cdot e^{-i\epsilon x}\varphi, S> + \delta_j$, where $\delta_j \to 0$, and $|<e^{-im_jx} \cdot e^{-i\epsilon x}\varphi, S>| = |<e^{-im_jx}, (e^{-i\epsilon x}\varphi) \cdot S>| = |T(m_j)| \to \infty$, where $T = (e^{-i\epsilon x}\varphi) \cdot S$ is a pseudomeasure, a contradiction.

If now $S \in PF$ but $S_t \notin PF$, then we can find, without loss of generality, $0 < n_0 < n_1 < \ldots$ and $\epsilon > 0$ such that $|S(tn_j)| \geq \epsilon$ and this leads to a contradiction as before. □

Lemma 4. Let $E \subseteq (0, 2\pi)$ be closed and let $t > 0$ be such that $tE = \{tx : x \in E\} \subseteq (0, 2\pi)$. If $S \in PM$ has support contained in E, then $S_t$ has support contained in tE. In particular if E is an $\mathcal{M}$-set, so is tE.

Proof. Choose h small so that $E_{2h} = \{x : \text{dist}(x, E) \leq 2h\} \subseteq (0, 2\pi)$ and $tE_{2h} \subseteq (0, 2\pi)$. Let

$$S^h(x) = \frac{\Delta^2 F_S(x, 2h)}{4h^2} = \sum S(n)\, u(nh) e^{inx}$$

where $u(x) = \left(\frac{\sin x}{x}\right)^2$. Then $S^h \in A$ and $\widehat{S^h}(n) = S(n) \cdot u(nh)$. Since $S^h(x) = 0$ for $x \notin E_{2h}$ (see the proof of (iv) $\Rightarrow$ (v) in 3.2) we have $\text{supp}(S^h) \subseteq E_{2h}$.

Now for each real y,

$$\widehat{S^h}(y) \to S(y), \text{ as } h \to 0.$$

This is because if $\varphi$ is real, $\varphi \in A$, $\varphi = 1$ in a nbhd of $E_{2h}$ and $\varphi = 0$ near 0 and $e^{-iyx}\varphi = \Sigma\, a_n e^{inx}$, we have

$$\widehat{S^h}(y) = \sum \widehat{S^h}(n) \cdot a_{-n} = \sum S(n) \cdot a_{-n} \cdot u(nh).$$

But $S(y) = \langle e^{-iyx}\varphi, S\rangle = \Sigma\, a_{-n} \cdot S(n)$. Thus by Lemma I.3.4 $\widehat{S^h}(y) \to S(y)$, as $h \to 0$.

The function

$$S_t^h = S^h(x/t), \text{ on } tE_{2h},$$
$$= 0 \quad , \text{ elsewhere}$$

has support contained in $tE_{2h}$ and

$$\widehat{S_t^h}(y) = t\widehat{S^h}(ty).$$

So $\widehat{S_t^h}(y) \to tS(ty)$ as $h \to 0$, and in particular $\widehat{S_t^h}(n) \to tS(tn)$ as $h \to 0$. Clearly $\widehat{S_t^h} \in PF$ and

$$\begin{aligned}|\widehat{S_t^h}(n)| &= |t\widehat{S^h}(tn)| \\ &= |t \cdot \langle e^{-itnx}\varphi, \widehat{S^h}\rangle|\end{aligned}$$

where $\varphi = 1$ in a nbhd of $E_{2h}$ and $\varphi = 0$ near 0. But for any $y > 0$

$$\begin{aligned}|\langle e^{-iyx}\varphi, \widehat{S^h}\rangle| &\le \|e^{-iyx}\varphi\|_A \cdot \|\widehat{S^h}\|_{PM} \\ &\le \|e^{-iyx}\varphi\|_A \cdot \|S\|_{PM}\end{aligned}$$

(since $|\widehat{S^h}(n)| \le |S(n)|$). Now we can write $y = n + \epsilon$ with $n \in \mathbf{Z}$ and $0 \le \epsilon < 1$ so that

$$\begin{aligned}\|e^{-iyx}\varphi\|_A &= \|e^{-inx} \cdot e^{-i\epsilon x}\varphi\|_A \\ &\le \|e^{-i\epsilon x}\varphi\|_A.\end{aligned}$$

By 2.1.

$$\begin{aligned}\|e^{-i\epsilon x}\varphi\|_A &\le |\widehat{e^{-i\epsilon x}\varphi}(0)| + C \cdot \|(e^{-i\epsilon x}\varphi)'\|_{L^2} \\ &\le K\end{aligned}$$

where K is a constant independent of $\epsilon$, but depending on $\varphi$. Since we can use the same $\varphi$ for all small enough h, we have for such h

$$|\langle e^{-iyx}\varphi, \widehat{S^h}\rangle| \leq M \cdot \|S\|_{PM}$$

where M is independent of y, h, i.e. $\|\widehat{S_t^h}\|_{PM}$ is bounded independently of h.

Let now $T = tS_t$ (i.e. $T(k) = tS_t(k)$). If $h_n \to 0$, then $\widehat{S_t^{h_n}}(k) \to T(k)$ and $\sup\|\widehat{S_t^{h_n}}\|_{PM} < \infty$, so $\widehat{S_t^{h_n}} \xrightarrow{w*} T$. But $\mathrm{supp}(\widehat{S_t^{h_n}}) = \mathrm{supp}(S_t^{h_n}) \subseteq tE_{2h_n}$, so $\mathrm{supp}(T) = \mathrm{supp}(S_t) \subseteq tE$.

Finally if E is an $\mathcal{M}$-set let $S \neq 0$ be a pseudofunction supported by E. Let $S(n_0) \neq 0$. Then if $S' = e^{-in_0x} \cdot S$, $S' \in PF$, $S'$ is also supported by E and $S'(0) = S(n_0) \neq 0$. Then $S'_t \in PF$, $S'_t$ is supported by tE and $S'_t \neq 0$, since $S'_t(0) = S'(0) \neq 0$. □

We can now complete the

<u>Proof of Theorem 2</u>. Say without loss of generality $0 < t < 1$. If $\mathrm{diam}(E) < 2\pi$ then for some small $\delta \in \mathbb{R}$, $\delta + E$ $(= \{\delta + x : x \in E\}) \subseteq (0, 2\pi)$ and $t\delta + tE \subseteq (0, 2\pi)$. Since $\mathcal{U}$-sets are closed under translation we are done by Lemma 4. If $\mathrm{diam}(E) = 2\pi$, split E into $E_1 = [0, \pi], \cap E$, $E_2 = [\pi, 2\pi] \cap E$. □

## §5. Rajchman measures and extended uniqueness sets

By 4.1 in order to show that a closed set E is an $\mathcal{M}$-set we have to produce a non-0 pseudofunction supported by E. The most common way of doing this is to actually construct a measure with this property.

Definition 1. A measure $\mu \in M(\mathbb{T})$ which is a pseudofunction (i.e. $\hat{\mu}(n) \to 0$ as $|n| \to \infty$) is called a Rajchman measure.

Closed sets which support non-0 Rajchman measures are said to be of restricted multiplicity. Most of the usual $\mathcal{M}$-sets turn out to be of restricted multiplicity and for quite a while it was speculated that this might always be the case. This was disproved in 1954 by Piatetski-Shapiro, who exhibited closed sets of multiplicity which support no non-0 Rajchman measure (see Chapter VII.3). Closed sets which are not of restricted multiplicity, i.e. support no non-0 Rajchman measure, are said to be of extended uniqueness. The study of their structure will be of central importance in this book. In fact we can define the concept of extended uniqueness for arbitrary sets in $\mathbb{T}$ as follows.

Definition 2. A set $P \subseteq \mathbb{T}$ is called a set of extended uniqueness if for every positive Rajchman measure $\mu$ we have $\mu(P) = 0$. Otherwise P is called a set of restricted multiplicity.

In particular $\lambda(P) = 0$ for all extended uniqueness P, as the Lebesgue measure $\lambda$ is a Rajchman measure.

We denote by $\mathcal{U}_0$ the class of sets of extended uniqueness and by $\mathcal{M}_0$ the class of sets of restricted multiplicity.

In a moment we will reconcile this definition for the case of closed sets with the one we gave earlier. But first let us note the following, where as usual a set P $\subseteq \mathbb{T}$ is called <u>universally measurable</u> if it is measurable for all positive Borel measures on $\mathbb{T}$. (In particular of course, all Borel sets are universally measurable).

<u>Proposition 3</u>. Every universally measurable set of uniqueness is of extended uniqueness, i.e. $\mathcal{U} \subseteq \mathcal{U}_0$ for universally measurable sets.

<u>Proof</u>. If $P \in \mathcal{U}$ but $P \in \mathcal{M}_0$ let $\mu \geq 0$ be a Rajchman measure, $\mu \neq 0$, with $\mu(P) > 0$. By regularity let $F \subseteq P$ be closed with $\mu(F) > 0$ and let $\nu = \mu|F$ (i.e. $\nu(B) = \mu(B \cap F)$). Then $\nu \neq 0$ and $\text{supp}(\nu) \subseteq F$, so it is enough to show that $\nu$ is a Rajchman measure. This is immediate from the following important closure property of Rajchman measures.

Recall that for measures $\rho$, $\sigma$ in $M(\mathbb{T})$ we define

$$\rho \ll \sigma \text{ iff } \rho \ll |\sigma|$$
$$\text{iff } \rho(B) = 0 \text{ for all Borel sets } B \text{ with } |\sigma|(B) = 0.$$

If $\rho \ll \sigma$ we say that $\rho$ is <u>absolutely continuous with respect to</u> $\sigma$. In particular, note that $|\sigma| \ll \sigma$ for all $\sigma \in M(\mathbb{T})$. By the Radon-Nikodym Theorem we have that

$$\rho \ll \sigma \text{ iff } d\rho = fd\sigma \text{ for some } f \in L^1(\mathbb{T}, |\sigma|) \ (\equiv L^1(|\sigma|))$$

<u>Lemma 4</u> (Milicer-Gruževska). Let $\sigma$ be a Rajchman measure and $\rho \ll \sigma$. Then $\rho$ is also a Rajchman measure. In particular, if $\sigma$ is a Rajchman measure, so is $|\sigma|$.

Proof. We have $d\rho = fd\sigma$ for some $f \in L^1(|\sigma|)$. Since the trigonometric polynomials are dense in $L^1(|\sigma|)$ we can find for each $\epsilon > 0$ $P(x) = \sum_{n=-N}^{N} c_n e^{inx}$ such that $\|f - P\|_{L^1(|\sigma|)} < \epsilon$. Let $d\rho_1 = Pd\sigma$. Then $\hat{\rho}_1(n) = \Sigma\, c_k\hat{\sigma}(n - k) \to 0$ as $|n| \to \infty$, so $\rho_1$ is a Rajchman measure. But $|\hat{\rho}(n) - \hat{\rho}_1(n)| \leq \|f - P\|_{L^1(|\sigma|)} < \epsilon$. So $\overline{\lim}\, |\hat{\rho}(n)| \leq \epsilon$, and thus $\hat{\rho}(n) \to 0$ as $|n| \to \infty$. □

Going back to our main proof now, we have clearly that $\nu \ll \mu$, so $\nu$ is a Rajchman measure and we are done. □

The following is obvious from the regularity of measures but worth keeping always in mind.

Proposition 5. A universally measurable set $P \subseteq \mathbb{T}$ is of extended uniqueness iff every closed subset of P is of extended uniqueness.

Thus there is no point in defining "interior extended uniqueness sets" (recall I.6.1).

Finally we provide a number of equivalent reformulations of the concept of extended uniqueness (or for convenience restricted multiplicity) for closed sets.

Proposition 6. Let $E \subseteq \mathbb{T}$ be a closed set. Then the following are equivalent:

(i) $E \in \mathcal{M}_0$,

(ii) There is a Rajchman measure $\mu$ with $\mu(E) \neq 0$,

(iii) E supports a non-0 positive Rajchman measure (i.e. there is such a $\mu$ with supp($\mu$) $\subseteq$ E),

(iv) E supports a non-0 Rajchman measure,

(v) There is a positive measure $\mu \neq 0$, with $\Sigma\ \hat{\mu}(n)e^{inx} = 0$ off E,

(vi) There is measure $\mu \neq 0$ with $\Sigma\ \hat{\mu}(n)e^{inx} = 0$ off E.

Proof. The implication (i) $\Rightarrow$ (ii) is obvious and (ii) $\Rightarrow$ (i) is immediate from the standard fact of measure theory which states that every measure $\mu$ is a linear combination of positive measures which are absolutely continuous with respect to $\mu$. That (iii) $\Leftrightarrow$ (v) and (iv) $\Leftrightarrow$ (vi) is immediate from 3.5. Also (i) $\Leftrightarrow$ (iii) follows as in Proposition 1. Since (iii) $\Rightarrow$ (iv) is obvious, it remains only to prove (iv) $\Rightarrow$ (iii): Say $\mu \neq 0$ is a Rajchman measure with supp($\mu$) $\subseteq$ E. Then supp ($|\mu|$) = supp($\mu$) $\subseteq$ E. Since $|\mu|$ is a Rajchman measure we are done. □

Remark. It can be also shown that these conditions are equivalent for arbitrary Borel sets E. That (i) $\Leftrightarrow$ (ii), (i) $\Leftrightarrow$ (iii), (iii) $\Rightarrow$ (v), (iii) $\Rightarrow$ (iv), (iv) $\Rightarrow$ (vi), (v) $\Rightarrow$ (vi) is exactly as before. Finally for the proof of (vi) $\Rightarrow$ (i) note that if $\lambda(E) = 1$ then certainly $E \in \mathcal{M}_0$, so we can assume that $\lambda(E) < 1$. Then by the Cantor-Lebesgue Lemma $\mu$ is a Rajchman measure. One can conclude now that $|\mu|$ (E) $> 0$. We will not however prove this here since we will not need it. (The proof can be found in Zygmund [1, Vol. II, p. 160; proof of 19i]). An alternative proof of (vi) $\Rightarrow$ (i) will be given in Chapter VIII, see remarks after Theorem VIII.3.4.

# Chapter III. Symmetric Perfect Sets and the Salem-Zygmund Theorem

Our main goal in this chapter is to prove the Salem-Zygmund Theorem which characterizes constant dissection symmetric (and more generally homogeneous) perfect sets of uniqueness. We discuss also an important class of $\mathcal{U}$-sets, the so-called $H^{(n)}$-sets.

## §1. $\underline{H}^{(n)}$-sets

We introduce here a generalization of the concept of H-set due to Piatetski-Shapiro.

Definition 1. A hyperplane in $\mathbf{Z}^n$ ($n = 1, 2, \ldots$) is the set of solutions (in $\mathbf{Z}^n$) of a linear equation of the form

$$\vec{m} \cdot \vec{x} = m_1 x_1 + \ldots + m_n x_n = p$$

where $\vec{m} \in \mathbf{Z}^n$, $\vec{m} \neq \vec{0}$ and $p \in \mathbf{Z}$.

Let $\mathfrak{I}_n$ be the ideal of subsets of $\mathbf{Z}^n$ generated by the hyperplanes, i.e. for $X \subseteq \mathbf{Z}^n$:

$$X \in \mathfrak{I}_n \Leftrightarrow X \text{ is contained in a finite union of hyperplanes in } \mathbf{Z}^n.$$

Note that for $n = 1$, $X \in \mathcal{I}_1 \Leftrightarrow X$ is finite.

Definition 2. A set $E \subseteq \mathbb{T}$ is an $H^{(n)}$-set $(n = 1, 2, \ldots)$ if there exists a nonempty open parallelepiped $I = \prod_{i=1}^{n} I_i$ in $\mathbb{T}^n$ (i.e. each $I_i$ is a nonempty open interval of $\mathbb{T}$) such that

$$\{\vec{m} \in \mathbb{Z}^n : \vec{m}E \cap I = \emptyset\} \notin \mathcal{I}_n$$

where (viewing E as a subset of $[0, 2\pi)$)

$$\vec{m}E = \{(m_1 x \ (\mathrm{mod}\ 2\pi), \ldots, m_n x \ (\mathrm{mod}\ 2\pi)) : x \in E\}.$$

(This definition is somewhat different but easily equivalent to the standard one, see e.g. Kahane-Salem [1]). Clearly $H = H^{(1)}$ and $H^{(1)} \subseteq H^{(2)} \subseteq \ldots$. Piatetski-Shapiro (see Piatetski-Shapiro [1] or Bary [2, Vol. II, p. 382]) has shown that there are closed sets in $H^{(n+1)}$ which are not even countable unions of $H^{(n)}$ sets. It is easy to check that if $E \in H^{(m)}$ and $F \in H^{(n)}$ then $E \cup F \in H^{(m+n)}$. Also a set $E \subseteq \mathbb{T}$ is an $H^{(n)}$-set iff $\bar{E}$ is an $H^{(n)}$-set.

We will verify now that every $H^{(n)}$-set is a $\mathcal{U}$-set. In order to do that let us first make explicit a sufficient criterion for a closed set to be in $\mathcal{U}$, which was implicit in the proof that H-sets are in $\mathcal{U}$ (I.4.6).

Since A as a Banach space (i.e. $\ell^1$) is the dual of PF ($= c_0$), we have on A the weak$^*$-topology generated by the maps $f \mapsto \langle f, S\rangle$ for $S \in$ PF. Again by standard functional analysis

$$f_n \xrightarrow{w^*} f \text{ iff } \forall S \in PF(<f_n, S> \to <f, S>)$$

$$\text{iff } (\sup \|f_n\|_A < \infty \text{ and } \hat{f}_n(k) \to \hat{f}(k), \forall k \in \mathbf{Z}).$$

We have now

Theorem 3 (Piatetski-Shapiro [2]). Let $E \subseteq \mathbf{T}$ be closed. If there is a sequence $f_n \in A$ with $\text{supp}(f_n) \cap E = \varnothing$ and $f_n \xrightarrow{w^*} 1$ (i.e. $\sup\|f_n\|_A < \infty$ and $\hat{f}_n(0) \to 1$, $\hat{f}_n(k) \to 0$ if $k \neq 0$), then $E \in \mathcal{U}$.

Proof. If not then let $S \in PF$, $S \neq 0$ have support contained in E. They by II.3.2 $<f_n, S> = 0$, so $<1, S> = \lim <f_n, S> = 0$. Thus $S(0) = 0$. Similarly using $e^{-inx}$ S instead of S we conclude that $S(n) = 0$ for all n, i.e. $S = 0$ a contradiction. □

The class of closed $\mathcal{U}$-sets which satisfy this criterion will play an important role in the sequel. (They are referred to as closed $\mathcal{U}$-sets of rank 1–see Chapter V.4).

Using Theorem 3 we can prove now

Theorem 4 (Piatetski-Shapiro [1]). Every $H^{(n)}$-set is a set of uniqueness.

Proof. It is enough to deal with closed sets. So let E be a closed $H^{(n)}$-set, let $I = \prod_{i=1}^n I_i$, where each $I_i$ is a non-empty open interval in $\mathbf{T}$, and let $\{\vec{m}^k\}$ be an infinite sequence in $\mathbf{Z}^n$ such that $\vec{m}^k E \cap I = \varnothing$ and

$$\vec{m}^k \text{ is not in the first k hyperplanes of } \mathbf{Z}^n$$

in some fixed enumeration of these hyperplanes.

Let $\varphi_i \in A$ have support contained in $I_i$ and $\hat{\varphi}_i(0) = 1$. Put $M = \prod_{i=1}^n \|\varphi_i\|_A$. Finally define

$$\psi_k(x) = \prod_{i=1}^n \varphi_i(m_i^k x)$$

where $\vec{m}^k = (m_1^k, \ldots, m_n^k)$. We will show that these $\psi_k$ satisfy the criterion of Theorem 3.

Since $\vec{m}^k E \cap I = \emptyset$ it follows that $\mathrm{supp}(\psi_k) \cap E = \emptyset$. Also $\|\psi_k\|_A \leq \prod_{i=1}^n \|\varphi_i\|_A = M$.

Finally consider

$$\hat{\psi}_k(q) = \sum_{\vec{m}^k \cdot \vec{p} = q} \prod_{i=1}^n \hat{\varphi}_i\,(p_i)$$

where $\vec{p} = (p_1, \ldots, p_n)$. For each N separate this sum into two pieces,

$$\sum\nolimits_1 = \sum_{\substack{\vec{m}^k \cdot \vec{p} = q \\ \forall i(|p_i| \leq N)}} \prod_{i=1}^n \hat{\varphi}_i(p_i)$$

and

$$\sum\nolimits_2 = \hat{\psi}_k(q) - \sum\nolimits_1.$$

Choose N so that $\sum_{|m|>N} |\hat{\varphi}_i(m)| \leq \epsilon$, for all $i = 1, \ldots, n$. Then clearly

$$\left|\sum\nolimits_2\right| \le n \cdot M^{n-1} \cdot \epsilon.$$

Choose now k so large that $\vec{m}_k, \vec{m}_{k+1}, \ldots$ are not in the hyperplanes $\vec{p} \cdot \vec{x} = q$ for any $\vec{p} \neq \vec{0}$ such that $\forall i(|p_i| \le N)$. (There are only finitely many of these). If $q \neq 0$, $\Sigma_1$ is the empty sum so $\Sigma_1 = 0$ and thus $|\hat{\psi}_k(q)| \le n \cdot M^{n-1} \cdot \epsilon$ for all large enough k. If $q = 0$, the only summand of $\Sigma_1$ comes from $\vec{p} = \vec{0}$, i.e.

$$\sum\nolimits_1 = \prod_{i=1}^{n} \hat{\varphi}_i(0) = 1$$

so $|\hat{\psi}_k(0) - 1| \le n \cdot M^{n-1} \cdot \epsilon$ for all large enough k and we are done. □

§2. Pisot numbers

Definition 1. An algebraic number $\theta$ is called an algebraic integer if $\theta$ is the root of some polynomial $P(x) \in \mathbb{Z}[x]$ with leading coefficient 1. An algebraic integer $\theta$ is called a Pisot number if $\theta > 1$ and all of its conjugates have absolute value $< 1$.

Examples. 2, 3, 4, ...; $\frac{1+\sqrt{5}}{2}$.

Some Results. (i) (Pisot) In any simple real algebraic extension of $\mathbb{Q}$ there is a Pisot number having the same degree as the field.

(ii) (Salem) The set of Pisot number is closed.

(For the proofs of these facts, that we will not use, see Salem [1], Kahane-Salem [1]).

Definition 2. For $x \in \mathbb{R}$ let

$$\{x\} = \text{dist}(x, \mathbb{Z})$$
$$= \text{distance of } x \text{ to the closest integer.}$$

Thus $0 \leq \{x\} \leq \frac{1}{2}$.

A Pisot number is loosely speaking a number whose powers are becoming "closer and closer" to integers. This is made precise by the following result in number theory.

Theorem 3 (Pisot). (i) Let $\theta$ be a Pisot number, $\gamma$ an algebraic integer in $\mathbb{Q}(\theta)$. Then $\{\gamma\theta^n\}$ decreases to 0 geometrically, so in particular $\sum_{n=0}^{\infty} \{\gamma\theta^n\} < \infty$.

(ii) Let $\gamma$, $\theta$ be reals such that $\gamma \neq 0$, $\theta > 1$ and $\sum_{n=0}^{\infty} \{\gamma\theta^n\}^2 < \infty$. Then $\theta$ is a Pisot number and $\gamma \in \mathbb{Q}(\theta)$.

We will only prove (i) and a special case of (ii). For the proof of the full Salem-Zygmund Theorem one needs (ii), but the application of this theorem in Chapter 4 needs only the special case. (The proof of (ii) can be found in Salem [1], Kahane-Salem [1]).

Proof of (i). Let n be the degree of the minimal polynomial P of $\theta = \theta^{(0)}$ and let $\theta^{(1)}, \ldots, \theta^{(n-1)}$ be the conjugates of $\theta$. Let $R(x) \in \mathbb{Q}[x]$ be such that $\gamma = R(\theta)$ and let $\gamma^{(i)} = R(\theta^{(i)})$. Then for each $p \in \mathbb{N}$,

$$\sum_{i=0}^{n-1} \gamma^{(i)} (\theta^{(i)})^p = \sum_{i=0}^{n-1} R(\theta^{(i)})^p (\theta^{(i)})^p$$

is a rational symmetric polynomial in the roots of P, so it is a rational number. But it is also an algebraic integer, since $\gamma^{(i)}$, $\theta^{(i)}$ are, so it is an integer. Therefore

$$\{\gamma\theta^p\} = \left\{\sum_{i=1}^{n-1} \gamma^{(i)} (\theta^{(i)})^p\right\}.$$

Since $|\theta^{(i)}| < 1$ for $i = 1, \ldots, n-1$, we have $\sum_{i=1}^{n-1} \gamma^{(i)} (\theta^{(i)})^p \to 0$ as $p \to \infty$, and so for p big enough,

$$\left\{\sum_{i=1}^{n-1} \gamma^{(i)}(\theta^{(i)})^p\right\} = \left|\sum_{i=1}^{n-1} \gamma^{(i)}(\theta^{(i)})^p\right|$$

$$\leq (n-1) \cdot \sup_{1\leq i\leq n-1} |\gamma^{(i)}| \cdot [\sup_{1\leq i\leq n-1} |\theta^{(i)}|]^p. \qquad \square$$

The following is a special case of 3.(ii) (with a weaker hypothesis).

<u>Proposition 4</u>. Let $\theta \in \mathbb{N}$, $\theta \geq 2$, $\gamma \in \mathbb{R}$. If $\overline{\lim_p} \{\gamma\theta^p\} < \frac{1}{2\theta}$, then $\gamma = \frac{n}{\theta^p}$ for some $n \in \mathbb{Z}$, $p \in \mathbb{N}$.

<u>Proof</u>. Assume the conclusion fails. Thus $\{\gamma\theta^p\} > 0$ for all p. Let $k_0$ be such that

$$k \geq k_0 \Rightarrow 0 < \{\gamma\theta^k\} < \frac{1}{2\theta}.$$

Then for some $\ell \geq 1$

$$\frac{1}{2\theta^{\ell+1}} \leq \{\gamma\theta^{k_0}\} < \frac{1}{2\theta^{\ell}}.$$

Let $n_{k_0}$ be the integer closest to $\gamma\theta^{k_0}$, so that

$$\{\gamma\theta^{k_0}\} = |\gamma\theta^{k_0} - n_{k_0}|.$$

Then

$$\frac{1}{2\theta^{\ell+1}} \le |\gamma\theta^{k_0} - n_{k_0}| < \frac{1}{2\theta^{\ell}}$$

so

$$\frac{1}{2\theta} \le |\gamma\theta^{k_0+\ell} - n_{k_0}\theta^{\ell}| < \frac{1}{2}$$

i.e.

$$\frac{1}{2\theta} \le \{\gamma\theta^{k_0+\ell}\}$$

a contradiction. □

Remark. It is not known if $\{\theta^p\} \to 0$ implies $\theta$ is Pisot. It is known that if $\theta > 1$ is an algebraic number and $\{\gamma\theta^p\} \to 0$ for some real $\gamma \neq 0$, then $\theta$ is Pisot and $\gamma \in \mathbb{Q}(\theta)$ (see Salem [1]).

## §3. Symmetric and homogeneous perfect sets

Let $\xi_1, \xi_2, \ldots$ be a sequence of numbers with $0 < \xi_i < \frac{1}{2}$. We construct the symmetric perfect set with dissection ratios $\xi_1, \xi_2, \ldots$, in symbols $E_{\xi_1,\xi_2,\ldots}$, as follows:

For each interval $[a, b]$ and number $0 < \xi < \frac{1}{2}$ consider the middle open interval $(a + \xi\ell, b - \xi\ell)$, where $\ell = b - a$ and let $E = [a, a + \xi\ell] \cup [b - \xi\ell, b]$ be the closed set remaining after removing this middle interval

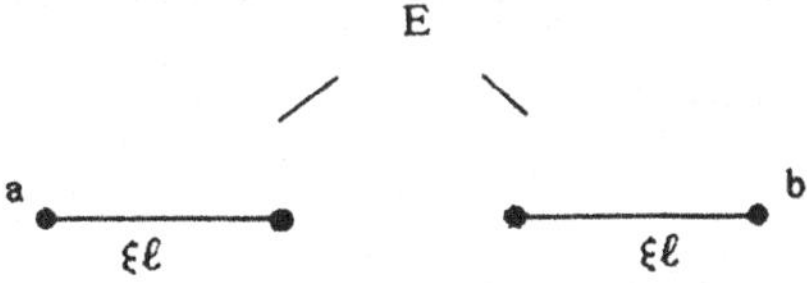

We say that E results from a dissection of [a, b] with ratio $\xi$. Now starting from the interval $[0, 2\pi]$ define the closed sets $E_1 \supseteq E_2 \supseteq \dots \supseteq E_k \supseteq \dots$ contained in $[0, 2\pi]$, where each $E_k$ is made up of $2^k$ closed disjoint intervals, as follows: $E_1$ results from $[0, 2\pi]$ by a dissection with ratio $\xi_1$, and $E_{k+1}$ results from $E_k$ after a dissection of each interval of $E_k$ with ratio $\xi_{k+1}$. Put

$$E_{\xi_1, \xi_2, \dots} = \cap_{k=1}^{\infty} E_k.$$

Then $E_{\xi_1, \xi_2, \dots}$ is a perfect nowhere dense set whose complement has as components the open intervals removed at each stage of the construction. Since the (normalized) measure of $E_k$ is $2^k \xi_1 \cdots \xi_k$, it follows that $\lambda(E) = 0$ iff $2^k \xi_1 \cdots \xi_k \to 0$. It is easy to check that the $2^k$ origins of the intervals that make up $E_k$ are given by the expression

$$2\pi[\epsilon_1(1 - \xi_1) + \epsilon_2\, \xi_1(1 - \xi_2) + \dots + \epsilon_k\, \xi_1 \cdots \xi_{k-1}\, (1 - \xi_k)]$$

where $\epsilon_i = 0$ or 1, and so

$$E_{\xi_1, \xi_2, \dots} = \Big\{2\pi \sum_{i=1}^{\infty} \epsilon_i\, \xi_1 \dots \xi_{i-1}\, (1 - \xi_i) : \epsilon_i = 0 \text{ or } 1\Big\}.$$

This gives a canonical homeomorphism of $2^{\mathbb{N}}$ $(= \{0, 1\}^{\mathbb{N}})$ onto $E_{\xi_1, \xi_2, \dots}$.

If all the $\xi_i$'s are equal, say to $0 < \xi < \frac{1}{2}$, then we obtain the <u>symmetric perfect set of constant ratio of dissection</u> $\xi$, in symbols $E_\xi$. For example $E_{1/3}$ is the classical Cantor set (on $[0, 2\pi]$).

We can generalize this as follows. Let $\eta_0 = 0, \eta_1, \ldots, \eta_k, \eta_{k+1} = 1$ be an increasing sequence and let $\xi = 1 - \eta_k$. Assume moreover that for $i < k$, $\xi < \eta_{i+1} - \eta_i$. We construct now the homogeneous perfect set associated to $(\xi; \eta_1 \ldots \eta_k)$, in symbols $E(\xi; \eta_1 \ldots \eta_k)$:

For each interval $[a, b]$, with $\ell = b - a$, consider the disjoint closed intervals $[a + \ell\eta_i, a + \ell\eta_i + \ell\xi]$, $i = 0 \ldots k$, and let E be their union. We say that

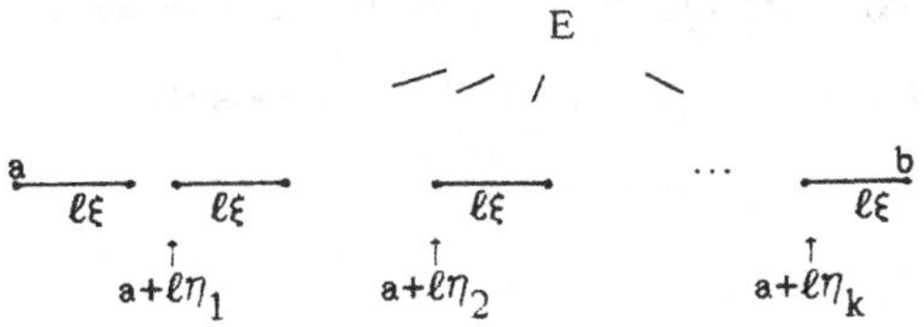

E

a $\ell\xi$ $\ell\xi$ $\ell\xi$ … $\ell\xi$ b

$a+\ell\eta_1$ $a+\ell\eta_2$ $a+\ell\eta_k$

E results from $[a, b]$ by a dissection of type $(\xi; \eta_1 \ldots \eta_k)$. Now starting from $[0, 2\pi]$ define closed sets $E_1 \supseteq E_2 \supseteq \ldots \supseteq E_n \supseteq \ldots$ by performing as before to each interval of $E_n$ a dissection of type $(\xi; \eta_1 \ldots \eta_k)$ to get $E_{n+1}$ (this is the same type for each n) and finally letting

$$E(\xi; \eta_1 \ldots \eta_k) = \cap_n E_n.$$

Again each $E_n$ is made up of $(k + 1)^n$ intervals of length $\xi^n$ so that, since $(k + 1)\xi < 1$, we have $(k + 1)^n \xi^n \to 0$ and $\lambda(E(\xi;\eta_1 \ldots \eta_k)) = 0$. Also the points of $E(\xi;\eta_1 \ldots \eta_k)$ are given by the formula

$$x = 2\pi \left(\sum_{i=0}^{\infty} \epsilon_i \xi^i\right)$$

where $\epsilon_i \in \{0, \eta_1 \dots \eta_k\}$. In case $k = 1$ we have $E(\xi; \eta_1) = E(\xi; 1 - \xi) = E_\xi$.

## §4. The Salem-Zygmund Theorem

We have now the following remarkable result proved in 1955 characterizing the homogeneous perfect sets of uniqueness $E(\xi; \eta_1 \dots \eta_k)$ in terms of the number theoretic properties of $\xi, \eta_1 \dots \eta_k$.

Theorem 1 (Salem-Zygmund). Let $E(\xi; \eta_1 \dots \eta_k)$ be the homogeneous perfect set of type $(\xi; \eta_1 \dots \eta_k)$. Then $E(\xi; \eta_1 \dots \eta_k)$ is a set of uniqueness iff

(i) $\theta = \frac{1}{\xi}$ is a Pisot number,

and

(ii) $\eta_1 \dots \eta_k \in \mathbb{Q}(\theta)$.

In particular $E_\xi \in \mathcal{U}$ iff $\theta = \frac{1}{\xi}$ is a Pisot number.

We will prove the two directions of this theorem by establishing actually quite a bit more in each case.

Theorem 2 (Salem-Zygmund). Suppose $\theta = \frac{1}{\xi}$ is a Pisot number of degree n and $\eta_1 \dots \eta_k \in \mathbb{Q}(\theta)$. Then $E(\xi; \eta_1 \dots \eta_k)$ is an $H^{(n)}$-set.

Proof. We prove first the special case when $n = 1$, i.e. $\theta$ is an integer. (This is what we will use in the next chapter).

We have $\eta_i \in \mathbb{Q}$ and $(k + 1)\,\xi < 1$, so that $\theta > k + 1$.

We will find an increasing sequence $n_m$ and an open nonempty interval I such that if $E = E(\xi; \eta_1 \dots \eta_k)$, then $n_m E \cap I = \varnothing$ for all m.

If $x \in E$ then

$$\frac{x}{2\pi} = \sum_{i=0}^{\infty} \epsilon_i \left(\frac{1}{\theta}\right)^i, \text{ with } \epsilon_i \in \{0, \eta_1 \dots \eta_k\}.$$

Choose $Q \in \mathbb{N}$ such that $Q\eta_i \in \mathbb{Z}$ for all i. Then for $N \in \mathbb{N}$, $p \in \mathbb{N}$

$$Q \cdot \theta^p \cdot \frac{x}{2\pi} = \sum_{i \le p} \epsilon_i \cdot Q \cdot \theta^{p-i} + \sum_{1 \le i \le N} \epsilon_{p+i} \cdot Q \cdot \theta^{-i} + \sum_{i > N} \epsilon_{p+i} \cdot Q \cdot \theta^{-i}.$$

Call these three sums $\Sigma_1$, $\Sigma_2$, $\Sigma_3$ respectively. Then $\Sigma_1 \in \mathbb{Z}$, so $\Sigma_1 \equiv 0 \pmod 1$. Also

$$\left|\sum\nolimits_3\right| \le Q \cdot \sum_{i > N} \left(\frac{1}{\theta}\right)^i = Q \cdot \frac{1}{\theta^N \cdot (\theta - 1)}.$$

Finally for $\Sigma_2$ there are at most $(k + 1)^N$ possibilities, as $\epsilon_{p+i}$ varies over $\{0, \eta_1, \dots, \eta_k\}$, say $x_1, \dots x_{(k+1)^N}$ (allowing repetitions). Put

$$F_N = 2\pi \left\{x \pmod 1 \colon \exists i \left(\operatorname{dist}(x, x_i) \le \frac{Q}{\theta^N \cdot (\theta - 1)}\right)\right\}.$$

Thus

$$\forall N \forall p[(Q \cdot \theta^p)E \subseteq F_N].$$

Also $F_N$ is a finite union of intervals of total length at most $2\pi$ times

$$2 \cdot \frac{Q}{\theta^N \cdot (\theta-1)} \cdot (k+1)^N = 2\left(\frac{k+1}{\theta}\right)^N \cdot \frac{Q}{\theta-1} \rightarrow 0$$

as $N \rightarrow \infty$. So for large N, some open interval I is disjoint from $F_N$ and thus $\forall p[(Q \cdot \theta^p)E \cap I = \emptyset]$. Therefore we can take $n_m = Q \cdot \theta^m$.

We deal now with the general case of arbitrary n.

First choose $Q \in \mathbb{N}$ so that $Q\eta_i$ is an algebraic integer in $\mathbb{Q}(\theta)$, $i = 1 \ldots k$. Let also $\gamma$ be any algebraic integer in $\mathbb{Q}(\theta)$. Then as before we have for $x \in E$,

$$Q \cdot \theta^p \cdot \gamma \cdot \frac{x}{2\pi} = \sum_{i \leq p} \epsilon_i \cdot Q \cdot \gamma \cdot \theta^{p-i} +$$

$$\sum_{1 \leq i \leq N} \epsilon_{p+i} \cdot Q \cdot \gamma \cdot \theta^{-i} +$$

$$\sum_{i > N} \epsilon_{p+i} \cdot Q \cdot \gamma \cdot \theta^{-i}$$

and again call these sums $\Sigma_1$, $\Sigma_2$, $\Sigma_3$ respectively.

We have, also as before, $|\Sigma_3| \leq \frac{Q \cdot |\gamma|}{\theta^N \cdot (\theta-1)}$. Moreover if $\theta^{(1)} \ldots \theta^{(n-1)}$ are the conjugates of $\theta$ ($= \theta^{(0)}$) and if we define $\gamma^{(j)}$ and $\epsilon_i^{(j)}$ as in the proof of Theorem 2.3(i) then (recalling that $Q \cdot \epsilon_i^{(j)}$ is an algebraic integer)

$$\sum\nolimits_1 + \sum_{1 \le j \le n-1} \sum_{i \le p} \epsilon_i^{(j)} \cdot Q \cdot \gamma^{(j)} \cdot (\theta^{(j)})^{p-i} \in \mathbf{Z}.$$

Note now that for $1 \le j \le n - 1$,

$$\left| \sum_{i \le p} \epsilon_i^{(j)} \cdot Q \cdot \gamma^{(j)} \cdot (\theta^{(j)})^{p-i} \right| \le M \cdot \frac{Q \cdot |\gamma^{(j)}|}{1-|\theta^{(j)}|}$$

where $M = \max_{\substack{j \le n-1 \\ i \le k}} |\eta_i^{(j)}|$.

<u>Claim</u>. We can choose $\gamma \ne 0$ and N so that

$$\frac{Q \cdot |\gamma|}{\theta^N \cdot (\theta - 1)} \le \frac{1}{2n(k+1)^{(N/n)+1}}.$$

and

$$M \cdot \frac{Q \cdot |\gamma^{(j)}|}{1-|\theta^{(j)}|} \le \frac{1}{2n(k+1)^{(N/n)+1}}.$$

We will assume this claim temporarily and proceed to complete the proof.

Since

$$\left| \sum_{1 \le j \le n-1} \sum_{i \le p} \epsilon_i^{(j)} \cdot Q \cdot \gamma^{(j)} \cdot (\theta^{(j)})^{p-i} \right| \le (n-1) \cdot \frac{1}{2n(k+1)^{(N/n)+1}}$$

it follows that each $Q \cdot \theta^p \cdot \gamma \cdot \frac{x}{2\pi}$ is closer than

$$\frac{1}{2} \cdot \frac{1}{(k+1)^{(N/n)+1}}$$

to one of at most $(k + 1)^N$ points $x_1, \ldots, x_{(k+1)^N}$, always of course (mod 1). These points as before come from letting in $\Sigma_2$ $(\epsilon_{p+1}, \ldots \epsilon_{p+N})$ take any values in $\{0, \eta_1, \ldots \eta_k\}$.

Consider now instead

$$Q \cdot \theta^{p+1} \cdot \gamma \cdot \frac{x}{2\pi}, \ldots, Q \cdot \theta^{p+n} \cdot \gamma \cdot \frac{x}{2\pi}.$$

Then in $\Sigma_2$, for each one of them, we have respectively

$$(\epsilon_{p+2}, \ldots, \epsilon_{p+N+1}), (\epsilon_{p+3}, \ldots, \epsilon_{p+N+2}), \ldots, (\epsilon_{p+n+1}, \ldots, \epsilon_{p+n+N})$$

and the indices of all these $\epsilon$'s are between $p + 2$ and $p + n + N$, i.e. there are $N + n - 1$ of them. So there are at most $(k + 1)^{N+n-1}$ points in $[0, 1]^n$, say

$$z_1, \ldots, z_{(k+1)^{N+n-1}}$$

such that each point

$$z = (Q \cdot \theta^{p+1} \cdot \gamma \cdot \frac{x}{2\pi} \text{ (mod 1)}, \ldots, Q \cdot \theta^{p+n} \cdot \gamma \cdot \frac{x}{2\pi} \text{ (mod 1)})$$

has the property that for some $1 \leq m \leq (k + 1)^{N+n-1}$

$$\sup_{1 \leq i \leq n} |z_m(i) - z(i)| \leq \frac{1}{2(k+1)^{(N/n)+1}}.$$

The volume of the set of points

$$\left\{y \in [0, 1]^n : \exists m\, [1 \leq m \leq (k+1)^{N+n-1} \;\&\; \left(\sup_{1 \leq i \leq n} |z_m(i) - y(i)| \leq \frac{1}{2(k+1)^{(N/n)+1}}\right)]\right\}$$

is at most

$$(k+1)^{N+n-1}\, [2 \cdot \frac{1}{2(k+1)^{(N/n)+1}}]^n = \frac{(k+1)^{N+n-1}}{(k+1)^{N+n}} < 1$$

so there is $I = \prod_{i=1}^n I_i \neq 0$ is $\mathbf{T}^n$ with

$$(Q \cdot \gamma \cdot \theta^{p+1}, \ldots, Q \cdot \gamma \cdot \theta^{p+n})\, E \cap I = \emptyset$$

for all p. Now although $Q \cdot \gamma \cdot \theta^{p+1}$ is not necessarily an integer we know by Theorem 2.3(i) that $\{Q \cdot \gamma \cdot \theta^j\} \to 0$, so we can choose $p_0$ big enough and $I' = \prod_{i=1}^n I_i' \subseteq I$, $I' \neq \emptyset$ so that

$$\vec{m}_p\, E \cap I' = \emptyset$$

for $p > p_0$, where

$$\vec{m}_p = (m_{p+1}, \ldots, m_{p+n})$$

with $m_j$ = closest integer to $Q \cdot \gamma \cdot \theta^j$.

It remains only to show that given any finitely many hyperplanes

$$\vec{v}_t \cdot \vec{x} = q_t \quad (t = 1, \ldots, \ell\,;\, \vec{v}_t \neq \vec{0})$$

we can choose $p > p_0$ large enough so that $\vec{m}_p$ is not contained in any of them. For that notice that if $\vec{v} \neq \vec{0}$ we have

$$\vec{v} \cdot \vec{m}_p = \sum_{1 \le i \le n} \gamma \cdot Q \cdot v(i) \cdot \theta^{p+i} - \sum_{1 \le i \le n} v(i) \cdot \pm \{Q \cdot \gamma \cdot \theta^{p+i}\}$$

$$= \gamma \cdot Q \cdot \theta^{p+1} \cdot \Big( \sum_{0 \le i \le n-1} v(i+1)\theta^i \Big) - \sum_{1 \le i \le n} v(i) \cdot \pm \{Q \cdot \gamma \cdot \theta^{p+i}\}$$

and since the degree of $\theta$ is n it follows that $\sum_{0 \le i \le n-1} v(i+1)\, \theta^i \neq 0$. So

$$\Big|\gamma \cdot Q \cdot \theta^{p+1} \cdot \Big( \sum_{0 \le i \le n-1} v(i+1)\theta^i \Big)\Big| \to \infty$$

and

$$\sum_{1 \le i \le n} v(i) \cdot \pm \{Q \cdot \gamma \cdot \theta^{p+i}\} \to 0$$

as $p \to \infty$, from which our previous assertion is evident.

It remains only to prove our claim. The proof is based on the following classical theorem of Minkowski in the geometry of numbers. (For a proof see Kahane-Salem [1]).

<u>Theorem 3</u> (Minkowski). Suppose we have n linear real forms

$$\gamma_p(\vec{x}) = \sum_{q=1}^{n} a_p^q \, x_q, \qquad a_p^q \in \mathbb{R}, \vec{x} \in \mathbb{R}^n, \quad p = 1, \dots, n$$

such that

$$\Delta = \det(a_p^q) \neq 0.$$

If $\delta_1, \delta_2, \ldots, \delta_n$ are positive numbers and

$$\prod_{i=1}^{n} \delta_i \geq |\Delta|$$

then there is $\vec{z} \neq \vec{0}$, $\vec{z} \in \mathbf{Z}^n$ such that

$$|\gamma_p(\vec{z})| \leq \delta_p, \quad p = 1, \ldots, n.$$

Moreover this is still valid if for some $p = 1, \ldots, n$ $(a_p^1, \ldots, a_p^n) \in \mathbb{C}^n$, provided that for each such p there is some $p' = 1, \ldots, n$ with $a_{p'}^q = \overline{a_p^q}$ and $\delta_{p'} = \delta_p$.

Let now

$$\gamma(\vec{x}) = x_1 + \theta x_2 + \ldots + \theta^{n-1} x_n$$

where $\vec{x} = (x_1, \ldots, x_n) \in \mathbf{R}^n$. The $\gamma$ which we will choose for our claim will be of that form for some $\vec{x} \in \mathbf{Z}^n$.

Consider the forms $\gamma_1, \ldots, \gamma_n$ given by

$$\gamma_1(\vec{x}) = \frac{Q \cdot \gamma(\vec{x})}{\theta^N \cdot (\theta - 1)},$$

$$\gamma_i(\vec{x}) = \frac{M \cdot Q \cdot \gamma^{(i-1)}(\vec{x})}{1 - |\theta^{(i-1)}|}, \quad i = 2, \ldots, n$$

where $\gamma^{(j)}(\vec{x})$ is defined by replacing $\theta$ by $\theta^{(j)}$ in $\gamma(\vec{x})$.

Note that these forms may be complex, but they certainly come in conjugate pairs. Let $\Delta$ be the determinant of the system. Then

$$\Delta = \frac{Q}{\theta^N \cdot (\theta-1)} \cdot \prod_{1 \le i \le n-1} \frac{M \cdot Q}{1 - |\theta^{(i)}|} \cdot \begin{vmatrix} 1 & \theta & \dots \theta^{n-1} \\ 1 & \theta^{(1)} & \dots (\theta^{(1)})^{n-1} \\ & \dots & \\ 1 & \theta^{(n-1)} & \dots (\theta^{(n-1)})^{n-1} \end{vmatrix}$$

and $\Delta \neq 0$, since $\theta^{(i)} \neq \theta^{(j)}$ for $0 \le i < j \le n-1$. Also clearly

$$\Delta = \frac{\Delta_0}{\theta^N}$$

where $\Delta_0 \neq 0$ is independent of N. Take $\delta_i = \delta = \frac{1}{2n(k+1)^{(N/n)+1}}$. By Minkowski's Theorem we will be done if we can choose N so that

$$\delta^n = \frac{1}{(2n)^n \cdot (k+1)^{N+n}} \ge \frac{|\Delta_0|}{\theta^N}.$$

But this is clearly possible, since $\theta > k + 1$, and our proof is complete. □

For the other direction of the theorem we will prove again the following stronger statement.

<u>Theorem 4</u> (Salem). If $\theta = \frac{1}{\xi}$ is not a Pisot number or if one of the $\eta_i$'s is not in $\mathbb{Q}(\theta)$, then $E(\xi; \eta_1 \dots \eta_k)$ is a set of restricted multiplicity.

<u>Proof.</u> If $k > 1$ then $E(\xi; \eta_1 \dots \eta_k)$ contains any $E(\xi; \eta_i, \eta_k)$ for $1 \le i < k$, so it is

enough to consider only the case of $E_\xi$ $(= E(\xi; 1 - \xi)$, and $E(\xi; \eta, 1 - \xi)$, where $\xi < \eta < 1 - 2\xi$. Call these two cases I and II respectively. In Case I, $\theta = \frac{1}{\xi}$ is not Pisot and in Case II, either $\theta$ is not Pisot or $\eta \notin \mathbb{Q}(\theta)$.

Consider the map

$$F : 2^{\mathbb{N}} \ (= \{0, 1\}^{\mathbb{N}}) \to \mathbb{T}$$

given by

$$F(x) = 2\pi \cdot \left[\sum_{i=0}^{\infty} x(i) \cdot \epsilon_i \cdot \xi^i\right]$$

where in Case I, $\epsilon_i = 1 - \xi$ and in Case II,

$$\epsilon_i = \begin{cases} \eta, \text{ if i is odd} \\ 1 - \xi, \text{ if i is even.} \end{cases}$$

Clearly F maps $2^{\mathbb{N}}$ continuously into the corresponding E $(= E_\xi$ or $E(\xi; \eta, 1 - \xi))$ in each case. Let $\sigma$ be the usual probability measure on $2^{\mathbb{N}}$ (i.e. the product measure coming from the measure $\tau$ on $\{0, 1\}$ where $\tau(\{0\}) = \tau(\{1\}) = \frac{1}{2}$). Let $\mu$ be the induced via F measure on $\mathbb{T}$, i.e.

$$\int g \, d\mu = \int (g \circ F) d\sigma.$$

Clearly $\text{supp}(\mu) \subseteq E$. We have now

$$\hat{\mu}(n) = \int e^{-inx} d\mu(x)$$

$$= \int e^{-inF(x)} d\sigma(x)$$

$$= \int e^{-in2\pi\left(\sum_{k=0}^{\infty} x(k)\, \epsilon_k\, \xi^k\right)} d\sigma(x)$$

$$= \lim_{N\to\infty} \int e^{-in2\pi\left(\sum_{k=0}^{N} x(k)\epsilon_k\xi^k\right)} d\sigma(x)$$

$$= \lim_{N\to\infty} \left[ \sum_{\vec{x}\in\{0,1\}^{N+1}} \frac{1}{2^{N+1}}\, e^{-2\pi in(\sum_{k=0}^{N} x(k)\epsilon_k\xi^k)} \right]$$

$$= \lim_{N\to\infty} \left[ \sum_{\vec{x}\in\{0,1\}^{N+1}} \frac{1}{2^{N+1}} \prod_{k=0}^{N} e^{-2\pi inx(k)\epsilon_k\xi^k} \right]$$

$$= \lim_{N\to\infty} \prod_{k=0}^{N} \left(\frac{e^{-2\pi in\epsilon_k\xi^k}+1}{2}\right)$$

$$= \prod_{k=0}^{\infty} \left(\frac{e^{-2\pi in\epsilon_k\xi^k}+1}{2}\right)$$

$$= \prod_{k=0}^{\infty} e^{-\pi in\epsilon_k\xi^k} \cdot \left(\frac{e^{\pi in\epsilon_k\xi^k}+e^{-\pi in\epsilon_k\xi^k}}{2}\right)$$

$$= e^{-\sum_{k=0}^{\infty}\pi in\epsilon_k\xi^k} \cdot \prod_{k=0}^{\infty} \cos(\pi n\, \epsilon_k\xi^k).$$

As the first factor has absolute value 1 it is enough to show that the second goes to 0 as $n \to \infty$.

If not, then we can find $n_1 < n_2 < \ldots$ such that for some $a > 0$

$$\prod_{k=0}^{\infty} \left|\cos(\pi n_j\, \epsilon_k\xi^k)\right| \geq a, \quad j = 1, 2, \ldots$$

or

$$\prod_{k=0}^{\infty} (1 - \sin^2(\pi n_j \epsilon_k \xi^k)) \geq a^2 > 0, \quad j = 1, 2, \dots .$$

Therefore

$$\sum_{k=0}^{\infty} \sin^2(\pi n_j \epsilon_k \xi^k) < C < \infty, \quad j = 1, 2, \dots .$$

Consider now separately the two cases:

Case I. Then $\epsilon_k = 1 - \xi$ and $n_j \epsilon_k \xi^k = \dfrac{(1-\xi)n_j}{\theta^k} \to 0$ as $k \to \infty$ so that we can find $k_j \to \infty$ nondecreasing such that

$$\frac{1}{\theta} \leq \frac{(1-\xi)n_j}{\theta^{k_j}} < 1.$$

By going to a subsequence we can assume that

$$\frac{(1-\xi)n_j}{\theta^{k_j}} \to \gamma \neq 0.$$

Now for $j < m$

$$\sum_{k=0}^{k_j} \sin^2\left(\frac{\pi n_m(1-\xi)}{\theta^{k_m}} \cdot \theta^k\right) < C$$

and letting $m \to \infty$ we have

$$\sum_{k=0}^{k_j} \sin^2(\pi \gamma \theta^k) \leq C$$

so that letting $j \to \infty$ we have

$$\sum_{k=0}^{\infty} \sin^2 (\pi \gamma \theta^k) < \infty$$

or

$$\sum_{k=0}^{\infty} \sin^2 (\pi \{\gamma \theta^k\}) < \infty,$$

and thus

$$\sum_{k=0}^{\infty} \{\gamma\theta^k\}^2 < \infty.$$

Then by Theorem 2.3(ii), $\theta$ is a Pisot number, a contradiction.

Case II. First by Case I we can assume $\theta$ is a Pisot number. Again if $\hat{\mu}(n)$ does not tend to 0 we have as before that

$$\sup_j \sum_{k=0}^{\infty} \sin^2 (\pi n_j(1 - \xi)\, \xi^{2k}) < \infty$$

and

$$\sup_j \sum_{k=0}^{\infty} \sin^2 (\pi n_j\, \eta\xi^{2k+1}) < \infty$$

for some $n_1 < n_2 < \dots$. Again choose $k_j \to \infty$ nondecreasing with

$$\frac{1}{\theta^2} \le \frac{\eta\xi n_j}{(\theta^2)^{k_j}} < 1$$

and $\frac{\eta\xi n_j}{(\theta^2)^{k_j}} \to \gamma_2 \neq 0$, so that also $\frac{(1-\xi)n_j}{(\theta^2)^{k_j}} \to \gamma_1 = \gamma_2 \cdot \frac{1-\xi}{\xi\eta}$. Then as before

$$\sum_{k=0}^{\infty} \sin^2(\pi\, \gamma_1\, \theta^{2k}) < \infty$$

and

$$\sum_{k=0}^{\infty} \sin^2 (\pi\, \gamma_2\, \theta^{2k}) < \infty,$$

so by Theorem 2.3(ii) $\gamma_1, \gamma_2 \in \mathbb{Q}(\theta)$, thus $\eta \in \mathbb{Q}(\theta)$, a contradiction, and our proof is complete. □

Remark. Note that in the proof of the preceding theorem, if $\theta$ is an integer we only needed the special case of 2.3(ii) that we proved in 2.4.

From Theorem 4 we obtain also immediately Menshov's Theorem on the existence of sets of restricted multiplicity that have measure 0.

Theorem 5 (Menshov). There exist closed sets of restricted multiplicity that have measure 0.

Proof. Consider $E = E_\xi$, where $\theta = \frac{1}{\xi}$ is not a Pisot number. □

We conclude with some comments on the general symmetric sets $E_{\xi_1,\xi_2,\cdots}$. Except for some special cases (see for example Kahane-Salem [1], Meyer [1]) the problem of characterizing which $E_{\xi_1,\xi_2,\cdots}$ are sets of uniqueness remains unsolved. An important result of Meyer (see Meyer [1]) asserts that the ultra-thin symmetric sets $E_{\xi_1,\xi_2,\cdots}$ with $\Sigma\xi_i^2 < \infty$ are all sets of uniqueness (in fact satisfy the criterion of Theorem 1.3 – see Lyons [4]). This rules out purely arithmetic or algebraic necessary conditions on the $\xi_i$ for $E_{\xi_1,\xi_2,\cdots}$ to be of uniqueness.

We will say a few more things on the $E_{\xi_1,\xi_2,\cdots}$ in the next chapter and Chapter VIII.2.

# Chapter IV. Classification of the Complexity of U

In this chapter we establish the first connections between the study of the structure of closed sets of uniqueness and descriptive set theory. After reviewing some basic facts from the theory of Borel, analytic and coanalytic sets in Polish spaces, we discuss the space K(E) of closed subsets of a compact, metrizable space E. We are primarily concerned here of course about the classes U, $U_0$ of closed sets of uniqueness, resp. extended uniqueness, in $\mathbb{T}$. The main result of this chapter, due to Solovay and (independently) Kaufman, classifies U, $U_0$ as being coanalytic but not Borel in $K(\mathbb{T})$. We discuss also the relevance of this theorem to the Characterization Problem. Finally, we present certain related results of the descriptive set theory of $\sigma$-ideals of closed sets in compact, metrizable spaces.

## §1. Some descriptive set theory

We review first some of the basic concepts and results of descriptive set theory in Polish spaces. References for the proofs of the theorems that we do not give are Kuratowski [1, Vol. I, §33–39] and Moschovakis [1, Ch. 1, 2].

i). A Polish space is a complete metrizable, separable space. The Borel sets in a Polish space X are obtained from the open sets by closing under the operations of complementation, countable union and intersection. The Borel sets can thus be classified into a transfinite hierarchy consisting of $\omega_1$ levels, where as usual $\omega_1$ is the first uncountable ordinal. We start with the class of open sets, which in modern (logically motivated) notation is denoted as

$$\underset{\sim}{\Sigma}^0_1 \equiv \text{open}.$$

We also let

$$\underset{\sim}{\Pi}^0_1 \equiv \text{closed}$$

and then inductively for each countable ordinal $\alpha < \omega_1$ we define $\underset{\sim}{\Sigma}^0_\alpha$, $\underset{\sim}{\Pi}^0_\alpha$ by

$$\underset{\sim}{\Sigma}^0_\alpha = \{\cup_n P_n : \text{Each } P_n \text{ is in } \underset{\sim}{\Pi}^0_{\alpha_n} \text{ for some } \alpha_n < \alpha\},$$

$$\underset{\sim}{\Pi}^0_\alpha = \{X - P : P \text{ is } \underset{\sim}{\Sigma}^0_\alpha\}.$$

We refer to $\underset{\sim}{\Sigma}^0_\alpha$ ($\underset{\sim}{\Pi}^0_\alpha$) as the dual class of $\underset{\sim}{\Pi}^0_\alpha$ ($\underset{\sim}{\Sigma}^0_\alpha$). When $\alpha = n$ is finite, the finite Borel classes $\underset{\sim}{\Sigma}^0_n$, $\underset{\sim}{\Pi}^0_n$ are then the following

$$\underset{\sim}{\Sigma}^0_2 = F_\sigma,$$

$$\underset{\sim}{\Pi}^0_2 = G_\delta,$$

$$\underset{\sim}{\Sigma}^0_3 = G_{\delta\sigma},$$

$$\underset{\sim}{\Pi}^0_3 = F_{\sigma\delta}, \text{ etc.}$$

A set is Borel iff it is in some $\underset{\sim}{\Sigma}^0_\alpha$, $\alpha < \omega_1$ (or in some $\underset{\sim}{\Pi}^0_\alpha$, $\alpha < \omega_1$; note that $\underset{\sim}{\Sigma}^0_\alpha \cup \underset{\sim}{\Pi}^0_\alpha \subseteq \underset{\sim}{\Sigma}^0_{\alpha+1} \cap \underset{\sim}{\Pi}^0_{\alpha+1}$). The __ambiguous__ classes $\underset{\sim}{\Delta}^0_\alpha$ are defined by

$$\underset{\sim}{\Delta}^0_\alpha = \underset{\sim}{\Sigma}^0_\alpha \cap \underset{\sim}{\Pi}^0_\alpha.$$

So a set is $\underset{\sim}{\Delta}^0_\alpha$ if it is both $\underset{\sim}{\Sigma}^0_\alpha$ and $\underset{\sim}{\Pi}^0_\alpha$, i.e. if both it and its complement are $\underset{\sim}{\Sigma}^0_\alpha$. Thus

$$\underset{\sim}{\Delta}^0_1 \equiv \text{clopen}$$

$$\underset{\sim}{\Delta}^0_2 \equiv F_\sigma \cap G_\delta, \text{ etc.}$$

A subset $P \subseteq X$ of a Polish space X is called __analytic__ or $\underset{\sim}{\Sigma}^1_1$,

$$\underset{\sim}{\Sigma}^1_1 \equiv \text{analytic (A)}$$

if it is the continuous image of a Borel set, i.e. there is $B \subseteq Y$ a Borel subset of a Polish space Y and $f : Y \to X$ continuous with $f[B] = P$.

The complements of analytic sets are called __coanalytic__ or $\underset{\sim}{\Pi}^1_1$ sets,

$$\underset{\sim}{\Pi}^1_1 \equiv \text{coanalytic (CA).}$$

The operations of complementation and taking continuous images can be iterated leading to the definition of the class of projective sets as follows:

Let

$$\underset{\sim}{\Sigma}^1_2 \equiv \text{PCA}$$

be the class of continuous images of $\underset{\sim}{\Pi}^1_1$ sets and

$$\underset{\sim}{\Pi}^1_2 \equiv \text{CPCA}$$

the class of complements of $\underset{\sim}{\Pi}^1_2$ sets, and in general

$$\underset{\sim}{\Sigma}^1_{n+1} = \text{the class of continuous images of } \underset{\sim}{\Pi}^1_n \text{ sets,}$$

$$\underset{\sim}{\Pi}^1_{n+1} = \text{the class of complements of } \underset{\sim}{\Sigma}^1_{n+1} \text{ sets.}$$

Again we refer to $\underset{\sim}{\Sigma}^1_n$ ($\underset{\sim}{\Pi}^1_n$) as the <u>dual</u> class of $\underset{\sim}{\Pi}^1_n$ ($\underset{\sim}{\Sigma}^1_n$). A set is <u>projective</u> if it belongs to some $\underset{\sim}{\Sigma}^1_n$ (or $\underset{\sim}{\Pi}^1_n$; again $\underset{\sim}{\Sigma}^1_n \cup \underset{\sim}{\Pi}^1_n \subseteq \underset{\sim}{\Sigma}^1_{n+1} \cap \underset{\sim}{\Pi}^1_{n+1}$). We define also the <u>ambiguous</u> classes $\underset{\sim}{\Delta}^1_n$ by

$$\underset{\sim}{\Delta}^1_n = \underset{\sim}{\Sigma}^1_n \cap \underset{\sim}{\Pi}^1_n.$$

ii). Let us review now some basic facts of the theory of Borel, analytic and coanalytic sets. (The higher level projective classes will not be used in this book).

First some simple (but very useful in computations) closure properties:

The analytic sets are closed under countable unions and intersections, and

Borel images and preimages (i.e. if $f : X \to Y$ is a Borel function between two Polish spaces X, Y and $P \subseteq X$, $Q \subseteq Y$ are $\underset{\sim}{\Sigma}^1_1$ then $f[P]$, $f^{-1}[Q]$ are also $\underset{\sim}{\Sigma}^1_1$). The coanalytic sets are closed under countable unions and intersections and Borel preimages. The Borel sets are closed under complements, countable unions and intersections, as well as Borel preimages.

Next we have hierarchy results. For each <u>uncountable</u> Polish space X we have that

$$\underset{\sim}{\Sigma}^0_\alpha \neq \underset{\sim}{\Pi}^0_\alpha, \quad \underset{\sim}{\Sigma}^0_\alpha \cup \underset{\sim}{\Pi}^0_\alpha \subsetneqq \underset{\sim}{\Delta}^0_{\alpha+1},$$

$$\underset{\sim}{\Sigma}^1_n \neq \underset{\sim}{\Pi}^1_n, \quad \underset{\sim}{\Sigma}^1_n \cup \underset{\sim}{\Pi}^1_n \subsetneqq \underset{\sim}{\Delta}^1_{n+1}.$$

In particular, of great importance for us is the fact that there are analytic (or coanalytic) non-Borel sets. We will see many examples of such sets later, but here are a couple of other ones occurring in analysis: The set of differentiable functions in $C([0, 1])$ is $\underset{\sim}{\Pi}^1_1$ but not Borel (see Mazurkiewicz [1]), and so is the set of functions in $C(\mathbb{T})$ with everywhere convergent Fourier series (see Ajtai–Kechris [1]).

Clearly every Borel set is $\underset{\sim}{\Delta}^1_1$, i.e. both $\underset{\sim}{\Sigma}^1_1$ and $\underset{\sim}{\Pi}^1_1$. In fact by a classical result of Suslin the converse is also true, i.e.

$$\text{Borel} = \underset{\sim}{\Delta}^1_1 \qquad (\underline{\text{Suslin's Theorem}}).$$

This, among other things, is a very useful tool for demonstrating that sets are Borel. For example it follows immediately that $f : X \to Y$ has Borel graph iff f is

Borel, since for $V \subseteq Y$ open we have

$$f(x) \in V \Leftrightarrow \exists y[f(x) = y \;\&\; y \in V]$$

$$\Leftrightarrow \forall y\, [f(x) = y \Rightarrow y \in V].$$

We can summarize now the above information on the containment relations of the Borel and projective classes in the following picture (for uncountable Polish spaces):

| | (open) | (F$_\sigma$) | (G$_{\delta\sigma}$) | | | | (A) | (PCA) | |
|---|---|---|---|---|---|---|---|---|---|
| | $\underset{\sim}{\Sigma}^0_1$ | $\underset{\sim}{\Sigma}^0_2$ | $\underset{\sim}{\Sigma}^0_3$ | | $\underset{\sim}{\Sigma}^0_\alpha$ | | $\underset{\sim}{\Sigma}^1_1$ | $\underset{\sim}{\Sigma}^1_2$ | |
| $\underset{\sim}{\Delta}^0_1$ | | $\underset{\sim}{\Delta}^0_2$ | $\underset{\sim}{\Delta}^0_3$ | $\cdots$ | $\underset{\sim}{\Delta}^0_{\alpha+1}$ | $\cdots$Borel $= \underset{\sim}{\Delta}^1_1$ | | $\underset{\sim}{\Delta}^1_2$ | $\cdots$ |
| | $\underset{\sim}{\Pi}^0_1$ | $\underset{\sim}{\Pi}^0_2$ | $\underset{\sim}{\Pi}^0_3$ | | $\underset{\sim}{\Pi}^0_\alpha$ | | $\underset{\sim}{\Pi}^1_1$ | $\underset{\sim}{\Pi}^1_2$ | |
| | (closed) | (G$_\delta$) | (F$_{\sigma\delta}$) | | | | (CA) | (CPCA) | |

Borel Hierarchy Projective Hierarchy

Suslin's Theorem has been strengthened by Lusin who proved the following separation property of $\underset{\sim}{\Sigma}^1_1$ sets:

If P, Q are disjoint $\underset{\sim}{\Sigma}^1_1$ sets in a Polish space X, then there is a Borel set $C \subseteq X$ with $P \subseteq C$, $C \cap Q = \emptyset$. (The Lusin Separation Theorem).

One of the many important consequences of the Suslin and Lusin Theorems is that the 1–1 image of a Borel set by a Borel function is also Borel, i.e.

If B is a Borel subset of a Polish space X, $f : X \to Y$ is a Borel map from X into a Polish space Y and f is 1–1 on B, then $C = f[B]$ is also Borel.

iii). We will be interested in the problem of classifying certain sets in the Borel or projective hierarchy.

Definition 1. If $\Gamma$ is any of the classes $\underset{\sim}{\Sigma}^0_\alpha$, $\underset{\sim}{\Pi}^0_\alpha$, $\underset{\sim}{\Sigma}^1_n$, $\underset{\sim}{\Pi}^1_n$ and $P \subseteq X$ is a subset of some Polish space we say that P is a true $\Gamma$ set if P is in $\Gamma$ but not in its dual class $\Gamma$.

For example the differentiable functions or the continuous functions with everywhere convergent Fourier series are true $\underset{\sim}{\Pi}^1_1$ sets.

If a given set $P \subseteq X$ in some Polish space is a true $\Gamma$ set, we view the class $\Gamma$ as providing a measure of the complexity of P. Such classification problems will play a prominent role in the sequel.

Many of the sets we will study turn out to be $\underset{\sim}{\Pi}^1_1$ and we are of course interested in methods for showing that they are not ($\underset{\sim}{\Sigma}^1_1$ or equivalently by Suslin's Theorem) Borel. There are two main such methods, the completeness method (to be discussed now) and the rank method (to be discussed in the next chapter). The completeness method applies as well to the other projective, and in an appropriate reformulation Borel, classes $\Gamma$ but we will not need this here.

The basic concept is that of a $\underset{\sim}{\Pi}^1_1$-complete set.

Definition 2. Let X be a Polish space. A subset P of X is called $\underset{\sim}{\Pi}^1_1$-complete if it is $\underset{\sim}{\Pi}^1_1$ and for any Polish space Y and any $\underset{\sim}{\Pi}^1_1$ subset Q of Y, there is a Borel function $f : Y \to X$ such that $Q = f^{-1}[P]$.

Since there are $\underset{\sim}{\Pi}^1_1$ but not Borel sets (in every uncountable Polish space) it follows that no $\underset{\sim}{\Pi}^1_1$-complete set is Borel. The completeness method for classifying a given set as being true $\underset{\sim}{\Pi}^1_1$ consists therefore of showing that it is actually $\underset{\sim}{\Pi}^1_1$-complete.

Let us first exhibit an elementary example of a $\underset{\sim}{\Pi}^1_1$-complete set. Let Y be a set and denote by

$$\text{Seq } Y$$

the set of all finite sequences from Y. If $s = (s(0),\ldots, s(n-1)) \in \text{Seq } Y$, we denote its length n by $\ell h(s)$. In particular the empty sequence $\emptyset$ has length 0. For $s = (s(0),\ldots, s(n-1))$, $t = (t(0),\ldots,t(k-1))$ the concatenation $s\hat{\ }t$ is defined by

$$s\hat{\ }t = (s(0),\ldots, s(n-1), t(0), \ldots, t(k-1)).$$

We also denote by

$$s \subseteq t$$

the <u>extension</u> order of finite sequences, i.e. $s \subseteq t$ iff $\ell h(s) \leq \ell h(t)$ and $\forall i \leq \ell h(s)$ $(s(i) = t(i))$. In this case we say also that s is an <u>initial segment</u> of t and write

$$s = t \restriction m$$

where $m = \ell h(s)$. Note that $\emptyset \subseteq s$, for all s. Similarly if $\epsilon \in Y^{\mathbb{N}}$ is an infinite sequence we write

$$s \subseteq \epsilon \text{ or } s = \epsilon \restriction m$$

where $m = \ell h(s)$, if $s = (\epsilon(0),\ldots, \epsilon(m-1))$.

Let T be a subset of Seq Y. We say that T is a <u>tree</u> if $t \in T$ and $s \subseteq t$ imply $s \in T$. A <u>branch</u> through T is an $\epsilon \in Y^{\mathbb{N}}$ such that for all n, $\epsilon \restriction n = (\epsilon(0),\ldots, \epsilon(n-1)) \in T$. We denote by

$$[T] = \{\epsilon \in Y^{\mathbb{N}} : \epsilon \text{ is a branch through } T\}$$

the set of all branches of T. This is a closed subset of the space $Y^{\mathbb{N}}$ (with the product topology, Y being discrete). Conversely every closed subset of $Y^{\mathbb{N}}$ is of the form [T] for some tree T. We call T <u>well founded</u> if $[T] = \emptyset$, i.e. T has no branches.

Let us specialize now to the case $Y = \mathbb{N}$. In this case the Polish space $\mathcal{N} = \mathbb{N}^{\mathbb{N}}$ is called the <u>Baire space</u> and is homeomorphic to the space of irrationals (as a subspace of $\mathbb{R}$). As a subset of Seq $\mathbb{N}$ a tree T can be identified with its characteristic function which is a member of the Polish space $\{0, 1\}^{\mathrm{Seq}\ \mathbb{N}} \equiv 2^{\mathrm{Seq}\ \mathbb{N}}$ homeomorphic to $2^{\mathbb{N}}$ (i.e. the Cantor set). The set of all trees is then easily a closed subset of $2^{\mathrm{Seq}\ \mathbb{N}}$. The following is a classical fact.

<u>Theorem 3</u>. Let WF be the set of <u>well founded</u> trees on Seq $\mathbb{N}$. Then WF is a $\underset{\sim}{\Pi}^1_1$-complete set (in $2^{\mathrm{Seq}\ \mathbb{N}}$).

<u>Proof</u>. For the proof we will need an important representation theorem for analytic sets which will be also useful many times in the sequel. (Its second form will only be used later on).

<u>Theorem 4</u> (Representation Theorem for $\underset{\sim}{\Sigma}^1_1$ sets). Let X be a Polish space and $P \subseteq X$ be $\underset{\sim}{\Sigma}^1_1$. Then there is a closed set $F \subseteq X \times \mathbb{N}^{\mathbb{N}}$ such that

$$x \in P \Leftrightarrow \exists \epsilon \in \mathbb{N}^{\mathbb{N}} (x, \epsilon) \in F$$

i.e. every $\underset{\sim}{\Sigma}^1_1$ set in X is the projection of a closed set in $X \times \mathbb{N}^{\mathbb{N}}$. (The converse is true by definition).

Also there is a $G_\delta$ set $G \subseteq X \times 2^{\mathbb{N}}$ such that

$$x \in P \Leftrightarrow \exists \epsilon \in 2^{\mathbb{N}} (x, \epsilon) \in G.$$

(One cannot replace here $G_\delta$ by closed, since $2^{\mathbb{N}}$ is compact and projections of compact sets are compact).

*Proof*. By a standard result in descriptive set theory,

every non-empty analytic set is the continuous image of $\mathbb{N}^{\mathbb{N}}$.

So let $f : \mathbb{N}^{\mathbb{N}} \to X$ be continuous with $f[\mathbb{N}^{\mathbb{N}}] = P$. (Provided $P \neq \varnothing$, since otherwise the result is trivial). Let then

$$(x, \epsilon) \in F \Leftrightarrow f(\epsilon) = x.$$

For the second statement notice that $\mathbb{N}^{\mathbb{N}}$ can be identified with a $G_\delta$ subset of $2^{\mathbb{N}}$ as follows. Let $< , > : \mathbb{N} \times \mathbb{N} \to \mathbb{N}$ be a 1–1 correspondence and assign to each $\epsilon \in \mathbb{N}^{\mathbb{N}}$ the element $\epsilon^* \in 2^{\mathbb{N}}$ given by

$$\epsilon^*(<n, m>) = 0 \text{ iff } \epsilon(n) = m.$$

The map $\epsilon \mapsto \epsilon^*$ is a homeomorphism between $\mathbb{N}^{\mathbb{N}}$ and the following $G_\delta$ subset of $2^{\mathbb{N}}$,

$$\{\delta \in 2^{\mathbb{N}}: \text{For all } n \text{ there is a unique } m \text{ with } \delta(<n, m>) = 0\}. \qquad \square$$

We return now to our main proof. Let $P \subseteq X$ be $\underset{\sim}{\Pi}^1_1$ and let $F \subseteq X \times \mathbb{N}^{\mathbb{N}}$ be closed such that

$$x \notin P \Leftrightarrow \exists \epsilon \in \mathbb{N}^{\mathbb{N}} (x, \epsilon) \in F.$$

For $s \in \text{Seq}\,\mathbb{N}$ let $\mathcal{N}_s = \{\epsilon \in \mathbb{N}^{\mathbb{N}} : s \subseteq \epsilon\}$. Assign now to each $x \in X$ the tree

$$T(x) = \{s \in \text{Seq}\,\mathbb{N} : \exists V \text{ open in } X\ (x \in V \text{ and diam}\,(V) \leq 2^{-\ell h(s)}$$
$$\text{and } (V \times \mathcal{N}_s) \cap F \neq \emptyset)\}.$$

Then $\{x : s \in T(x)\}$ is open in $X$ for all $s \in \text{Seq}\,\mathbb{N}$, so the function $x \mapsto T(x)$ is Borel from $X$ into $2^{\text{Seq}\,\mathbb{N}}$. We claim now that

$$(*) \qquad x \notin P \Leftrightarrow T(x) \notin WF$$

thus $P = f^{-1}[WF]$, where $f(x) = T(x)$, so WF is $\underset{\sim}{\Pi}^1_1$-complete.

Indeed if $x \notin P$ let $\epsilon$ be such that $(x, \epsilon) \in F$. Then $\epsilon$ is a branch through $T(x)$. Conversely if $\epsilon$ is a branch through $T(x)$ let for each $n$, $\epsilon_n \in \mathbb{N}^{\mathbb{N}}$ with $\epsilon_n \restriction n = \epsilon \restriction n$ and $x_n \in X$ be such that $(x_n, \epsilon_n) \in F$ and $\text{dist}(x_n, x) \leq 2^{-n}$. Then $x_n \to x$ and $\epsilon_n \to \epsilon$, so $(x, \epsilon) \in F$ and therefore $x \notin P$. □

It is clear that if $X$, $X'$ are Polish spaces, $g : X \to X'$ is a Borel function, $P \subseteq X$ is $\underset{\sim}{\Pi}^1_1$-complete and $P' \subseteq X'$ is $\underset{\sim}{\Pi}^1_1$ and such that $P = g^{-1}[P']$, then $P'$ is also $\underset{\sim}{\Pi}^1_1$-complete. When $P = g^{-1}[P']$ as above we say that $P$ is <u>reduced</u> to $P'$ via $g$. The terminology comes from the fact that $x \in P \Leftrightarrow g(x) \in P'$, which means that the question of membership in $P$ is reduced via $g$ to that of $P'$. So a standard procedure for showing the $\underset{\sim}{\Pi}^1_1$-completeness of a set is to reduce one of the already known $\underset{\sim}{\Pi}^1_1$-complete sets to it. One can use this method to show that the set of differentiable functions and the set of continuous functions with everywhere convergent Fourier series are $\underset{\sim}{\Pi}^1_1$-complete.

It is clear that since every $\underset{\sim}{\Pi}_1^1$ set can be reduced to any $\underset{\sim}{\Pi}_1^1$-complete set, the $\underset{\sim}{\Pi}_1^1$-complete sets are in some sense the "most" complicated true $\underset{\sim}{\Pi}_1^1$ sets. One naturally wonders whether there are "intermediate" sets, which are true $\underset{\sim}{\Pi}_1^1$ but not complete. The answer is no (Martin, Wadge), i.e. indeed every true $\underset{\sim}{\Pi}_1^1$ set is actually complete. However (Harrington) this fact cannot be proved from the usual classical axioms of ZFC set theory and requires the so-called strong axioms of set theory (determinacy or large cardinal hypotheses) for its proof. For more on this fascinating story, see for example Martin-Kechris [1].

One final remark before we leave the subject of completeness. We defined this notion in terms of Borel reducing maps. Finer notions, where these maps are more restricted, like of Baire class 1, continuous, etc., can be studied, but we will have no use for them here.

iv). We conclude our brief survey with some facts concerning regularity properties of analytic and coanalytic sets.

First, it is a result of Suslin that every uncountable $\underset{\sim}{\Sigma}_1^1$ set contains a subset homeomorphic to $2^{\mathbb{N}}$ (i.e. the Cantor set). We have used the special case of that result for Borel sets, which was proved by Alexandroff and Hausdorff, in the proof of I.5.6. A similar result is true for $\underset{\sim}{\Pi}_1^1$ sets, but cannot be proved in ZFC and requires the use of strong axioms of set theory (Solovay; see Moschovakis [1]).

Also (Lusin-Sierpinski) every $\underset{\sim}{\Sigma}_1^1$ or $\underset{\sim}{\Pi}_1^1$ set has the property of Baire (i.e. it is equal to an open set modulo a set of the first category), and (Lusin) is universally measurable, i.e. for any given positive (Borel) measure $\mu$, it is equal to a Borel set modulo a set of $\mu$-measure 0.

## §2. The theorem of Solovay and Kaufman

For each compact, metrizable (therefore Polish) space E we denote by K(E) the space of closed ($\equiv$ compact) subsets of E with the Hausdorff topology, generated by the sets $\{K \in K(E) : K \cap V \neq \emptyset\}$, $\{K \in K(E) : K \subseteq V\}$, where V is open in E. So the basic open sets in K(E) are of the form

$$\{K \in K(E) : K \subseteq V_0 \;\&\; K \cap V_1 \neq \emptyset \;\&\ldots\&\; K \cap V_n \neq 0\}$$

where $V_0, V_1, \ldots, V_n$ are open sets in E. This is a compact, metrizable space with the following metric

$$\delta(K,L) = \sup\{\max(\text{dist}(x, K), \text{dist}(y, L)) : x \in L \;\&\; y \in K\}, \quad \text{if } K, L \neq \emptyset,$$

$$= \text{diam}(E), \text{ otherwise.}$$

We will use freely various simple facts about this topology (see Kuratowski [1, Vol. 2]), like for example:

i) The set of finite subsets of E is dense in K(E), and in fact the set of finite subsets of some fixed countable dense subset of E is a countable dense subset of K(E).

ii) The set of perfect subsets of E is a $G_\delta$ set in K(E).

iii) The union function $\cup : K(K(E)) \to K(E)$ given by $\cup(L) \equiv \cup L = \cup\{K :$

$K \in L\}$ is continuous. Also the function $\cup : K(E) \times K(E) \to K(E)$ given by $\cup(K, L) = K \cup L$ is continuous. The intersection function $\cap : K(E) \times K(E) \to K(E)$ given by $\cap(K, L) = K \cap L$ is Borel (but not necessarily continuous). If however L is clopen in E, the map $K \mapsto K \cap L$ is continuous.

iv) Finally, if $f : E \to E'$ is continuous, then $f'' : K(E) \to K(E')$ given by $f''(K) = f[K] = \{f(x) : x \in K\}$ is also continuous.

Our main interest will be of course in the space $K(\mathbb{T})$ (and $K(E)$ for $E \subseteq \mathbb{T}$ closed). Let us denote by U the set of closed $\mathcal{U}$-sets, i.e.

$$U = K(\mathbb{T}) \cap \mathcal{U}$$

and similarly for $U_0$,

$$U_0 = K(\mathbb{T}) \cap \mathcal{U}_0.$$

Let also $M = K(\mathbb{T}) - U$, $M_0 = K(\mathbb{T}) - U_0$.

We will prove here the result of Solovay and (independently) Kaufman which classifies U, $U_0$ as $\underset{\sim}{\Pi}^1_1$-complete sets. The problem of the classification of U had been raised in 1982 by Kechris, whose motivation came from a discussion with S. Pichorides on the Characterization Problem. Kechris noted the (easy) fact that U is $\underset{\sim}{\Pi}^1_1$ in $K(\mathbb{T})$ and conjectured that U is not Borel. (An interesting logical point: The definition of U in terms of trigonometric series is far too complex, i.e. $\underset{\sim}{\Pi}^1_2$. One needs a reformulation of the concept of closed $\mathcal{U}$-sets, as for example that in Chapter II, to see that it is actually $\underset{\sim}{\Pi}^1_1$). He mentioned this problem to Solovay in the fall of 1983, and Solovay proved the conjecture to be true in December 1983 (private communication, Solovay [1]), but has not published his proof yet. Independently of all this R. Kaufman who was studying the problem of representing bounded $\underset{\sim}{\Sigma}^1_1$ subsets of $\mathbb{C}$ as point spectra of operators on Banach spaces (see Kaufman [5], [7]), arrived at the same result a few months later in 1984, and published his result in Kaufman [6] (see also Kaufman [8]).

The proof of the result of Solovay and Kaufman that we give below is a simplification of the original Solovay proof, which makes use of some ideas from Kechris-Louveau-Woodin [1].

In order to show that U, $U_0$ are $\underset{\sim}{\Pi}^1_1$-complete we will reduce some previously known $\underset{\sim}{\Pi}^1_1$-complete set to them. It is most convenient to use the following classical example of Hurewicz [1], a proof of which is presented here in the context of modern descriptive set theoretic ideas, which we will also use later.

<u>Theorem 1</u> (Hurewicz [1]). Let $\mathbb{Q}'$ be the set of rationals in [0, 1]. Then

$$K(\mathbb{Q}') = \{K \in K([0, 1]) : K \subseteq \mathbb{Q}'\}$$

is $\underset{\sim}{\Pi}^1_1$-complete.

<u>Proof</u>. Work first in $2^{\mathbb{N}}$ and let Q be a countable dense subset of $2^{\mathbb{N}}$. For each $F_\sigma$ set $B \subseteq 2^{\mathbb{N}}$ consider the following infinite, two-person, perfect information game (an instance of the so-called <u>Wadge</u> games—see Moschovakis [1])

| I | II | |
|---|---|---|
| $\delta(0)$ | | In each run of the game, Players I, II take turns (with I starting first) |
| | $\epsilon(0)$ | choosing successively $\delta(0)$, $\epsilon(0)$, |
| $\delta(1)$ | | $\delta(1)$, $\epsilon(1)$, ...; $\delta(i)$, $\epsilon(i) \in \{0, 1\}$. |
| | $\epsilon(1)$ | So each player has produced |
| $\vdots$ | | at the end $\delta, \epsilon \in 2^{\mathbb{N}}$. Player I wins this |
| $\delta$ | $\epsilon$ | run of the game if $\delta \notin B \Leftrightarrow \epsilon \in Q$, otherwise II wins. |

A <u>strategy</u> for player I is a rule that tells him how to play each $\delta(n)$ when he sees $\epsilon(0) \ldots \epsilon(n-1)$, i.e. it is a function $\sigma : \mathrm{Seq}\,\{0, 1\} \to \{0, 1\}$, from the set of finite sequences of 0's and 1's into $\{0, 1\}$. If in a run of the game II plays $\epsilon \in 2^{\mathbb{N}}$, then I plays $\delta = \sigma^*(\epsilon)$ following $\sigma$, where

$$\sigma^*(\epsilon)(n) = \sigma(\epsilon \restriction n).$$

A strategy $\sigma$ is a <u>winning strategy</u> for I if he always wins any run of the game in which he follows $\sigma$, i.e. for all $\epsilon \in 2^{\mathbb{N}}$, the run $(\sigma^*(\epsilon), \epsilon)$ is a win for I. Similarly we define the concept of a winning strategy for II. Note the obvious but important fact that the function $\sigma^* : 2^{\mathbb{N}} \to 2^{\mathbb{N}}$ corresponding to any strategy $\sigma$ is continuous.

Now since the payoff set of this game $\{(\delta, \epsilon) \in 2^{\mathbb{N}} \times 2^{\mathbb{N}} : \delta \notin B \Leftrightarrow \epsilon \in Q\}$ is clearly a Boolean combination of $F_\sigma$ sets, it follows by a result of Morton Davis [1] (a special case of the Borel Determinacy Theorem of Martin; see Moschovakis [1], Martin [1]) that this game is <u>determined</u>, i.e. one of the two players has a winning strategy. If I had one, say $\sigma$, then we would have for each $\epsilon \in 2^{\mathbb{N}}$

$$\sigma^*(\epsilon) \notin B \Leftrightarrow \epsilon \in Q$$

i.e. letting $f = \sigma^*$,

$$Q = f^{-1}[2^{\mathbb{N}} - B].$$

Thus, since B is $F_\sigma$ and f continuous, Q would be a dense $G_\delta$. But Q is countable, so its complement is also a dense $G_\delta$ violating the Baire Category Theorem.

So II must have a winning strategy and thus as above there is a continuous function $F : 2^{\mathbb{N}} \to 2^{\mathbb{N}}$ with $\delta \in B \Leftrightarrow F(\delta) \in Q$, i.e. $B = F^{-1}[Q]$.

Now it turns out that we can prove in our special game that II has indeed a winning strategy without using Davis' Theorem. (The reason we went through all that was to bring out a bit of the general context of determinacy in which these ideas naturally fit, and as preparation for a later result in §3). Indeed since B is $F_\sigma$, we can write it as $B = \bigcup_n B_n$, where each $B_n$ is closed, thus of the form $B_n = [T_n]$, with $T_n$ a tree on Seq {0, 1} (recall here §1). Enumerate also Q in a sequence $q_0, q_1, \ldots$ .

As I plays $\delta$ ($= \delta(0), \delta(1), \ldots$) bit-by-bit, II plays $\epsilon$ ($= \epsilon(0), \epsilon(1), \ldots$) bit-by-bit as follows: As long as $\delta$ stays within $T_0$ (i.e. $\delta \upharpoonright n \in T_0$) II plays $\epsilon(0) = q_0(0)$, $\epsilon(1) = q_0(1)$, ... i.e. follows $q_0 \equiv q_0'$. If $\delta$ ever gets out of $T_0$ let $n_0 + 1$ be least such that $\delta \upharpoonright (n_0 + 1) \notin T_0$. Then II plays $\epsilon(n_0) \neq q_0(n_0)$, and chooses (by the density of Q) some $q'_1 \in Q$ with $\epsilon \restriction (n_0 + 1) \subseteq q'_1$. From then on as long as $\delta$ stays within $T_1$ II follows $q'_1$. If $\delta$ ever gets out of $T_1$, let $n_1$ be least with $n_1 > n_0$ such that $\delta \restriction (n_1 + 1) \notin T_1$. Then II plays $\epsilon(n_1) \neq q_1(n_1)$, and chooses $q'_2 \in Q$ with $\epsilon \restriction (n_1 + 1) \subseteq q'_2$, etc.

If $\delta \in B = \bigcup_n B_n$ let n be least with $\delta \in [T_n]$. Then clearly $\epsilon = q'_n \in Q$. If however $\delta \notin B$ then $\delta \notin [T_n]$ for all n, so $\epsilon \neq q_n$ for all n (since at the time $\delta$ gets out of $T_n$ we made sure that II avoided $q_n$). Therefore this is a winning strategy for II.

The fact that any $F_\sigma$ subset of $2^{\mathbb{N}}$ can be reduced to Q via a continuous function implies easily that

$$K(Q) = \{K \in K(2^{\mathbb{N}}) : K \subseteq Q\}$$

is $\underset{\sim}{\Pi}^1_1$-complete. First, it is clearly $\underset{\sim}{\Pi}^1_1$. Let now $P \subseteq 2^{\mathbb{N}}$ be $\underset{\sim}{\Pi}^1_1$-complete (for example P could be WF as in 1.3). Then by the Representation Theorem 1.4 there is $B \subseteq 2^{\mathbb{N}} \times 2^{\mathbb{N}}$, B an $F_\sigma$ with

$$x \in P \Leftrightarrow \forall \epsilon \in 2^{\mathbb{N}} (x, \epsilon) \in B.$$

Since $2^{\mathbb{N}} \times 2^{\mathbb{N}}$ is homeomorphic to $2^{\mathbb{N}}$, let F be continuous such that $B = F^{-1}[Q]$. Let then $F_1 : 2^{\mathbb{N}} \to K(2^{\mathbb{N}})$ be defined by

$$F_1(x) = F[\{x\} \times 2^{\mathbb{N}}].$$

Then $F_1$ is continuous and

$$x \in P \Leftrightarrow F_1(x) \in K(Q).$$

Thus K(Q) is $\underset{\sim}{\Pi}_1^1$-complete as well.

Choose now Q to be the set of all eventually periodic sequences in $2^{\mathbb{N}}$. We will reduce this K(Q) to K($\mathbb{Q}'$), thereby showing that K($\mathbb{Q}'$) is $\underset{\sim}{\Pi}_1^1$-complete. Let $g : 2^{\mathbb{N}} \to [0, 1]$ be given by

$$g(\epsilon) = \sum_{n=0}^{\infty} \epsilon(n)2^{-n}.$$

Then g is continuous and

$$\epsilon \in Q \Leftrightarrow g(\epsilon) \in \mathbb{Q}'.$$

Let $G : K(2^{\mathbb{N}}) \to K([0, 1])$ be given by

$$G(K) = g[K] = \{g(x) : x \in K\}$$

Then G is continuous and

$$K \in K(Q) \Leftrightarrow G(K) \in K(\mathbb{Q}'). \qquad \square$$

We are now ready to give the proof of

<u>Theorem</u> $\underset{\sim}{2}$ (Solovay [1], Kaufman [6]). The classes U, $U_0$ of closed sets of

uniqueness, resp. closed sets of extended uniqueness, are $\underset{\sim}{\Pi}^1_1$-complete. (In particular they are $\underset{\sim}{\Pi}^1_1$ non-Borel sets).

Proof. First we check that U and $U_0$ are $\underset{\sim}{\Pi}^1_1$ sets in K(T).

For U we use II.4.1, i.e. for $K \in K(\mathbb{T})$

$$(*) \quad K \in M \Leftrightarrow \exists S \in PF(S \neq 0 \;\&\; \mathrm{supp}(S) \subseteq K).$$

Now recall the classical result of Banach that the unit ball $B_1 \equiv B_1(PM) = \{S \in PM : \|S\|_{PM} \leq 1\}$ of PM with the weak*-topology is compact, metrizable (as PM is the dual of the separable Banach space A). Note then the following simple fact which extends our remarks in II.3.

Lemma 3. The set

$$\{(S, E) : S \in B_1(PM) \;\&\; E \in K(\mathbb{T}) \;\&\; \mathrm{supp}(S) \subseteq E\}$$

is closed in $B_1(PM) \times K(\mathbb{T})$, where $B_1(PM)$ is equipped with the weak*-topology.

Proof. Since $B_1$ is metrizable it is enough to show that if $S_n \xrightarrow{w*} S$, $E_n \to E$ and $\mathrm{supp}(S_n) \subseteq E_n$, then $\mathrm{supp}(S) \subseteq E$. But this is clear, since if $f \in A$ has support disjoint from E, i.e. $E \subseteq \mathbb{T} - \mathrm{supp}(f)$, then $E_n \subseteq \mathbb{T} - \mathrm{supp}(f)$ for all large enough n, so $\langle f, S_n\rangle = 0$, thus $\langle f, S\rangle = 0$. □

Thus rewriting (*) as

$$K \in M \Leftrightarrow \exists S \in B_1 \, [S \neq 0 \ \& \ (S(n) \to 0, \text{ as } |n| \to \infty) \ \& \ \text{supp}(S) \subseteq K]$$

we have that M is $\underset{\sim}{\Sigma}^1_1$, and U is $\underset{\sim}{\Pi}^1_1$.

For $U_0$, let

$$M_1(\mathbb{T}) = \{\mu \in M(\mathbb{T}) : \|\mu\|_M \leq 1\}$$

with the weak*-topology (coming from $C(\mathbb{T})^* = M(\mathbb{T})$), for which it is again compact, metrizable by Banach's Theorem. Then using II.5.6 we have for $K \in K(\mathbb{T})$,

$$K \in M_0 \Leftrightarrow \exists \mu \in M_1(\mathbb{T}) \, [\mu \neq 0 \ \& \ (\hat{\mu}(n) \to 0, \text{ as } |n| \to \infty) \ \& \ \text{supp}(\mu) \subseteq K].$$

Since $\mu \to \hat{\mu}(n)$, for each $n \in \mathbb{Z}$, is continuous in the weak*-topology of $M_1(\mathbb{T})$ we have as before that $M_0$ is $\underset{\sim}{\Sigma}^1_1$ and $U_0$ is $\underset{\sim}{\Pi}^1_1$.

To prove the $\underset{\sim}{\Pi}^1_1$-completeness of U, $U_0$ we are going to use (a special case of) the Salem-Zygmund Theorem. For each $x \in [0, 1]$ let

$$f(x) = E\left(\tfrac{1}{4}; \tfrac{3}{8} + \tfrac{x}{9}, \tfrac{3}{4}\right)$$

where $E(\xi; \eta_1 \ldots \eta_k)$ is the homogeneous perfect set of III.3. It is easy to check that f is continuous from [0, 1] into $K(\mathbb{T})$. Since $\left(\frac{1}{4}\right)^{-1} \in \mathbb{N}$ and $\mathbb{Q}\left(\frac{1}{4}\right) = \mathbb{Q}$ it follows from III.4.2 and III.4.4 that

$$x \in \mathbb{Q} \Rightarrow f(x) \in H \text{ (i.e. } f(x) \text{ is an H-set)},$$

$$x \notin Q \Rightarrow f(x) \in M_0.$$

(Notice here that we use only the particular case of III.4.4. for $\theta \in \mathbb{N}$, and thus by the remark following its proof we need only the special case of III.2.3(ii) that was proved in III.2.4).

Finally define $F : K([0, 1]) \to K(\mathbb{T})$ by

$$F(K) = \bigcup f[K]$$
$$= \bigcup \{f(x) : x \in K\}.$$

Then F is continuous and

$$K \subseteq Q \Rightarrow F(K) \text{ is a countable union of H-sets}$$
$$\Rightarrow F(K) \in U$$

since a closed set which is a countable union of closed $\mathcal{U}$-sets is also in $\mathcal{U}$ (a special case of Bary's Theorem I.5.1, which we also proved in II.4). Also

$$K \not\subseteq Q \Rightarrow F(K) \in M_0$$

since a set which contains an $M_0$-set is also $M_0$. Putting these two things together we obtain, letting $Q' = Q \cap [0, 1]$,

$$K(Q') = F^{-1}[U] = F^{-1}[U_0]$$

so by Theorem 1, U and $U_0$ are $\underset{\sim}{\Pi}^1_1$-complete. □

Note also that since the H-sets used in the above proofs are of course all perfect, it follows that <u>the set of perfect U-sets is $\underset{\sim}{\Pi}^1_1$-complete, and similarly for</u> $U_0$.

The preceding arguments give actually quite a bit more. Let us introduce first the following concepts.

<u>Definition 4</u>. Let E be compact, metrizable and let $B \subseteq K(E)$. We denote by $B_\sigma$ the set

$$B_\sigma = \{K \in K(E)\colon \text{There is a sequence } K_i \in B \text{ with } K = \cup_i K_i\}$$

i.e. $B_\sigma$ consists of all closed sets which are unions of sequences of closed sets from B.

<u>Definition 5</u>. A set P in some Polish space X is called $\underset{\sim}{\Pi}^1_1$-<u>hard</u> if for every Polish space Y and every $\underset{\sim}{\Pi}^1_1$ set Q in Y there is a Borel map $f : Y \to X$ with $Q = f^{-1}[P]$, i.e. every $\underset{\sim}{\Pi}^1_1$ set can be reduced to P. However P itself is not required to be $\underset{\sim}{\Pi}^1_1$. If it is, then P is $\underset{\sim}{\Pi}^1_1$-complete. Clearly no $\underset{\sim}{\Pi}^1_1$-hard set is $\underset{\sim}{\Sigma}^1_1$.

We have now the following corollary of the proof of Theorem 2.

<u>Theorem 6</u> (Solovay, Kaufman). Let H be the class of closed H-sets. Then if P is such that

$$H_\sigma \subseteq P \subseteq U_0$$

P is $\underset{\sim}{\Pi}^1_1$-hard, therefore not $\underset{\sim}{\Sigma}^1_1$. In particular any $\underset{\sim}{\Pi}^1_1$ such set P is $\underset{\sim}{\Pi}^1_1$-complete.

<u>Proof</u>. If F is as in the proof of Theorem 2, then clearly $K(\mathbb{Q}') = F^{-1}[P]$. Since $K(\mathbb{Q}')$ is $\underset{\sim}{\Pi}^1_1$-complete we are done. □

In contrast with the above results classes of "simple" closed $\mathcal{U}$-sets can be seen to be Borel.

For example denoting again by $H^{(n)}$ the set of closed $H^{(n)}$-sets, let us note that $H^{(n)}$ is Borel, in fact $\underset{\sim}{\Sigma}^0_3$ ($\equiv G_{\delta\sigma}$) and thus so is $\cup_n H^{(n)}$.

<u>Proposition 7</u>. For each n, $H^{(n)}$ is $\underset{\sim}{\Sigma}^0_3$ in $K(\mathbb{T})$ and thus so is $\cup_n H^{(n)}$.

<u>Proof</u>. Recall from III.1 that for $K \in K(\mathbb{T})$,

$$K \in H^{(n)} \Leftrightarrow \text{there is a basic open set } V \text{ of } \mathbb{T}^n \text{ with } \{\vec{m} \in \mathbb{Z}^n : \vec{m}\, K \cap \bar{V} = \emptyset\} \notin \mathcal{I}_n$$

where $\mathcal{I}_n$ is the ideal of subsets of $\mathbb{Z}^n$ generated by the hyperplanes. For each basic open V let

$$K \in R_V \Leftrightarrow \{\vec{m} \in \mathbb{Z}^n : \vec{m}\, K \cap \bar{V} = \emptyset\} \in \mathcal{I}_n$$

$\Leftrightarrow$ There are hyperplanes $H_1, \ldots, H_k$ such that if $\vec{m} \in \mathbb{Z}^n$ and $\vec{m}\, K \subseteq \mathbb{T}^n - \bar{V}$, then $\vec{m} \in H_1 \cup \ldots \cup H_k$.

Since there are only countably many hyperplanes and $K \mapsto \vec{m}K$ is continuous from $K(\mathbb{T})$ into $K(\mathbb{T}^n)$, for each $\vec{m} \in \mathbb{Z}^n$, it follows that $R_V$ is $\underset{\sim}{\Sigma}^0_2$ in $K(\mathbb{T})$, so (since there are only countably many such V's) $H^{(n)}$ is $\underset{\sim}{\Sigma}^0_3$. □

We can see also that the set U′ of closed $\mathcal{U}$-sets which satisfy the Piatetski-Shapiro criterion III.1.3 (and which contains $\cup_n H^{(n)}$ by III.1.4) is also $\underset{\sim}{\Sigma}^0_3$.

Proposition 8. Let

$$U' = \{K \in K(\mathbb{T}) : \exists \{f_n\} \ (f_n \in A \text{ and } \mathrm{supp}(f_n) \cap K = \emptyset \text{ and } f_n \xrightarrow{w*} 1\} \subseteq U.$$

Then U′ is $\underset{\sim}{\Sigma}^0_3$ (i.e. $G_{\delta\sigma}$) in $K(\mathbb{T})$.

Proof. We have, denoting by $B_r = B_r(A)$ the ball $\{f \in A : \|f\|_A \leq r\}$ which is compact, metrizable in the weak*-topology,

$$K \in U' \Leftrightarrow \exists n \in \mathbb{N} \ \forall m \in \mathbb{N} \ \exists f \in B_n$$
$$[\mathrm{supp}\,(f) \cap K = \emptyset \text{ and } \forall |k| \leq m \ (|\hat{f}(k) - \hat{1}(k)| < \tfrac{1}{m+1})].$$

Thus U′ is $\underset{\sim}{\Sigma}^0_3$. □

It follows also trivially that the set of homogeneous U- or $U_0$- sets is Borel, being countable. However this question is open for the symmetric sets. Since the map

$$(\xi_1, \xi_2, \ldots) \mapsto E_{\xi_1, \xi_2, \ldots}$$

with domain $(0, \frac{1}{2})^{\mathbb{N}-\{0\}}$ into $K(\mathbb{T})$ is clearly continuous and 1-1, it follows (see §1) that the question of whether the set of symmetric perfect U-sets is Borel in $K(\mathbb{T})$ is equivalent to the following

<u>Problem 9</u>. Is the set

$$\{(\xi_1,\xi_2,\ldots) \in (0, \tfrac{1}{2})^{\mathbb{N}-\{0\}} : E_{\xi_1,\xi_2,\ldots} \in U\}$$

Borel in $(0, \frac{1}{2})^{\mathbb{N}-\{0\}}$?

We do not state the corresponding problem for $U_0$ because in Chapter VIII (see VIII.3.6) we will obtain the affirmative answer that if U is replaced by $U_0$, this is indeed a Borel set.

We will finish this section with some remarks concerning the relevance of the non-Borelness of U, $U_0$ to the Characterization Problem. This clearly rules out any kind of analytic, geometric or arithmetic structural characterizations which are "too simple" or "explicit enough" so they can be expressed in terms of countable operations from any standard specification of a closed set, like e.g. its contiguous intervals. This is because any such description would lead to a Borel definition of U or $U_0$.

On the other hand there are other situations, where a class of small sets although of high complexity can still be analyzed in terms of simpler components. A typical instance of that is the class $K_\omega(E)$ of the countable closed sets of a compact, metrizable space E. If E is uncountable, this is a $\utilde{\Pi}^1_1$-complete set as well.

(We will prove this shortly in §3). So although in general we do not have an explicit criterion of when a closed set is countable, we can still decompose every such set in a countable union of very simple sets, i.e. singletons. So one can ask whether a similar situation might exist with the U- or $U_0$-sets. This is the "Borel Basis Problem" that will occupy our attention in Chapters VI, VII, VIII. Surprisingly the answer is different for U and $U_0$. It is negative for U-sets, thereby establishing an even more negative implication for the Characterization Problem. On the other hand it is positive for $U_0$-sets, which leaves open some further possibilities for positive results, for instance in the case of symmetric perfect sets. We will discuss these issues further in the relevant chapters.

## §3. On $\sigma$-ideals of closed sets in compact, metrizable spaces

Let us recall a general set-theoretic terminology concerning classes of sets.

Let $\mathfrak{F}$ be a family of sets and let $\mathcal{C} \subseteq \mathfrak{F}$. Then $\mathcal{C}$ is an ideal if $\mathcal{C}$ is nonempty hereditary, i.e. $P \in \mathcal{C}$, $Q \in \mathfrak{F}$ and $Q \subseteq P \Rightarrow Q \in \mathcal{C}$, and closed under finite unions which are in $\mathfrak{F}$, i.e. if $P \in \mathcal{C}$, $Q \in \mathcal{C}$ and $P \cup Q \in \mathfrak{F}$, then $P \cup Q \in \mathcal{C}$. We call $\mathcal{C}$ a $\sigma$-ideal if it is also closed under countable unions which are in $\mathfrak{F}$, i.e. if $P_0, P_1, \ldots \in \mathcal{C}$ and $P = P_0 \cup P_1 \cup \ldots$ is in $\mathfrak{F}$, then $P \in \mathcal{C}$.

We will be mainly interested in classes of closed subsets of some compact, metrizable set E, i.e. in subsets of K(E). Specializing the above terminology, a set I $\subseteq$ K(E) is an ideal if it is nonempty hereditary and closed under finite unions and it is a $\sigma$-ideal if it is also closed under countable unions which are closed. The basic closure and definability properties of U, $U_0$ can then be summarized as follows:

The classes U, $U_0$ of closed $\mathcal{U}$-sets, resp. $\mathcal{U}_0$-sets, are $\sigma$-ideals in K(E). Moreover they are both $\underset{\sim}{\Pi}^1_1$-complete.

Of course $\sigma$-ideals of closed sets occur very frequently in various parts of analysis as "smallness" notions or "exceptional sets". Some examples are: the nowhere dense closed sets, the $\mu$-measure 0 closed sets for some positive measure $\mu$ (and various types of similar negligible sets for other types of set functions, like (subadditive) capacities, etc.), the countable closed sets, and of course the above examples U, $U_0$.

Most of the interesting examples of such $\sigma$-ideals of closed sets are of course explicitly definable in low levels of the projective hierarchy. Indeed their vast majority are easily seen to be $\underset{\sim}{\Pi}^1_1$. Motivated by the result of Solovay and Kaufman, a systematic study of the descriptive set theoretic properties of $\underset{\sim}{\Pi}^1_1$ $\sigma$-ideals of closed sets in compact, metrizable spaces has been undertaken in Kechris-Louveau-Woodin [1]. In this section we will state and prove one of the main results of that paper, which concerns the complexity of such $\sigma$-ideals. This result, the so-called Dichotomy Theorem, will be used often in the sequel. We will return to another major theme of Kechris-Louveau-Woodin [1], namely the Borel Basis Problem for $\underset{\sim}{\Pi}^1_1$ $\sigma$-ideals, in Chapter VI.

Let us state the main result now

<u>Theorem 1</u> (The Dichotomy Theorem; Kechris-Louveau-Woodin [1]). Let E be a compact, metrizable space. Then every $\underset{\sim}{\Pi}^1_1$ $\sigma$-ideal of closed subsets of E is either $\underset{\sim}{\Pi}^0_2$ (i.e. $G_\delta$) or else $\underset{\sim}{\Pi}^1_1$-complete.

In particular every true $\underset{\sim}{\Pi}^1_1$ such $\sigma$-ideal is $\underset{\sim}{\Pi}^1_1$-complete.

This shows that the structural requirement of being a $\sigma$-ideal imposes strong definability restrictions. (Recall here that the second conclusion above – "true $\underset{\sim}{\Pi}^1_1$" implies "$\underset{\sim}{\Pi}^1_1$-complete" – is valid for any set using strong axioms of set theory; see §1. But of course the proof above for $\sigma$-ideals is carried within the usual classical set theory ZFC).

We will derive an even stronger version of Theorem 1 from a Hurewicz-type theorem. The original Hurewicz Theorem states the following

Theorem 2 (Hurewicz [1]). Let B be a $\underset{\sim}{\Pi}^1_1$ subset of a compact, metrizable space E. Then either B is $G_\delta$ or else there is a homeomorphism $f : 2^{\mathbb{N}} \to E$ (not necessarily onto E) and a countable dense set $Q \subseteq 2^{\mathbb{N}}$ with $f[Q] \subseteq B$, $f[2^{\mathbb{N}} - Q] \subseteq E - B$. In other words if B is not $G_\delta$, one can construct in E a homeomorphic copy F of $2^{\mathbb{N}}$ such that $F \cap B$ is a countable dense subset of F.

This is enough (by the argument in Theorem 4 below) to prove Theorem 1, but one can obtain a better result using the following strengthening of Hurewicz's Theorem.

Theorem 3 (Kechris-Louveau-Woodin [1]). Let E be a compact, metrizable space. Let P, B be two disjoint subsets of E with P in $\underset{\sim}{\Sigma}^1_1$. If there is no $F_\sigma$ set C separating P from B (i.e. $P \subseteq C$, $B \cap C = \emptyset$), then there is a homeomorphic copy F of $2^{\mathbb{N}}$ with $F \subseteq P \cup B$, and $F \cap B$ countable dense in F.

By taking $B = E - P$ we obtain Hurewicz's Theorem.

Let us see first how Theorem 3 implies the promised stronger version of Theorem 1. Indeed from Theorem 3 we can derive

Theorem 4 (Kechris-Louveau-Woodin [1]). Let I be a $\underset{\sim}{\Pi}^1_1$ $\sigma$-ideal of closed sets in a compact, metrizable space E. Let $B \subseteq I$, and let $B_\sigma$ be the class of closed sets which are countable unions of sets in B. (Thus $B_\sigma \subseteq I$). If there is no $G_\delta$ set G with $B \subseteq G \subseteq I$, then $B_\sigma$ is $\underset{\sim}{\Pi}^1_1$-hard.

Proof. If no such $G_\delta$ set G exists apply Theorem 3 to $P = K(E) - I$ and B. Then we obtain a homeomorphic copy F of $2^{\mathbb{N}}$ with $F \subseteq (K(E) - I) \cup B$ and $F \cap B = Q$ countable dense in F. Consider the continuous map $g : K(F) \to K(E)$ given by $g(L) = \cup L = \cup\{K : K \in L\}$. Then for $L \in K(F)$,

$$L \subseteq F \cap B \Leftrightarrow \cup L \in B_\sigma \Leftrightarrow \cup L \in I$$

so $g^{-1}[B_\sigma] = g^{-1}[I] = K(Q) = \{K \in K(F) : K \subseteq Q\}$. But in the proof of 2.1 we showed that K(Q) is $\underset{\sim}{\Pi}^1_1$-complete, so $B_\sigma$ is $\underset{\sim}{\Pi}^1_1$-hard and we are done. □

Letting $B = I$ in Theorem 4 we obtain Theorem 1.

It remains only to give the

Proof of Theorem 3. By a standard fact (see Kuratowski [1]), since E is compact, metrizable there is a continuous surjection $f : 2^{\mathbb{N}} \twoheadrightarrow E$. Let $P' = f^{-1}[P]$, $B' =$

$f^{-1}[B]$.

Let Q be a countable dense subset of $2^{\mathbb{N}}$ and consider the following Wadge-type game, where for basic explanations about games see the proof of 2.1:

| I | II | |
|---|---|---|
| $\delta$ | $\epsilon$ | I plays bit-by-bit $\delta \in 2^{\mathbb{N}}$ and II plays bit-by-bit $\epsilon \in 2^{\mathbb{N}}$; II wins this run of the game iff |

$$(\delta \in Q \Rightarrow \epsilon \in B') \text{ and } (\delta \notin Q \Rightarrow \epsilon \in P').$$

If player I has a winning strategy, this strategy gives a continuous function $g : 2^{\mathbb{N}} \to 2^{\mathbb{N}}$ with $C' = g^{-1}[Q]$ a $F_\sigma$ set separating $P'$ from $B'$. Then $C = f[C']$ is a $F_\sigma$ set (since continuous images of closed sets are closed by compactness) separating P from B, a contradiction. So if this game is determined, II has a winning strategy, and there is a continuous function $h : 2^{\mathbb{N}} \to 2^{\mathbb{N}}$ with $h[2^{\mathbb{N}}] \subseteq P' \cup B'$ and $h^{-1}[B'] = Q$. Composing this with f we obtain a continuous $p : 2^{\mathbb{N}} \to E$ with $p[2^{\mathbb{N}}] \subseteq P \cup B$ and $p^{-1}[B] = Q$. Then $F' = p[2^{\mathbb{N}}]$ is compact in E and $p[Q]$, $p[2^{\mathbb{N}} - Q]$ are disjoint dense subsets of $F'$, hence $F'$ is perfect. Then by a standard Cantor-type splitting construction we can find inside $F'$ a homeomorphic copy F of $2^{\mathbb{N}}$ with $F \cap p[Q] = F \cap B$ dense in F, so we are done.

It remains to show that the above game is determined. As B is arbitrary, the payoff is arbitrary too, and we cannot appeal to determinacy results. However, due to the particular nature of the game, we can argue as follows:

Let $G \subseteq 2^{\mathbb{N}} \times 2^{\mathbb{N}}$ be $G_\delta$ and project to P′, by the representation theorem 1.4. Consider the largest open set V whose projection proj $(V \cap G)$ is $F_\sigma$-separable from B′. Then $G_0 = G - V \neq \emptyset$, since P′ cannot be separated from B′ by a $F_\sigma$ set. Now $G_0$ being a $G_\delta$ in a Polish space is also a Polish space (in the relative topology it inherits from $2^{\mathbb{N}} \times 2^{\mathbb{N}}$) - this is a standard fact about Polish spaces, see Kuratowski [1]. Let $\{W_n\}$ be a basis for the topology of $G_0$, consisting of non-empty open sets. By the maximality of V, $\overline{\mathrm{proj}(W_n)} \cap B' \neq \emptyset$, so choose $x_n \in \overline{\mathrm{proj}(W_n)} \cap B'$. Let $B_0 = \{x_0, x_1, \ldots\}$, and consider the pair $G_0$, $B_0 \times 2^{\mathbb{N}}$ in $2^{\mathbb{N}} \times 2^{\mathbb{N}}$. Now $G_0$ cannot be separated from $B_0 \times 2^{\mathbb{N}}$ by a $F_\sigma$ set, because if $\cup_i K_i$ is such a set, $K_i$ closed, then by the Baire Category Theorem (applied to $G_0$), for some i, n $W_n \subseteq K_i$ and so $\overline{\mathrm{proj}(W_n)} \subseteq \overline{\mathrm{proj}(K_i)} = \mathrm{proj}(K_i)$, since $\mathrm{proj}(K_i)$ is closed. So $x_n \in \mathrm{proj}(K_i)$ and $K_i \cap (B_0 \times 2^{\mathbb{N}}) \neq \emptyset$, a contradiction. We can play now the Wadge-type game as above for $G_0$, $B_0 \times 2^{\mathbb{N}}$ instead of P′, B′ (identifying here $2^{\mathbb{N}} \times 2^{\mathbb{N}}$ with $2^{\mathbb{N}}$). This game is a Boolean combination of $G_\delta$ sets, so it is determined by the Morton Davis result, a particular case of Martin's Borel Determinacy Theorem which asserts that all Borel games are determined (see Moschovakis [1], Martin [1]). As above player II must have a winning strategy in this game. Composing the continuous function coming from his strategy with the projection we obtain the function h as before. □

Theorem 1 separates the $\underset{\sim}{\Pi}^1_1$ $\sigma$-ideals into two categories

(A) The "simple" ones which are $\underset{\sim}{\Pi}^0_2$ (i.e. $G_\delta$). Typical examples are the nowhere dense closed sets and the $\mu$-measure 0 closed sets for any positive measure $\mu$.

(B) The "complicated" ones, which are $\underset{\sim}{\Pi}^1_1$-complete. Typical examples are the countable closed sets in an uncountable space and also U, $U_0$. For completeness let us give also a proof for the countable closed sets, a classical result of Hurewicz [1], using Theorem 3.

<u>Theorem</u> <u>5</u> (Hurewicz [1]). Let E be a compact, metrizable space and assume E is uncountable. Then the $\sigma$-ideal $K_\omega(E)$ of countable closed subsets of E is $\underset{\sim}{\Pi}^1_1$-complete.

<u>Proof</u>. First notice that $K_\omega(E)$ is $\underset{\sim}{\Pi}^1_1$ since

$$K \in K_\omega(E) \Leftrightarrow \forall L \in K(E) \, (L \subseteq K \ \& \ L \neq \emptyset \Rightarrow L \text{ is not perfect}).$$

By Theorem 1 it is enough now to show that $K_\omega(E)$ is not $G_\delta$.

Since every uncountable Polish space contains a homeomorphic copy of $2^{\mathbb{N}}$ (see §1) it suffices to prove this for $E = 2^{\mathbb{N}}$.

Let $D = \{\epsilon \in 2^{\mathbb{N}} : \epsilon$ has only finitely many 1's$\}$. Then D is countable and dense in $2^{\mathbb{N}}$, so it cannot be a $G_\delta$ by the Baire Category Theorem. Define $f : 2^{\mathbb{N}} \to K(2^{\mathbb{N}})$ by

$$f(\epsilon) = \{\epsilon' : \forall n(\epsilon'(n) \leq \epsilon(n))\}.$$

Clearly f is continuous, and $f(\epsilon)$ is finite if $\epsilon \in D$, while $f(\epsilon)$ is perfect if $\epsilon \notin D$. So

$$\epsilon \in D \Leftrightarrow f(\epsilon) \in K_\omega(2^{\mathbb{N}})$$

and therefore $K_\omega(2^{\mathbb{N}})$ cannot be a $G_\delta$ set. By Theorem 3 it must be $\underset{\sim}{\Pi}^1_1$-complete. □

As a final remark, let us point out that a similar proof shows that Theorem 2.1 holds for arbitrary perfect compact, metrizable E, i.e. if $Q \subseteq E$ is countable dense, then K(Q) is $\underset{\sim}{\Pi}^1_1$-complete. This is because Q is not $G_\delta$ and $x \in Q \Leftrightarrow \{x\} \in K(Q)$, so K(Q) cannot be a $G_\delta$, thus it has to be $\underset{\sim}{\Pi}^1_1$-complete by Theorem 1.

# Chapter V. The Piatetski-Shapiro Hierarchy of U-sets

In [2], Piatetski-Shapiro introduced a rank on U-sets, i.e. a function assigning to each closed set of uniqueness a countable ordinal number, in order to prove a decomposition theorem that we will discuss in Chapter VI. This rank can be viewed as a classification of the U-sets according to how hard it is to verify that they belong to the class U. For the analyst, such a rank will be "natural" if it is nicely related to natural objects associated with the closed set E under consideration. Similar ideas of ranking arise in descriptive set theory, in the study of $\underset{\sim}{\Pi}_1^1$ sets. They were introduced at the early stages of the theory, by Lusin and Sierpinski. In this context, the notion of naturalness for a rank is replaced by definability conditions on the associated ordering, leading to the modern notion of $\underset{\sim}{\Pi}_1^1$-rank, due to Moschovakis. We study $\underset{\sim}{\Pi}_1^1$-ranks on $\underset{\sim}{\Pi}_1^1$ sets in §1, establishing their existence and main properties. We introduce in §2 the ideas from functional analysis that form the background in Piatetski-Shapiro's definition of a rank for U. The main concept, due to Banach, is that of the order (a countable ordinal number) of a subspace of the dual of a separable Banach space. In §3, we give two other ways of defining ranks for such subspaces and prove their equivalence with that of §2. These ideas will be also useful in later chapters. In §4 the Piatetski-Shapiro rank of U-sets is introduced and the equivalence between the ranks introduced in §2, §3 is used to prove Solovay's result that this rank is indeed a $\underset{\sim}{\Pi}_1^1$-rank on U. In the final §5, we discuss U-sets of rank 1, i.e. the "simplest" ones in the hierarchy, showing in particular that $H^{(n)}$-sets, and countable closed sets are of rank 1.

## §1. $\underset{\sim}{\Pi}^1_1$-ranks on $\underset{\sim}{\Pi}^1_1$ sets

A rank on a set P is just a function $\varphi$ from P into the set of countable ordinals. Associated with $\varphi$ is a relation $\leq_\varphi$ on the elements of P, defined by

$$x \leq_\varphi y \Leftrightarrow \varphi(x) \leq \varphi(y).$$

When P is a subset of some space X, we will extend $\varphi$ to all of X by letting $\varphi(x) = \omega_1$ (= the first uncountable ordinal) for $x \notin P$, and hence extend $\leq_\varphi$ to $\leq^*_\varphi$ defined by

$$\begin{aligned} x \leq^*_\varphi y &\Leftrightarrow x \in P \text{ and } \varphi(x) \leq \varphi(y) \\ &\Leftrightarrow x \in P \text{ and } [y \notin P \text{ or } (y \in P \text{ and } \varphi(x) \leq \varphi(y))]. \end{aligned}$$

We will also consider $<^*_\varphi$ defined by

$$\begin{aligned} x <^*_\varphi y &\Leftrightarrow x \in P \text{ and } \varphi(x) < \varphi(y) \\ &\Leftrightarrow \varphi(x) < \varphi(y) \\ &\Leftrightarrow x \in P \text{ and } [y \notin P \text{ or } (y \in P \text{ and } \varphi(x) < \varphi(y))]. \end{aligned}$$

Definition 1. Let X be a Polish space and $P \subseteq X$ be $\underset{\sim}{\Pi}^1_1$. A rank $\varphi$ on P is a $\underset{\sim}{\Pi}^1_1$-rank if the relations

$$x \leq^*_\varphi y,\ x <^*_\varphi y$$

are both $\underset{\sim}{\Pi}^1_1$, as subsets of $X^2$.

In descriptive set theory this notion, except for our added (harmless) requirement that $\varphi$ maps into $\omega_1$ as opposed to arbitrary ordinals, is referred to as a $\underset{\sim}{\Pi}^1_1$-norm (see Moschovakis [1]). We chose to change this standard terminology here as it seemed less appropriate in the current context.

Example. The case of the set of well founded trees WF (see IV.1.3). To define a rank on WF, we define by induction for any well founded tree T on Seq Y the height of s in T, ht(s, T), by

$$\mathrm{ht}(s, T) = 0, \text{ if } s \notin T,$$

$$\mathrm{ht}(s, T) = \sup\{\mathrm{ht}(s^\frown(y), T) + 1 : y \in Y\}, \text{ for } s \in T$$

$$(= \sup\{\mathrm{ht}(t, T) + 1 : s \subsetneqq t\}).$$

We define finally ht(T), the height of T, as ht(∅, T). Note that this is a countable ordinal for each $T \in \mathrm{WF}$, i.e. each well founded tree on Seq $\mathbb{N}$.

Lemma 2. The function ht is a $\underset{\sim}{\Pi}^1_1$-rank on the $\underset{\sim}{\Pi}^1_1$ set WF.

Proof. Extend as usual ht to all trees by letting $\mathrm{ht}(T) = \omega_1$ if T is not well founded. Say that a function $f : \mathrm{Seq}\,\mathbb{N} \to \mathrm{Seq}\,\mathbb{N}$ respects inclusion if $s \subsetneqq t \Rightarrow f(s) \subsetneqq f(t)$. We claim that for trees $T_1$, $T_2$

$$\mathrm{ht}(T_1) \leq \mathrm{ht}(T_2) \Leftrightarrow \exists f \text{ respecting inclusion with } f[T_1] \subseteq T_2$$

To see this, suppose first $f : \mathrm{Seq}\,\mathbb{N} \to \mathrm{Seq}\,\mathbb{N}$ respects inclusion and sends $T_1$ into $T_2$. If $T_2$ is well founded, $T_1$ must be well founded too, as the image by f of a branch through $T_1$ would yield a branch through $T_2$. One then proves by induction on $ht(s, T_1)$ that $ht(s, T_1) \leq ht(f(s), T_2)$. This is clear if $s \notin T_1$. Assuming the result for all $s^\frown(n)$, $n \in \mathbb{N}$, we get for $s \in T_1$,

$$\begin{aligned} ht(s, T_1) &= \sup\{ht(s^\frown(n), T_1) + 1 : n \in \mathbb{N}\} \\ &\leq \sup\{ht(f(s^\frown(n)), T_2) + 1 : n \in \mathbb{N}\} \\ &\leq \sup\{ht(t, T_2) + 1 : f(s) \subsetneq t\} \\ &= ht(f(s), T_2). \end{aligned}$$

This proves direction $\Leftarrow$.

Direction $\Rightarrow$ is easy if $T_2$ has a branch $\epsilon$: Define then $f(s) = (\epsilon(0),\ldots, \epsilon(\ell h(s) - 1))$. If now $T_2$ is well founded, one proves $\Rightarrow$ by induction on the ordinal $\alpha = ht(T_2)$. Suppose the result is true for all trees $T_1'$, $T_2'$ with $ht(T_1') \leq ht(T_2') < \alpha$. Let $T_1$ be a tree with $ht(T_1) \leq ht(T_2)$. For each n, consider the tree $T_1^n = \{s \in \mathrm{Seq}\ \mathbb{N} : (n)^\frown s \in T_1\}$, and similarly for $T_2^n$. One easily checks that $T_1^n$ is well founded, and $ht(T_1^n) = ht(\langle n\rangle, T_1) < ht(T_2)$. As $ht(T_2) = \sup\{ht(T_2^m) + 1 : m \in \mathbb{N}\}$, one can define $g(n) = \text{least } m(ht(T_1^n) \leq ht(T_2^m))$. Then by the induction hypothesis, there exists $f_n$ respecting inclusion with $f_n[T_1^n] \subseteq T_2^{g(n)}$. Define then f by $f(\emptyset) = \emptyset$, and $f((n)^\frown s) = (g(n))^\frown f_n(s)$. One easily checks that f works. This establishes our claim.

Using this claim, one gets (for trees $T_1$, $T_2$):

$$\begin{aligned} &T_1 \in WF \text{ and } ht(T_1) < ht(T_2) \\ \Leftrightarrow &T_1 \in WF \text{ and } ht(T_2) \not\leq ht(T_1) \end{aligned}$$

$$\Leftrightarrow T_1 \in WF \text{ and for all inclusion respecting } f: \text{Seq } \mathbb{N} \to \text{Seq } \mathbb{N}$$
$$f[T_2] \underset{\neq}{\not\subseteq} T_1$$

and similarly,

$$T_1 \in WF \text{ and } ht(T_1) < ht(T_2)$$
$$\Leftrightarrow T_1 \in WF \text{ and } ht(T_2) \not\leq ht(T_1)$$
$$\Leftrightarrow T_1 \in WF \text{ and } \forall m\ (ht(T_2) \not\leq ht(T_1^m))$$
$$\Leftrightarrow T_1 \in WF \text{ and for all inclusion respecting } f: \text{Seq } \mathbb{N} \to \text{Seq } \mathbb{N}$$
$$\text{and all } m,\ f[T_2] \underset{\neq}{\not\subseteq} T_1^m.$$

This easily gives that both relations are $\underset{\sim}{\Pi}^1_1$, hence ht is a $\underset{\sim}{\Pi}^1_1$-rank on WF. For example to see that the relation

$$(T_1, T_2) \in P \Leftrightarrow \text{for all inclusion respecting}$$
$$f: \text{Seq } \mathbb{N} \to \text{Seq } \mathbb{N},\ f[T_2] \underset{\neq}{\not\subseteq} T_1$$

is $\underset{\sim}{\Pi}^1_1$ (as a subset of $2^{\text{Seq } \mathbb{N}} \times 2^{\text{Seq } \mathbb{N}}$), let $\rho : \mathbb{N} \to \text{Seq } \mathbb{N}$ be a 1–1 correspondence and note that

$$(T_1, T_2) \in P \Leftrightarrow \forall g \in \mathbb{N}^{\mathbb{N}}[(\forall m, n \in \mathbb{N}(\rho(m) \subsetneq \rho(n) \Rightarrow \rho(g(m)) \subsetneq \rho(g(n)))) \Rightarrow$$
$$\exists m\ (\rho(m) \in T_2\ \&\ \rho(f(m)) \notin T_1)].$$

The relation of g, $T_1$, $T_2$ defined by the expression in [...] is clearly Borel (in $\mathbb{N}^{\mathbb{N}} \times 2^{\text{Seq } \mathbb{N}} \times 2^{\text{Seq } \mathbb{N}}$) and thus P is $\underset{\sim}{\Pi}^1_1$. □

The following is obvious from the closure properties of $\underset{\sim}{\Pi}^1_1$-sets (see IV.1).

<u>Lemma</u> 3. Let X, Y be Polish spaces, and $f : X \to Y$ Borel. Let $P \subseteq X$ and $Q \subseteq Y$ be $\underset{\sim}{\Pi}_1^1$ and assume $f^{-1}[Q] = P$. If $\varphi$ is a $\underset{\sim}{\Pi}_1^1$-rank on Q, then $\varphi \circ f$ is a $\underset{\sim}{\Pi}_1^1$-rank on P.

<u>Theorem</u> 4. Any $\underset{\sim}{\Pi}_1^1$ set admits a $\underset{\sim}{\Pi}_1^1$-rank.

<u>Proof</u>. Immediate from Lemmas 2, 3 and the fact that WF is $\underset{\sim}{\Pi}_1^1$-complete; see IV.1.3. □

The existence of $\underset{\sim}{\Pi}_1^1$-ranks on $\underset{\sim}{\Pi}_1^1$ sets has many applications in the structure theory of these sets. We will concentrate here on some of them which will be needed in the sequel. For more information, the reader may consult Moschovakis [1].

First, $\underset{\sim}{\Pi}_1^1$-ranks give nice approximations of $\underset{\sim}{\Pi}_1^1$ sets by Borel ones:

<u>Lemma</u> 5. Let $\varphi$ be a $\underset{\sim}{\Pi}_1^1$-rank on the $\underset{\sim}{\Pi}_1^1$ set P. Then for each $\alpha < \omega_1$, $P_\alpha = \{x \in P : \varphi(x) \leq \alpha\}$ is a Borel subset of P. In particular if P is not Borel, the $\underset{\sim}{\Pi}_1^1$-rank $\varphi$ is unbounded on P.

<u>Proof</u>. The second assertion comes immediately from the first, so it is enough to show that for $x_0 \in P$

$$C = \{x : \varphi(x) \leq \varphi(x_0)\}$$

is Borel. Now

$$\varphi(x) \leq \varphi(x_0) \Leftrightarrow x \leq^*_\varphi x_0$$

hence C is $\underset{\sim}{\Pi}^1_1$, and

$$\varphi(x) \leq \varphi(x_0) \Leftrightarrow \text{not } (x_0 <^*_\varphi x)$$

hence C is $\underset{\sim}{\Sigma}^1_1$. By Suslin's Theorem (see IV.1) C is Borel. □

One of the most useful results concerning $\underset{\sim}{\Pi}^1_1$-ranks is a converse to this lemma. Although it can be proved directly, we prefer to derive it from a more general result about well founded $\underset{\sim}{\Sigma}^1_1$ relations. It is a particular case of the Kunen-Martin Theorem, although it was known much earlier than that; see Moschovakis [1].

Let X be a Polish space. A binary relation $\prec$ on X is <u>well founded</u> if there is no sequence $x_0, x_1, x_2, \ldots, x_n, \ldots$ in X such that for all n $x_{n+1} \prec x_n$. If $\prec$ is well founded, one can define inductively $\ell h(x, \prec)$ by $\ell h(x, \prec) = 1$ if there is no $y \prec x$ (i.e. x is minimal for $\prec$), and if x is not minimal $\ell h(x, \prec) = \sup\{\ell h(y, \prec) + 1 : y \prec x\}$. The <u>length</u> of the well founded relation $\prec$ is the ordinal $\ell h(\prec) = \sup\{\ell h(x, \prec) + 1 : x \in X\}$.

<u>Theorem</u> <u>6</u>. Let X be a Polish space. Any $\underset{\sim}{\Sigma}^1_1$ well founded relation on X has countable length.

<u>Proof</u>. Instead of working directly with the well founded relation $\prec$, consider the tree T on Seq X defined by $s = (x_0, \ldots x_{n-1}) \in T \Leftrightarrow x_{n-1} \prec x_{n-2} \cdots \prec x_1 \prec x_0$. Clearly as $\prec$ is well founded, T is well founded too. One easily checks by induction on $\ell h(x, \prec)$, that for $(x_0, \ldots, x_{n-1}, x) \in T$

$$ht((x_0, \ldots, x_{n-1}, x), T) = \ell h(x, \prec)$$

so that $\ell h(\prec) = \sup\{ht((x), T) + 1 : x \in X\} = ht(T)$.

Thus it is enough to prove that ht(T) is countable. To do this, we construct a well founded tree $T^*$ on Seq $\mathbb{N}$, and a function $f : T \to T^*$ such that $u \subsetneq v \Rightarrow f(u) \subsetneq f(v)$, (i.e. f respects inclusion). As in Lemma 2, this implies that $ht(T) \leq ht(T^*)$. But as $T^*$ is countable, $ht(T^*)$ is a countable ordinal and we are done.

To construct $T^*$ and f, let $\{V_n\}_{n \in \mathbb{N}}$ be a basis for the topology on $X^2$, and let $F \subseteq X^2 \times \mathbb{N}^{\mathbb{N}}$ be closed so that $y \prec x \Leftrightarrow \exists\, \epsilon \in \mathbb{N}^{\mathbb{N}}\ (x, y, \epsilon) \in F$, by the representation theorem IV.1.4. For each pair (x, y) with $y \prec x$, pick for each m an integer $n = n_{x,y}(m)$ such that $(x,y) \in V_n$ and $\mathrm{diam}(V_n) \leq 2^{-m}$, and an $\epsilon_{x,y}$ such that $(x, y, \epsilon_{x,y}) \in F$. Define $g_{x,y} : \mathbb{N} \to \mathbb{N}$ by:

$$g_{x,y}(2m) = \epsilon_{x,y}(m)$$

$$g_{x,y}(2m + 1) = n_{x,y}(m).$$

The functions $g_{x,y}$ have the following property:

(*) if $x_n$, $y_n$ are sequences of points in X, if for all n, $y_n \prec x_n$, and if for each $m \in \mathbb{N}$ the sequence $g_{x_n,y_n}(m)$ is eventually constant as $n \to \infty$, i.e. $g_{x_n,y_n}$ converges in $\mathbb{N}^{\mathbb{N}}$, then for some x, y in X, $x_n \to x$, $y_n \to y$ and $y \prec x$.

To see this, let $g(m) = \lim g_{x_n,y_n}(m)$. For each m and n big enough, $(x_n,y_n) \in$

$V_{g(2m+1)}$ which is of diameter $\leq 2^{-m}$. This shows that $x_n \to x$ and $y_n \to y$ for some x,y in X. Let $\epsilon(m) = g(2m)$, $\epsilon_n = \epsilon_{x_n,y_n}$. By our assumption, $\epsilon_n \to \epsilon$ as $n \to \infty$, hence, as $(x_n, y_n, \epsilon_n) \in F$, $(x, y, \epsilon) \in F$. This shows $y \prec x$ and proves (*).

We now define $f : T \to \mathrm{Seq}\ \mathbb{N}$ as follows:

$$f(\emptyset) = \emptyset,\ f(x_0) = (0),\ f((x_0, x_1)) = (0, g_{x_0,x_1}(0)),$$

$$f(x_0, x_1, x_2) = f((x_0, x_1))\ \hat{}\ (g_{x_1,x_2}(0), g_{x_1,x_2}(1), g_{x_0,x_1}(1)),$$

and more generally for $n > 2$,

$$\begin{aligned} f(x_0, x_1,\dots, x_n) = f(x_0, \dots, x_{n-1})\hat{}(&g_{x_{n-1},x_n}(0), g_{x_{n-1},x_n}(1), \dots,\\ &g_{x_{n-1},x_n}(n-1), g_{x_{n-2},x_{n-1}}(n-1), g_{x_{n-3},x_{n-2}}(n-1),\\ &\dots, g_{x_0,x_1}(n-1)). \end{aligned}$$

Let $T^* = \{s \in \mathrm{Seq}\ \mathbb{N} : \exists u \in T(s \subseteq f(u))\}$. Clearly $T^*$ is a tree on Seq $\mathbb{N}$ and $f : T \to T^*$ satisfies $u \subsetneq v \Rightarrow f(u) \subsetneq f(v)$. So it is enough to check that $T^*$ is well founded.

Assume not, towards a contradiction. Let $v_n$ be in $T^*$ with $lh(v_n) = n$ and $v_n \subseteq v_{n+1}$, and let $u_n = (x_0^n,\dots, x_{n-1}^n)$ in T be such that $v_n \subseteq f(u_n)$. Then easily for each p and m, the sequence $g_{x_p^n,x_{p+1}^n}(m)$ is eventually constant (equal to $v_{q+1}(q)$, for some q depending on p and m). Using (*), this implies that for each p, $x_p^n$ converges with n to some $x_p \in X$, and for all p $x_{p+1} \prec x_p$, contradicting the well foundedness of the relation $\prec$. This finishes the proof. □

As a corollary, one gets the following result (proved originally for a specific rank by Lusin).

<u>Theorem</u> <u>7</u> (The Boundedness Theorem). Let P be a $\underset{\sim}{\Pi}^1_1$ set and $\varphi$ a $\underset{\sim}{\Pi}^1_1$-rank on P. If $Q \subseteq P$ is $\underset{\sim}{\Sigma}^1_1$, then $\varphi$ is bounded on Q, i.e. $\sup\{\varphi(x) : x \in Q\} < \omega_1$. In particular

$$P \text{ is Borel} \Leftrightarrow \varphi \text{ is bounded on } P.$$

<u>Proof</u>. The second statement follows from the first and Lemma 5. To prove the first, consider the relation

$$x \prec y \Leftrightarrow x \in Q \text{ and } y \in Q \text{ and } \varphi(x) < \varphi(y).$$

It is clearly a well founded relation on X and is $\underset{\sim}{\Sigma}^1_1$, since

$$x \prec y \Leftrightarrow x \in Q \text{ and } y \in Q \text{ and not } (y \leq^*_\varphi x).$$

By Theorem 6, $\prec$ has countable length, hence $\{\varphi(x) : x \in Q\}$ is countable and its sup is less than $\omega_1$. □

This theorem gives an alternative way of proving that a given $\underset{\sim}{\Pi}^1_1$ set P is not Borel, quite different from the completeness method discussed in IV.1. It is enough to be able to find a $\underset{\sim}{\Pi}^1_1$-rank on the set P for which one can prove it is unbounded on P. For obvious reasons we call this the <u>rank</u> <u>method</u>.

For example, it is easy to construct trees in Seq $\mathbb{N}$ which are well founded and of arbitrary countable height. By the Boundedness Theorem and Lemma 2, this gives that WF is true $\underset{\sim}{\Pi}^1_1$. Note that by IV.1.3 WF is in fact $\underset{\sim}{\Pi}^1_1$-complete. We will use in the sequel this rank method to prove that certain $\sigma$-ideals of closed sets are true $\underset{\sim}{\Pi}^1_1$, and by combining it with Theorem IV.3.1, that they are $\underset{\sim}{\Pi}^1_1$-complete.

There is another interesting way of using the Boundedness Theorem, sometimes called the <u>overspill method</u>: Suppose P is a $\underset{\sim}{\Pi}^1_1$ set, Q is a $\underset{\sim}{\Sigma}^1_1$ set and one wants to find some x in Q–P. By the Boundedness Theorem, it is enough to find a $\underset{\sim}{\Pi}^1_1$-rank on P, and construct points in P ∩ Q of unbounded rank. This replaces the construction of one point <u>off</u> P (and in Q) by that of $\omega_1$ points <u>in</u> P (and in Q), which is of course done by induction on the countable ordinals. This is sometimes a very powerful way of getting existence results.

The existence of $\underset{\sim}{\Pi}^1_1$-ranks can also be used to prove a great variety of separation properties. We give here one of them, sometimes called the $\underset{\sim}{\Delta}^1_1$-Selection Theorem, sometimes the Novikov Separation Theorem.

<u>Theorem 8</u> (Novikov). Let P be a $\underset{\sim}{\Pi}^1_1$ set in $X \times \mathbb{N}$ (X Polish, $\mathbb{N}$ discrete), and Q a Borel subset of X such that $\forall x \in Q\ \exists n\ (x, n) \in P$. Then there is a Borel function $f : Q \to \mathbb{N}$ such that $\forall x \in Q\ (x, f(x)) \in P$.

<u>Proof</u>. Let $\varphi$ be a $\underset{\sim}{\Pi}^1_1$-rank on P. For $x \in Q$, define $\alpha_x$ = least $\alpha$ ($\exists n\ (x, n) \in P$ and $\varphi(x, n) = \alpha$) and let f(x) = least n ($(x, n) \in P$ and $\varphi(x, n) = \alpha_x$). The only point to check is that f is Borel, and for this by Suslin's Theorem (see IV.1), it is enough to check that the relation

$$x \in Q \text{ and } f(x) = n$$

is both $\underset{\sim}{\Sigma}^1_1$ and $\underset{\sim}{\Pi}^1_1$. But

$$x \in Q \text{ and } f(x) = n \Leftrightarrow x \in Q \text{ and } (x, n) \in P \text{ and } \forall m[(x, n) <^*_\varphi (x,m) \text{ or } [(x, n) \leq^*_\varphi (x, m) \text{ and } n \leq m]]$$

so it is $\underset{\sim}{\Pi}^1_1$, and

$$x \in Q \text{ and } f(x) \neq n \Leftrightarrow x \in Q \text{ and } \exists m(m \neq n \text{ and } x \in Q \text{ and } f(x) = m)$$

so it is also $\underset{\sim}{\Sigma}^1_1$, and we are done. □

By considering $Q_n = Q - \{x \in X : (x, n) \in P\}$, one can state the previous result as : For every sequence $\{Q_n\}$ of $\underset{\sim}{\Sigma}^1_1$ sets with $\cap_n Q_n = \emptyset$, there is a sequence $P_n$ of Borel sets with $Q_n \subseteq P_n$ and $\cap_n P_n = \emptyset$. This is the classical statement of Novikov's Theorem.

## §2. Ranks for subspaces of Banach spaces

In this section X denotes a (complex) Banach space with norm $\| \ \|$, and $X^*$, $X^{**}$ its dual and second dual respectively with norms $\| \ \|_*$, $\| \ \|_{**}$. The application of $x^* \in X^*$ to $x \in X$ is denoted as usual by $\langle x, x^* \rangle$. Identifying $x \in X$ with $x^{**} \in X^{**}$ given by $\langle x^*, x^{**} \rangle = \langle x, x^* \rangle$ we can view X as a closed subspace of $X^{**}$. Thus for $x \in X$, $\langle x, x^* \rangle = \langle x^*, x \rangle$.

For each Banach space Y we use B(Y; x, a) to denote the closed ball of Y with center x and radius a and $B^0(Y; x, a)$ for the open ball. Any of Y, x, a will be omitted if there is no danger of confusion. In the particular case $x = 0$ we will also write $B_a(Y)$, $B_a^0(Y)$ for $B(Y; 0, a)$, $B^0(Y; 0, a)$.

The _weak topology_ of $X$ is the topology of duality with $X^*$, i.e. the smallest topology making continuous all the maps $x \mapsto \langle x, x^* \rangle$ for $x^* \in X^*$. The _weak*(w*)-topology_ on $X^*$ is the topology of duality with $X$, i.e. the smallest topology making continuous the maps $x^* \mapsto \langle x, x^* \rangle$ for all $x \in X$. We denote by $\overline{P}$, $\overline{P}^w$, $\overline{P}^{w*}$ closures in the norm-topology, weak topology, and weak*-topology respectively.

Basic open nbhds of 0 in the weak topology of X are the form

$$\{x \in X : |\langle x, x_1^* \rangle| < \epsilon, \ldots, |\langle x, x_n^* \rangle| < \epsilon\}$$

for $x_1^*, \ldots, x_n^* \in X^*$ and $\epsilon > 0$, and similarly for the weak*-topology of $X^*$ they are of the form

$$\{x^* \in X^* : | \langle x_1, x^* \rangle | < \epsilon, \ldots, | \langle x_n, x^* \rangle | < \epsilon\}$$

where $x_1, \ldots, x_n \in X$ and $\epsilon > 0$. With this topology $X^*$ is a locally convex topological vector space with dual $X$. We state the following standard theorem without proof (see Rudin [2]).

_Theorem 1_ (Banach, Alaoglu). Let $X$ be a Banach space. Then any closed ball $B_a(X^*)$ of $X^*$ is w*-compact (i.e. compact is the weak*-topology). Moreover if $X$ is separable, and $\{x_n\}$ is a countable dense subset in the unit ball $B_1(X)$, the metric

$$d(x^*, y^*) = \sum 2^{-n} \left| \langle x_n, x^* \rangle - \langle x_n, y^* \rangle \right|$$

gives the weak$^*$-topology in each $B_a(X^*)$. Thus $B_a(X^*)$ is compact, metrizable in the weak$^*$-topology.

In general however the weak$^*$-topology is not metrizable. This is certainly the case for $X = PF$ ($X^* = A$, $X^{**} = PM$), which is our primary interest. In particular we cannot expect to obtain closures in the weak$^*$-topology by closing under w$^*$-limits of <u>sequences</u> only. This will be the underlying cause for many important phenomena on the structure of U-sets, as we will see later. At the moment let us notice that for a separable Banach space X, and a sequence $\{x_i\}$ dense in $B_1(X)$ we have

$$x_n^* \xrightarrow{w^*} x^* \text{ iff } \forall x \in X(\langle x, x_n^* \rangle \to \langle x, x^* \rangle)$$
$$\text{iff } \sup_n \|x_n^*\|_* < \infty \text{ and } \forall i(\langle x_i, x_n^* \rangle \to \langle x_i, x^* \rangle)$$

(The second equivalence follows from the Uniform Boundedness Principle of Banach-Steinhaus).

So in the case $X = PF$ recall that for $f_n$, f in A ($= X^*$)

$$f_n \xrightarrow{w^*} f \quad \text{iff } (\sup_n \|f_n\|_A < \infty \text{ and } \forall k(\hat{f}_n(k) \to \hat{f}(k))).$$

Also for $S_n$, $S \in PM(= X^{**})$

$$S_n \xrightarrow{w^*} S \text{ iff } (\sup_n \|S_n\|_{PM} < \infty \text{ and } \forall k(S_n(k) \to S(k))).$$

Finally one can also easily check that for $x_n$, $x$ in PF( $=$ X)

$$x_n \xrightarrow{w} x \quad \text{iff } \left[\sup_n \|x_n\|_{PF} < \infty \text{ and } \forall k\ (x_n(k) \to x(k))\right]$$

$$\text{iff } x_n \xrightarrow{w^*} x \text{ (in } X^{**}).$$

Let now Y be a subspace (= linear subspace) of $X^*$. We denote by $Y^{(1)}$ the $w^*$-<u>sequential</u> <u>closure</u> of Y, i.e. the set of $w^*$-limits of <u>sequences</u> in Y. One can define a transfinite sequence $Y^{(\alpha)}$, $\alpha$ an ordinal, by setting:

$$Y^{(0)} = Y,$$

$$Y^{(\alpha+1)} = (Y^{(\alpha)})^{(1)},$$

$$Y^{(\lambda)} = \bigcup_{\alpha<\lambda} Y^{(\alpha)} \text{ for limit } \lambda.$$

This gives an increasing family of subspaces of $X^*$, which must terminate at some ordinal

$$\alpha(Y) = \text{least } \alpha(Y^{(\alpha+1)} = Y^{(\alpha)}).$$

<u>Theorem</u> <u>2</u> (Banach). Let X be a separable Banach space. Let $Y \subseteq X^*$ be a subspace of $X^*$. Then the following are equivalent:

(i) Y is $w^*$-closed

(ii) Y is $w^*$-sequentially closed, i.e. Y is closed under $w^*$-limits of sequences in Y. In particular, for each subspace Y of $X^*$, $Y^{(\alpha(Y))}$ is the $w^*$-closure of Y.

Proof. (i) ⇒ (ii) is clear, and the last statement is trivial from the first one and the definition of $Y^{(\alpha(Y))}$. To prove (ii) ⇒ (i), we shall need the following result:

Theorem 3 (Banach). Let X be a separable Banach space. If $Y \subseteq X^*$ is a subspace and $Y \cap B_1(X^*)$ is $w^*$-closed, Y is $w^*$-closed in $X^*$.

Granting this result, we finish the proof of Theorem 2: We know that Y is $w^*$-sequentially closed, i.e. $Y = Y^{(1)}$. By Theorem 3, it is enough to show that $Y \cap B_1(X^*)$ is $w^*$-closed. To see this, let $x^* \in \overline{Y \cap B_1(X^*)}^{w^*} \subseteq B_1(X^*)$ be given. By the metrizability of $B_1(X^*)$, $x^*$ is the $w^*$-limit of some sequence in $Y \cap B_1(X^*)$, so $x^* \in Y^{(1)} = Y$. As $x^* \in B_1(X^*)$, $x^* \in B_1(X^*) \cap Y$ and we are done. □

Proof of Theorem 3. First note that as $Y \cap B_1(X^*)$ is $w^*$-closed, it is norm-closed, hence Y is norm-closed. Let $x_0^* \notin Y$. We need to show that there is an $x \in X$ with $\langle x, x_0^*\rangle \neq 0$ but $\langle x, y^*\rangle = 0$ for all $y^* \in Y$. By choosing $\epsilon$ small enough, we can assume $B(x_0^*, 2\epsilon) \cap Y = \emptyset$. We need the following lemma:

Lemma 4. If there is a sequence $\{x_n\}$ in X with $x_n \to 0$ such that $Y \cap (\bigcap_n \{y^* \in X^* : |\langle x_n, x_0^* - y^*\rangle| \leq \epsilon\}) = \emptyset$, then there is an $x \in X$ such that $\langle x, x_0^*\rangle \neq 0$ but $\langle x, y^*\rangle = 0$ for all $y^* \in Y$.

Proof. Fix $\{x_n\}$ satisfying the hypothesis of the lemma. Consider the map T :

$X^* \to c_0(\mathbb{N})$, given by $T(x^*) = \{\langle x_n, x^*\rangle\}$. The space $T[Y]$ is a subspace of $c_0(\mathbb{N})$, and by the hypothesis, $\operatorname{dist}(T(x_0^*), T[Y]) \geq \epsilon$. By the Hahn-Banach Theorem, there is a sequence $\{c_n\} \in \ell^1(\mathbb{N})$ such that $\langle T(x_0^*), \{c_n\}\rangle \neq 0$ but $\langle T(y^*), \{c_n\}\rangle = 0$ for all $y^* \in Y$. Let $x = \sum_n c_n x_n$. Then

$$\langle x, x_0^*\rangle = \sum_n c_n \langle x_n, x_0^*\rangle = \langle T(x_0^*), \{c_n\}\rangle \neq 0$$

while

$$\langle x, y^*\rangle = \sum_n c_n \langle x_n, y^*\rangle = \langle T(y^*), \{c_n\}\rangle = 0$$

for all $y^* \in Y$ and we are done. □

To complete the proof of Theorem 3, we construct a sequence $x_n$ satisfying the hypotheses of Lemma 4:

Consider the set $E = \frac{1}{\epsilon}(Y - x_0^*)$. Then $E \cap B_r(X^*)$ is $w^*$-closed for all r. Moreover $E \cap B_1(X^*) = \varnothing$. We want to find $x_n \to 0$ with $E \cap (\bigcap_n \{y^* : |\langle x_n, y^*\rangle| \leq 1\}) = \varnothing$. To find the $x_n$'s, we inductively define a sequence $\{F_k\}_{k \in \mathbb{N}}$ of finite subsets of X, such that $F_0 = \{0\}$, $F_k \subseteq B_{1/k}(X)$, and

$$P(F_0) \cap \ldots \cap P(F_{k-1}) \cap E \cap B_k(X^*) = \varnothing$$

where

$$P(F) = \{y^* : |\langle x, y^*\rangle| \leq 1 \text{ for all } x \in F\}.$$

If this can be done, $E \cap \bigcap_{k=0}^{\infty} P(F_k) = \varnothing$, and if $\bigcup_k F_k$ is enumerated as $\{x_n\}_{n \in \mathbb{N}}$, $x_n \to 0$ as desired.

So it is enough to construct the $F_k$'s. Assume that $F_0, \ldots, F_{k-1}$ have already been chosen, satisfying the properties above. Let $Q = P(F_0) \cap \ldots \cap P(F_{k-1}) \cap E \cap B_{k+1}(X^*)$. If for all finite $F \subseteq B_{1/k}(X)$, $P(F) \cap Q \neq \emptyset$, then as $\cap \{P(F) : F$ finite subset of $B_{1/k}(X)\} = B_k(X^*)$, $P(F) \cap P(F') = P(F \cup F')$, and $Q$ is $w^*$-compact, we get that $Q \cap B_k(X^*) \neq \emptyset$, contradicting the induction hypothesis. So for some finite $F_k \subseteq B_{1/k}(X)$, $P(F_k) \cap Q = \emptyset$ and we can take this $F_k$ to complete the induction step of the construction. This finishes the proof. □

Proposition 5 (Banach). Let X be a separable Banach space, and Y a subspace of $X^*$. The closure ordinal $\alpha(Y)$ for the iteration of $w^*$-sequential closure is countable.

Proof. Consider for every ordinal $\alpha$, $Z^\alpha = Y^{(\alpha)} \cap B_1(X^*)$. Let $\alpha_1$ be such that $Z^{\alpha_1} = Z^\beta$ for all $\beta \geq \alpha_1$. Then $Z^{\alpha_1}$ is $w^*$-closed, and by Theorem 3 $Y^{(\alpha_1)}$ is too, i.e. $\alpha(Y) \leq \alpha_1$. So it is enough to prove that $\alpha_1 < \omega_1$. Now $Z^\alpha \subseteq \overline{Z^\alpha}^{w^*} \subseteq Z^{\alpha+1}$, and $\overline{Z^\alpha}^{w^*}$ is an increasing sequence of closed sets in $B_1(X^*)$ with the weak$^*$-topology, which is compact metrizable by Theorem 1. So the sequence $\overline{Z^\alpha}^{w^*}$ stabilizes at some ordinal $< \omega_1$, and so does $Z^\alpha$. □

We shall be mainly interested in the case where the subspace Y of $X^*$ is $w^*$-dense in $X^*$:

Definition 6. Let X be a separable Banach space and Y a $w^*$-dense subspace of $X^*$. We define the order of Y in $X^*$, ord(Y), to be the least ordinal $\alpha$ such that $Y^{(\alpha)} = X^*$. We extend ord to non-$w^*$-dense subspaces by setting ord(Y) $= \omega_1$. So the function ord is a rank on the family of $w^*$-dense subspaces of $X^*$.

Note that ord(Y) is always a successor ordinal, unless $Y = X^*$ for which it is 0. This is because if $X^* = Y^{(\lambda)} = \cup_{\alpha<\lambda} Y^{(\alpha)}$ for some countable limit $\lambda$ then $X^* = \cup_{\alpha<\lambda} \overline{Y^{(\alpha)}}$, so by the Baire Category Theorem some $\overline{Y^{(\alpha)}}$, $\alpha < \lambda$, contains an open ball and thus $\overline{Y^{(\alpha)}} = X^*$. Since $\overline{Y^{(\alpha)}} \subseteq Y^{(\alpha+1)}$ it follows that $\text{ord}(Y) \leq \alpha + 1 < \lambda$.

Particularly important in the sequel will be subspaces Y which are <u>sequentially</u> $w^*$-<u>dense</u> in $X^*$, i.e. Y with order less than or equal to 1, $Y^{(1)} = X^*$. For these subspaces there is an important characterization due to Banach that we will explain now.

<u>Definition 7</u>. Let X be a Banach space. For Y a subspace of $X^*$ and $Z \subseteq X$ define

$$s(Z, Y) = \inf\left\{\frac{\|x\|_Y}{\|x\|} : x \in Z,\ x \neq 0\right\},$$

where $\|x\|_Y$ is the norm of x as a functional on Y, i.e.

$$\|x\|_Y = \sup\{|\langle x, y^*\rangle| : y^* \in Y,\ \|y^*\|_* \leq 1\}.$$

We let

$$s(Y) = s(X, Y).$$

<u>Theorem 8</u>. Let X be a separable Banach space and let Y be a subspace of $X^*$. Then

(i) (Dixmier [1]) $s(Y) = \sup\{a \geq 0 : B_a(X^*) \subseteq \overline{Y \cap B_1(X^*)}^{w^*}\}$,

(ii) (Banach) $Y^{(1)} = X^*$ iff $s(Y) > 0$.

Proof. (i) Let

$$r = \sup(a \geq 0 : B_a(X^*) \subseteq \overline{Y \cap B_1(X^*)}^{w^*}).$$

Clearly this supremum is attained. We first show that $r \leq s(Y)$, i.e. for $x \in X$, $\|x\|_Y \geq r \cdot \|x\|$. We may assume $r > 0$ and $\|x\| = 1$. Then for any $y^* \in Y \cap B_1(X^*)$, $|\langle x, y^*\rangle| \leq \|x\|_Y$, hence also for $y^* \in \overline{Y \cap B_1(X^*)}^{w^*}$, so for $y^* \in B_r(X^*)$. But then $1 = \|x\| = \sup\{\frac{1}{r} |\langle x, y^*\rangle| : y^* \in B_r(X^*)\} \leq \frac{1}{r} \cdot \|x\|_Y$ and we are done. Conversely, we show that $B_a(X^*) \subseteq \overline{Y \cap B_1(X^*)}^{w^*}$ for all $a < s(Y)$. (If $s(Y) = 0$, there is nothing to prove). So let $x^* \in X^*$ be such that $0 < \|x^*\|_* \leq a < s(Y)$. For each $x \in X$, $x \neq 0$, $\|x\|_Y \geq s(Y) \cdot \|x\|$, so for some $y^* \in Y \cap B_1(X^*)$ $|\langle x, y^*\rangle| \geq \|x^*\|_* \ \|x\| \geq |\langle x, x^*\rangle|$. Thus $z^* = \dfrac{\langle x, x^*\rangle}{\langle x, y^*\rangle} \cdot y^*$ satisfies $z^* \in Y \cap B_1(X^*)$ and $\langle x, x^*\rangle = \langle x, z^*\rangle$. Thus $x^*$ cannot be separated from $Y \cap B_1(X^*)$ by any $x \in X$, i.e. $x^* \in \overline{Y \cap B_1(X^*)}^{w^*}$ and we are done.

(ii) If for some $r > 0$, $Y \cap B_1(X^*)$ is $w^*$-dense in $B_r(X^*)$, then $Y \cap B_1(X^*)$ is sequentially $w^*$-dense in $B_r(X^*)$ and by scalar multiplication $Y$ is sequentially $w^*$-dense in $X^*$. So direction $\Leftarrow$ follows from part (i).

Suppose now $Y^{(1)} = X^*$. This means that for any $x^* \in X^*$ there is a sequence $y_n^* \in Y$ $w^*$-converging to $x^*$, and this sequence is of course bounded in norm. In other words, if $K = \overline{Y \cap B_1(X^*)}^{w^*}$, $X^* = \cup_{n \in \mathbb{N}} nK$. But $K$ is $w^*$ (hence norm)-closed in $X^*$, so by the Baire Category Theorem some $nK$ contains a ball $B(x^*, \epsilon)$ for some $\epsilon > 0$. But $K$ is convex and symmetric, hence $B(-x^*, \epsilon) \subseteq nK$ and finally $B_\epsilon(X^*) \subseteq \frac{1}{2}B(x^*, \epsilon) + \frac{1}{2}B(-x^*, \epsilon) \subseteq nK$, i.e. $s(Y) \geq \epsilon/n > 0$ by part (i). □

We will identify, using duality, the quantity $s(Y)$ with still another quantity. For that we need the following basic standard fact.

Proposition 9. Let $X$ be a Banach space, $Y \subseteq X$ a subspace and $x^* \in X^*$. Then if $\|x^*\|_Y = \sup\{|\langle x, x^*\rangle| : x \in Y, \|x\| \leq 1\}$,

$$\|x^*\|_Y = \operatorname{dist}(x^*, Y^{\perp})$$

where $Y^{\perp}$ is the annihilator of $Y$, i.e.

$$Y^{\perp} = \{y^* \in X^* : \forall y \in Y(\langle y, y^*\rangle = 0)\}.$$

Proof. Since $Y^{\perp} = (\bar{Y})^{\perp}$, $\|x^*\|_Y = \|x^*\|_{\bar{Y}}$, we can assume $Y$ is closed.

(i) For $\|x^*\|_Y \leq \operatorname{dist}(x^*, Y^{\perp})$ : Let $y \in Y$, $\|y\| \leq 1$ and $y^* \in Y^{\perp}$. Then $|\langle y, x^*\rangle| = |\langle y, x^* - y^*\rangle| \leq \|x^* - y^*\|_*$.

(ii) For $\operatorname{dist}(x^*, Y^{\perp}) \leq \|x^*\|_Y$: Let $\epsilon < \operatorname{dist}(x^*, Y^{\perp})$. We will show that $\epsilon \leq \|x^*\|_Y$. By Hahn-Banach there is $x_0 \in X$, $\|x_0\| = 1$ with $x_0 \in {}^{\perp}(Y^{\perp}) = \{x \in X : \forall y^* \in Y^{\perp}(\langle x, y^*\rangle = 0)\} = Y$ (by Hahn-Banach again since $Y$ is closed) and $|\langle x_0, x^*\rangle| \geq \epsilon$. So $\|x^*\|_Y \geq \epsilon$. □

Applying this to the context of Theorem 8 we have (recalling that $X \subseteq X^{**}$)

$$s(Y) = \inf\{\frac{\operatorname{dist}(x, Y^{\perp})}{\|x\|} : x \in X, x \neq 0\}$$

$$= \inf\{\frac{\|x + y^{**}\|_{**}}{\|x\|} : x \in X, y^{**} \in Y^{\perp}, x \neq 0\}$$

$$=^{\text{def}} t(Y^{\perp}).$$

One can think of $t(Y^{\perp})$ as the (sine of the ) angle between X, $Y^{\perp}$. In this general context we can define also the (sine of the) angle between $Y^{\perp}$, X (for $\bar{Y} \neq X^*$, so that $Y^{\perp} \neq \{0\}$) by

$$\eta(Y^{\perp}) = \inf\{\frac{\text{dist}(y^{**}, X)}{\|y^{**}\|_{**}} : y^{**} \in Y^{\perp}, y^{**} \neq 0\}$$

$$= \inf\{\frac{\|x + y^{**}\|_{**}}{\|y^{**}\|_{**}} : y^{**} \in Y^{\perp}, y^{**} \neq 0, x \in X\}.$$

One can easily check that

$$t \geq \frac{1}{1 + \eta^{-1}} = \frac{\eta}{\eta + 1},$$

and symmetrically

$$\eta \geq \frac{t}{1 + t}$$

so that $t\eta \geq |\eta - t|$ and thus $t = 0$ iff $\eta = 0$ (a fact which can be checked directly as well). We will see in §5.5 that for X = PF we actually have $t = \frac{\eta}{1 + \eta}$.

Before we proceed to apply the preceding ideas of rank to the structure of

U-sets, we will introduce two other ranks on $w^*$-dense subspaces of $X^*$.

## §3. The tree-rank and the R-rank

The results in this section are due to Kechris and Louveau, unless otherwise stated.

### The tree-rank

Let here X be a separable Banach space, and fix a countable set D dense in the open unit ball of X, and closed under multiplication by elements $q \in \mathbb{Q} + i\mathbb{Q}$ (provided the product is still in the open unit ball).

Given a subspace Y of $X^*$ and $\epsilon \in \mathbb{Q} \cap (0, 1)$, define a tree $T_Y^\epsilon$ on Seq D by

$$T_Y^\epsilon = \{\emptyset\} \cup \{(x_0,\ldots,x_n) : \forall j \le n (x_j \in D \text{ and } \|x_j\| \ge \epsilon) \\ \& \ \forall j<n \ (\|x_j - x_{j+1}\| \le 2^{-(j+3)}) \ \& \ \forall j \le n \ (\|x_j\|_Y \le 2^{-(j+1)})\}$$

where as usual $\|x\|_Y$ is the norm of x viewed as a linear functional on the subspace Y, i.e. $\|x\|_Y = \sup\{|\langle x, y\rangle| : y \in Y, \|y\|_* \le 1\}$.

Proposition 1. The following are equivalent, for Y a subspace of $X^*$:

(i) Y is $w^*$-dense in $X^*$,

(ii) for all $\epsilon \in \mathbb{Q} \cap (0, 1)$, $T_Y^\epsilon$ is well founded,

(iii) for some $\epsilon \in \mathbb{Q} \cap (0, 1)$, $T_Y^\epsilon$ is well founded.

<u>Proof</u>. If $Y$ is not $w^*$-dense in $X^*$, there is an element $x \in X$ of norm 1 in the anihilator $Y^\perp$ of $Y$, i.e. $\langle x, y^* \rangle = 0$ for all $y^* \in Y$. If $\{x_n\}$ are sequences in $D$ converging fast enough to $x$, they will give branches through all the trees $T^\epsilon_Y$. Conversely, a branch through some tree $T^\epsilon_Y$ is a Cauchy sequence in $X$ converging to some $x \in Y^\perp$ of norm $\geq \epsilon$. □

The preceding proposition allows us to associate to each $w^*$-dense $Y \subseteq X^*$ an ordinal $\beta(Y)$ by

$$\begin{aligned}\beta(Y) &= \sup\{ht(T^\epsilon_Y) + 1 : \epsilon \in \mathbb{Q} \cap (0, 1)\} \\ &= \lim_{\epsilon \to 0} (ht(T^\epsilon_Y) + 1)\end{aligned}$$

(for clearly if $\epsilon \leq \epsilon'$, $T^{\epsilon'}_Y \subseteq T^\epsilon_Y$, hence $ht(T^{\epsilon'}_Y) \leq ht(T^\epsilon_Y)$). We proceed now to show that $\beta(Y)$ is always a successor limit ordinal, i.e. of the form $\omega \cdot \alpha$ for some $\alpha$ a successor ordinal.

<u>Lemma</u> <u>2</u>. Let $Y$ be $w^*$-dense in $X^*$, and $\epsilon$, $\epsilon'$ in $\mathbb{Q} \cap (0, 1)$. Then $ht(T^\epsilon_Y)$ and $ht(T^{\epsilon'}_Y)$ differ at most by an integer, i.e. for all $\alpha$ $ht(T^\epsilon_Y) \geq \omega \cdot \alpha$ implies $ht(T^{\epsilon'}_Y) \geq \omega \cdot \alpha$.

<u>Proof</u>. The case $\epsilon' \leq \epsilon$ is clear from the remark above. So assume $\epsilon' \geq \epsilon$. For $\alpha = 0$ there is nothing to prove, so let $\alpha > 0$. For each positive integer $N$, let

$$T^\epsilon_{Y,N} = \{(x_0,\ldots,x_n) : x_j \in D \text{ and } \|x_j\| \geq \epsilon \text{ and }$$
$$\|x_j - x_{j+1}\| \leq 2^{-(N+j+3)} \text{ and } \|x_j\|_Y \leq 2^{-(N+j+1)}\} \cup \{\emptyset\}.$$

Notice that if $s = (x_0,\ldots, x_{N-1})$ is in $T^\epsilon_Y$, then the tree $(T^\epsilon_Y)_s = \{t : s^\frown t \in T^\epsilon_Y\}$ is a subtree of $T^\epsilon_{Y,N}$. This implies

$$\begin{aligned} ht(T^\epsilon_{Y,N}) &\geq \sup\{ht((T^\epsilon_Y)_s) : s \in T^\epsilon_Y, \ell h(s) = N\} \\ &\geq \omega \cdot \alpha, \end{aligned}$$

as $ht(T^\epsilon_Y) \geq \omega \cdot \alpha$.

Fix now $\eta$ rational $> 0$ with $\epsilon' + 3\eta < 1$, and N large enough so that $2^{-N} \cdot p < \eta$, where $p = \frac{1 - 2\eta}{\epsilon}$. To each $u = (x_0,\ldots,x_n) \in T^\epsilon_{Y,N}$ we associate $u' = (x'_0,\ldots,x'_n)$ as follows:

Choose $r \in \mathbb{Q}^+$, $r \leq p$ such that $1 - 2\eta \geq r\ \|x_0\| \geq \epsilon' + \eta$, which is possible by our choice of $\eta$, and let for $j \leq n$, $x'_j = rx_j$. Clearly, as the choice of r depends only on $x_0$, $u \subsetneq v$ implies $u' \subsetneq v'$. So in order to get $ht(T^{\epsilon'}_Y) \geq \omega \cdot \alpha$, it is enough to prove that for $u \in T^\epsilon_{Y,N}$, one has $u' \in T^{\epsilon'}_Y$ (recall here the proof of 1.2). First $r\|x_0\| < 1$ and $\|x_0 - x_j\| \leq 2^{-(N+2)}$, so $\|x'_0 - x'_j\| \leq p \cdot 2^{-(N+2)} < \eta$, thus $\|x'_j\| < r \cdot \|x_0\| + \eta \leq 1 - \eta < 1$, so that $x'_j \in D$, and $\|x'_j\| \geq r \cdot \|x_0\| - \eta \geq \epsilon'$. Also $\|x'_j - x'_{j+1}\| = r \cdot \|x_j - x_{j+1}\| \leq r \cdot 2^{-(N+j+3)} \leq 2^{-(j+3)}$ and $\|x'_j\|_Y = r \cdot \|x_j\|_Y \leq r \cdot 2^{-(N+j+1)} \leq 2^{-(j+1)}$. This shows that $u' \in T^{\epsilon'}_Y$ and completes the proof. □

<u>Lemma 3</u>. For Y a w*-dense subspace of X*, $\beta(Y) = \sup_\epsilon(ht(T^\epsilon_Y) + 1)$ is of the form $\omega \cdot \alpha$ for some successor ordinal $\alpha$.

<u>Proof</u>. By Lemma 2, as $\beta(Y)$ is the supremum of a sequence of ordinals differing

only by an integer, $\beta(Y)$ cannot be of form $\omega \cdot \alpha$ for some limit $\alpha$. So it is enough to show that $\beta(Y) \geq \omega$, and if $\beta(Y) > \omega \cdot \alpha$ for some $\alpha > 0$, then $\beta(Y) \geq \omega \cdot (\alpha + 1)$.

Fix $N \in \mathbb{N}$, and let $x \in D$ satisfy $2^{-(N+2)} \leq \|x\| \leq 2^{-(N+1)}$. Then $s = (x,\dots, x)$ ($N + 1$ times) is in $T_Y^\epsilon$, with $\epsilon = 2^{-(N+2)}$, hence $ht(T_Y^\epsilon) \geq N + 1$ and $\beta(Y) \geq \omega$.

Next let $\alpha > 0$ and suppose $\beta(Y) > \omega\cdot\alpha$, so that for some $\epsilon > 0$ $ht(T_Y^\epsilon) \geq \omega\cdot\alpha$. Let $N \in \mathbb{N}$ be given, and let $\epsilon' = \epsilon \cdot 2^{-(N+4)}$. The tree $T' = 2^{-(N+4)} \cdot T_{Y,N}^\epsilon$ is a subtree of $T_Y^{\epsilon'}$. And if $(x) \in T'$, $s = (x_0,\dots, x_n) \in T'$, then $\|x - x_0\| \leq 2^{-(N+3)}$, hence $(\underbrace{x,\dots, x}_{N+1 \text{ times}}, x_0,\dots,x_n) \in T_Y^{\epsilon'}$. As $ht(T') = ht(T_{Y,N}^\epsilon) \geq \omega \cdot \alpha$, $ht(T_Y^{\epsilon'}) \geq \omega \cdot \alpha + N + 1$. This shows that $\beta(Y) \geq \omega \cdot \alpha + \omega = \omega \cdot (\alpha + 1)$ and we are done. □

__Definition 4__. We define the __tree-rank__ $rk_T(Y)$, for Y a $w^*$-dense subspace of $X^*$, by the equality $\beta(Y) = \omega \cdot rk_T(Y)$ (using Lemma 3). As usual we extend this by assigning the value $rk_T(Y) = \omega_1$, for Y's which are not $w^*$-dense.

Our next goal is to show that the two ranks ord(Y) and $rk_T(Y)$ coincide.

__Theorem 5__. Let X be a separable Banach space, and $Y \neq X^*$ a subspace of $X^*$. Then $\text{ord}(Y) = rk_T(Y)$.

__Proof__. By definition, ord (Y) and $rk_T(Y)$ are both $\omega_1$ if Y is not $w^*$-dense in $X^*$. So we may assume Y is $w^*$-dense. We shall use the following two lemmas:

Lemma 6. Suppose $u \in T^\epsilon_Y$, and $ht(u, T^\epsilon_Y) \geq \omega \cdot \alpha$. Then $u \in T^\epsilon_{Y^{(\alpha)}}$. (Recall (§2) that $Y^{(\alpha)}$ is the $\alpha^{th}$ iterate of Y in the operation of sequential w*-closure).

Lemma 7. Let $u = (x_0,\ldots, x_n) \in T^\epsilon_{Y^{(\alpha)}}$, and assume moreover that $\|x_n\| \cdot (1 + 2^{-(n+3)}) < 1$, $\|x_n\| \cdot (1 - 2^{-(n+3)}) > \epsilon$ and $\|x_n\|_{Y^{(\alpha)}} \leq 2^{-(n+3)}$. Then $ht(u, T^\epsilon_Y) \geq \omega \cdot \alpha$.

From these two lemmas, we can complete the proof of Theorem 5 (recall from the remarks following 2.6 that ord(Y) is a successor ordinal, since $Y \neq X^*$):

(i) Suppose $\beta(Y) > \omega \cdot (\beta + 1)$, so that for some $\epsilon > 0$ and $u = (x_0,\ldots, x_n) \in T^\epsilon_Y$, $ht(u, T^\epsilon_Y) \geq \omega \cdot (\beta + 1)$. Then for all $N \in \mathbb{N}$ there is $v_N = (x^N_{n+1},\ldots, x^N_{n+N})$ with $u^\frown v_N \in T^\epsilon_Y$ and $ht(u^\frown v_N, T^\epsilon_Y) \geq \omega \cdot \beta$. By Lemma 6, for all N, $u^\frown v_N \in T^\epsilon_{Y^{(\beta)}}$. But then the sequence $(x^N_{n+N})_{N\in\mathbb{N}}$ is such that $\|x^N_{n+N}\| \geq \epsilon$ and $\|x^N_{n+N}\|_{Y^{(\beta)}} \leq 2^{-(n+N+1)}$, so it witnesses that $s(Y^{(\beta)}) = 0$. By Theorem 2.8, $Y^{(\beta)}$ is not w*-sequentially dense in $X^*$, i.e. $ord(Y) > \beta + 1$. This proves that $\beta(Y) \leq \omega \cdot ord(Y)$.

(ii) Conversely, suppose $Y^{(\beta+1)} \neq X^*$, hence by Theorem 2.8 again $s(Y^{(\beta)}) = 0$. Then we can find a sequence $\{x_n\}$ – that we can choose in D by the density of D – with $\|x_n\| \cdot (1 + 2^{-(n+3)}) < 1$, $\|x_n\| \cdot (1 - 2^{-(n+3)}) > \frac{1}{2}$ and $\|x_n\|_{Y^{(\beta)}} \leq 2^{-(n+3)}$. Let $u_n = (x_n,\ldots, x_n)$ (n times). By the choice of $x_n$, $u_n \in T^{1/2}_{Y^{(\beta)}}$ and satisfies the hypotheses of Lemma 7, hence $ht(u_n, T^{1/2}_Y) \geq \omega \cdot \beta$. But this implies that $ht(T^{1/2}_Y) \geq \omega \cdot \beta + n$ for all n, i.e. $ht(T^{1/2}_Y) \geq \omega \cdot (\beta + 1)$. Thus $\beta(Y) > \omega \cdot (\beta + 1)$. So $\beta(Y) \geq \omega \cdot ord(Y)$ and we are done. □

Proof of Lemma 6. We want to show that for all $\alpha$, $ht(u, T^\epsilon_Y) \geq \omega \cdot \alpha \Rightarrow u \in$

$T^\epsilon_{Y^{(\alpha)}}$. We prove it by induction on $\alpha$. It is trivial for $\alpha = 0$. If $\alpha$ is limit, ht(u, $T^\epsilon_Y) \geq \omega \cdot \alpha$ implies ht(u, $T^\epsilon_Y) \geq \omega \cdot \beta$ for all $\beta < \alpha$, hence by the induction hypothesis $u \in T^\epsilon_{Y^{(\beta)}}$ for all $\beta < \alpha$. But as $Y^{(\alpha)} = \bigcup_{\beta<\alpha} Y^{(\beta)}$, $\| \ \|_{Y^{(\alpha)}} = \sup_{\beta<\alpha} \| \ \|_{Y^{(\beta)}}$, hence $T^\epsilon_{Y^{(\alpha)}} = \bigcap_{\beta<\alpha} T^\epsilon_{Y^{(\beta)}}$ and $u \in T^\epsilon_{Y^{(\alpha)}}$.

So assume the result is known for $\alpha$; we prove it for $\alpha + 1$. Let $u = (x_0, \ldots, x_n) \in T^\epsilon_Y$ with ht(u, $T^\epsilon_Y) \geq \omega \cdot (\alpha + 1)$, so that for all k there exists $v_k = (x^k_{n+1}, \ldots, x^k_{n+k})$ with $u^\frown v_k \in T^\epsilon_Y$ and ht($u^\frown v_k$, $T^\epsilon_Y) \geq \omega \cdot \alpha$. By the induction hypothesis $u^\frown v_k \in T^\epsilon_{Y^{(\alpha)}}$ for all k.

In order to verify that u is in $T^\epsilon_{Y^{(\alpha+1)}}$, it is enough to see that for all $y^* \in Y^{(\alpha+1)}$ and all $j \leq n$, $|\langle x_j, y^*\rangle| \leq 2^{-(j+1)} \cdot \|y^*\|_*$. Fix $y^* \in Y^{(\alpha+1)}$, and let $\{y^*_p\}$ be a sequence in $Y^{(\alpha)}$ with $\|y^*_p\|_* \leq M$ and $y^*_p \xrightarrow{w^*} y^*$. Now

$$(*) \qquad |\langle x_j, y^*\rangle| \leq |\langle x_j, y^* - y^*_p\rangle| + |\langle x_j - x^k_{n+k}, y^*_p\rangle| + |\langle x^k_{n+k}, y^*_p\rangle|$$

for all j, k, p. As $u^\frown v_k \in T^\epsilon_{Y^{(\alpha)}}$, $|\langle x^k_{n+k}, y^*_p\rangle| \leq M \cdot 2^{-(n+k+1)}$, so by choosing k large enough we can insure that $|\langle x^k_{n+k}, y^*_p\rangle| \leq 2^{-(j+2)} \cdot \|y^*\|_*$ for all p. Fix such a k and let $p \to \infty$ in (*). The first term goes to 0, and the second to $|\langle x_j - x^k_{n+k}, y^*\rangle| \leq \|x_j - x^k_{n+k}\| \cdot \|y^*\|_* \leq 2^{-(j+2)} \cdot \|y^*\|_*$ by the definition of $T^\epsilon_{Y^{(\alpha)}}$. This finally gives $|\langle x_j, y^*\rangle| \leq 2^{-(j+1)} \cdot \|y^*\|_*$ and we are done. □

In order to prove Lemma 7, we need another lemma, of interest in its own right.

<u>Lemma</u> <u>8</u>. Let X be a separable Banach space and Y a subspace of $X^*$. Let $x_0 \neq 0$, $x_0 \in X$. If $\|x_0\| > a > 0$ and $s(B(X; x_0, a), Y) > 0$, then $\|x_0\|_{Y^{(1)}} \geq a$.

Using this lemma we can complete the

<u>Proof</u> <u>of</u> <u>Lemma</u> <u>7</u>. The proof is by induction on $\alpha$. There is nothing to prove for $\alpha = 0$, and it is immediate for $\alpha$ limit. So suppose the result is known for $\alpha$. Let $u = (x_0, \ldots, x_n) \in T^{\epsilon}_{Y^{(\alpha+1)}}$, with $\|x_n\| \cdot (1 + 2^{-(n+3)}) < 1$, $\|x_n\| \cdot (1 - 2^{-(n+3)}) > \epsilon$ and $\|x_n\|_{Y^{(\alpha+1)}} \leq 2^{-(n+3)}$. Let $0 < b < \|x_n\|$, and $a = b \cdot \|x_n\|_{Y^{(\alpha+1)}}$. We claim that $s(B(X; x_n, a), Y^{(\alpha)}) = 0$. For if not, as $a < \|x_n\|$, Lemma 8 gives $\|x_n\|_{Y^{(\alpha+1)}} \geq a$, a contradiction since $b < 1$.

By definition of $s(B(X; x_n, a), Y^{(\alpha)})$, there is a sequence $\{y_k\}_{k \in \mathbb{N}}$ in $B(X; x_n, a)$ with $\lim \dfrac{\|y_k\|_{Y^{(\alpha)}}}{\|y_k\|} = 0$, and by density we can choose the $y_k$ in D. (Note here that $\|x_n\| + a < 1$). Now for $y \in B(X; x_n, a)$

$$\|y\| \geq \|x_n\| - \|x_n - y\|$$

$$\geq \|x_n\| - \|x_n\| \cdot \|x_n\|_{Y^{(\alpha+1)}}$$

$$\geq \|x_n\| \cdot (1 - 2^{-(n+3)}) > \epsilon.$$

So by considering a subsequence if necessary, we can choose $k_0$ such that if $k \geq k_0$ $\|y_k\|_{Y^{(\alpha)}} \leq 2^{-(n+k+3)}$, $\|x_n\| \cdot (1 + 2^{-(n+3)}) \cdot (1 + 2^{-(n+k+3)}) < 1$ and $\|x_n\| \cdot (1 - 2^{-(n+3)}) \cdot (1 - 2^{-(n+k+3)}) > \epsilon$. But then for $k \geq k_0$, $v_k = u^\frown(y_k, \ldots, y_k)$ (k times) is in $T^{\epsilon}_{Y^{(\alpha)}}$, and easily $v_k$ satisfies the hypotheses for $\alpha$ (note that $\|y_k\| \leq$

$\|x_n\| \cdot (1 + 2^{-(n+3)})$). So we get $ht(v_k, T^{\epsilon}_Y) \geq \omega \cdot \alpha$, hence $ht(u, T^{\epsilon}_Y) \geq \omega \cdot \alpha + k$ for all $k \geq k_0$, i.e. $ht(u, T^{\epsilon}_Y) \geq \omega \cdot (\alpha + 1)$ and we are done. □

For the proof of Lemma 8, we need the following standard result:

Theorem 9 (Goldstine). Let X be a Banach space. The unit ball of X is $w^*$-dense in the unit ball of $X^{**}$. (Recall that X is naturally identified with a closed subspace of $X^{**}$).

Proof. Let $x^{**} \notin \overline{B_1(X)}^{w^*} = E$. Since E is $w^*$-closed and convex in $X^{**}$, there is by Hahn-Banach, a $y^* \in X^*$ of norm 1 with $\mathrm{Re}(\langle x^{**}, y^* \rangle) > \sup_{x \in E} \mathrm{Re}(\langle x, y^* \rangle) \geq \|y^*\|_* = 1$, as E contains $B_1(X)$. This gives $\|x^{**}\|_{**} > 1$. □

Proof of Lemma 8. We start with $x_0 \in X$, $a > 0$ such that $\|x_0\| > a$, $\|x_0\|_{Y^{(1)}} < a$ and we want to prove that $s(B(X; x_0, a), Y) = 0$, i.e. we want to find a sequence $x_n \in B(X; x_0, a)$ with $\|x_n\|_Y \to 0$.

In $X^*$, consider

$$E = \overline{Y \cap B_1(X^*)}^{w^*} \subseteq B_1(X^*) \cap Y^{(1)}.$$

With the weak*-topology, E is compact, metrizable. Morever E is clearly convex. Let $C \subseteq C(E)$ ($= \{f : E \to \mathbb{C} : f \text{ continuous}\}$) be defined by

$$C = \{x \restriction E : x \in B(X; x_0, a)\}$$

where $x \restriction E$ is the function $x^* \mapsto \langle x, x^*\rangle$ for $x^* \in E$. The family C is a convex set of continuous functions on E, and it is enough to prove that $0 \in \bar{C}$ = the norm-closure of C in the Banach space C(E). For then there are $x_n \in B(X; x_0, a)$ with $\|x_n \restriction E\|_{C(E)} \to 0$, and since $Y \cap B_1(X^*) \subseteq E$, a fortiori, $\|x_n\|_Y \to 0$.

Recall here Mazur's Theorem that in a Banach space the weak-closure of a convex set is the same as its norm-closure. So it is enough to show that $0 \in \bar{C}^w$ = the weak-closure of C in C(E), i.e for the topology of duality with the space $C(E)^*$ = M(E) = the space of (complex, Borel) measures on E. A typical nbhd of 0 in the weak topology of C(E) is given by

$$V = \{f \in C(E) : |\langle f, \nu_1\rangle| < \epsilon \ \&\ldots\&\ |\langle f, \nu_n\rangle| < \epsilon\}$$

where $\nu_1,\ldots, \nu_n$ are measures on E. Write each $\nu_j$ as a linear combination $\nu_j = \Sigma a_m^j \mu_m^j$ with $\mu_m^j$ probability measures on E. Now every probability measure $\mu$ on E has a <u>barycenter,</u> i.e. a point $b(\mu) \in E$ satisfying $\langle f, \mu\rangle = f(b(\mu))$ for all linear continuous f on E, that is all f of the form $f = x \restriction E$ where $x \in X$. (The subset of all probabilities which have barycenters is $w^*$-closed in M(E) as E is compact, and contains all probabilities with finite support as E is convex, hence is identical with the set of all probability measures on E).

Let $y_m^j \in E$ be the barycenter of $\mu_m^j$. As each $f = x \restriction E \in C$ is linear continuous we get for $x \in B(x; x_0, a)$

$$\langle x \restriction E, \nu_j\rangle = \sum a_m^j \, x \restriction E(y_m^j)$$

$$= \sum a_m^j \langle x, y_m^j \rangle$$

$$= \langle x, y_j \rangle$$

where

$$y_j = \sum a_m^j\, y_m^j \in Y^{(1)},$$

since $E \subseteq Y^{(1)}$.

As $\|x_0\|_{Y^{(1)}} < a$, there is by 2.9 some $x^{**} \in X^{**}$, $x^{**} \in (Y^{(1)})^{\perp}$ with $\|x^{**} - x_0\|_{**} = a' < a$. Applying (a trivial variant of) Goldstine's Theorem to the balls $B(X; x_0, a')$ and $B(X^{**}; x_0, a')$ and the w*-nbhd in $X^{**}$

$$\{z^{**} \in X^{**} : |\langle y_j, z^{**} - x^{**} \rangle| < \epsilon,\ j = 1,\ldots, n\}$$

which clearly contains $x^{**}$, we get some $x \in B(x; x_0, a')$ with

$$|\langle y_j, x - x^{**} \rangle| < \epsilon,\ j = 1,\ldots, n.$$

But since $y_j \in Y^{(1)}$ and $x^{**} \in (Y^{(1)})^{\perp}$ we obtain

$$|\langle x \restriction E, \nu_j \rangle| = |\langle x, y_j \rangle| = |\langle y_j, x \rangle| < \epsilon$$

for $j = 1,\ldots, n$. Thus $C \cap V \neq \emptyset$ and we are done. □

Notice that Theorem 5 provides an alternative definition for ord (Y) in terms of the space X (as opposed to the space $X^*$). Its main use will be in providing simple computations for the complexity of ranks.

We turn now to a third rank, which this time deals with the space $X^{**}$.

## The R-rank

We put now a further restriction on the Banach space X: we assume that X has separable dual, i.e. $X^*$ is separable for the norm topology. (In the applications to the U-sets, where $X = PF$ and $X^* = A$, this condition is fulfilled). In this case, any point in $B_1(X^{**})$ is a $w^*$-limit of a sequence of points in $B_1(X)$. [Note that the hypothesis that X has separable dual is stronger: By a theorem of Odell-Rosenthal, see Diestel [1], it is enough to assume that $\ell^1$ does not embed in X—but we do not need this level of generality here). Let Y be a subspace of $X^*$, with associated sequence $Y^{(\alpha)}$. The idea of the R-rank is to define the sequence of annihilators $Z_{(\alpha)} = (Y^{(\alpha)})^{\perp}$ in $X^{**}$ a priori (i.e. without reference to the subspace Y). We will use this idea again in Chapter VIII.

So let Z be any subset of $X^{**}$. We define a derivative operation by

$$Z_{(1)} = \{x^{**} \in Z\text{: There is a sequence } x_n^{**} \in Z \text{ with } x_n^{**} \xrightarrow{w^*} x^{**} \text{ and } R(x_n^{**}) \to 0\}$$

where $R(x^{**})$ is the distance from $x^{**}$ to X, i.e.

$$R(x^{**}) = \inf_{x \in X} \|x^{**} - x\|_{**}.$$

We define inductively then a sequence $Z_{(\alpha)}$ by

$$Z_{(0)} = Z$$
$$Z_{(\alpha+1)} = (Z_{(\alpha)})_{(1)},$$
$$Z_{(\lambda)} = \cap_{\alpha<\lambda} Z_{(\alpha)}, \text{ for } \lambda \text{ limit.}$$

<u>Definition</u> <u>10</u>. We define the R-<u>rank</u> of $Z \subseteq X^{**}$ by

$$\text{rk}_R(Z) = \text{least } \alpha \ (Z_{(\alpha)} = \{0\})$$

if for some countable ordinal $\alpha$ $Z_{(\alpha)} = \{0\}$, and otherwise we let $\text{rk}_R(Z) = \omega_1$.

<u>Theorem</u> <u>11</u>. Let X be a Banach space with separable dual, and Y a subspace of $X^*$. Then for every ordinal $\alpha$, $(Y^{\perp})_{(\alpha)} = (Y^{(\alpha)})^{\perp}$. In particular, $\text{rk}_R(Y^{\perp})$ is countable iff Y is $w^*$-dense in $X^*$, and $\text{rk}_R(Y^{\perp}) \leq \text{ord}(Y) \leq \text{rk}_R(Y^{\perp}) + 1$.

<u>Proof</u>. The proof reduces to the case $\alpha = 1$, i.e. $(Y^{\perp})_{(1)} = (Y^{(1)})^{\perp}$. For assuming by induction the equality $(Y^{\perp})_{(\alpha)} = (Y^{(\alpha)})^{\perp}$, we get

$$(Y^{\perp})_{(\alpha+1)} = (Y^{\perp}_{(\alpha)})_{(1)} = ((Y^{(\alpha)})^{\perp})_{(1)} = (Y^{(\alpha+1)})^{\perp}$$

and for limit $\lambda$

$$(Y^{\perp})_{(\lambda)} = \cap_{\alpha<\lambda}(Y^{\perp})_{(\alpha)} = \cap_{\alpha<\lambda}(Y^{(\alpha)})^{\perp} = (\cup_{\alpha<\lambda}Y^{(\alpha)})^{\perp} = (Y^{(\lambda)})^{\perp}.$$

As moreover $(Y^{(\alpha)})^{\perp} = \{0\}$ iff $Y^{(\alpha)}$ is (norm-) dense in $X^*$, we get immediately the second assertion.

So we have only to prove $(Y^{\perp})_{(1)} = (Y^{(1)})^{\perp}$.

(i) Suppose first $x^{**} \in (Y^{\perp})_{(1)}$, so that $x_n^{**} \xrightarrow{w^*} x^{**}$, where $x_n^{**} \in Y^{\perp}$ and $R(x_n^{**}) \to 0$. Let $y^* \in Y^{(1)}$. We want to show $<y^*, x^{**}> = 0$. Let $y_p^* \in Y$, $y_p^* \xrightarrow{w^*} y^*$, and as $R(x_n^{**}) \to 0$, let $x_n \in X$ be such that $\|x_n - x_n^{**}\|_{**} \to 0$. Then $<y^*, x^{**}> = <y^*, x^{**} - x_n^{**}> + <y^* - y_p^*, x_n^{**}>$ as $<y_p^*, x_n^{**}> = 0$. The first term goes to 0 with n, as $x_n^{**} \xrightarrow{w^*} x^{**}$. Write the second as $<y^* - y_p^*, x_n^{**} - x_n> + <y^* - y_p^*, x_n>$. Choosing n large enough, we can make the first term above as small as we want, uniformly in p, since $\|x_n^{**} - x_n\|_{**} \to 0$ and $y^* - y_p^*$ is bounded in norm. Choosing then p large enough, we can make $<y^* - y_p^*, x_n>$ as small as we want, since $y_p^* \xrightarrow{w^*} y^*$. This shows that $<y^*, x^{**}> = 0$.

(ii) Conversely, let $x^{**} \in (Y^{(1)})^{\perp}$, $\|x^{**}\|_{**} = 1$. Since $B_1(X^{**})$ is metrizable in the weak$^*$-topology and $B_1(X)$ is $w^*$-dense in $B_1(X^{**})$ by Goldstine's Theorem, we can find a sequence $x_n$ in $B_1(X)$ with $x_n \xrightarrow{w^*} x^{**}$. Let $E = \overline{Y \cap B_1(X^*)}^{w^*} \subseteq B_1(X^*) \cap Y^{(1)}$, so that E is compact, metrizable. The sequence of continuous functions $x_n \restriction E$ is uniformly bounded on E and converges pointwise to $x^{**} \restriction E \equiv 0$, i.e. this sequence of functions converges weakly to 0 in $C(E)$. So by Mazur's Theorem there is a sequence $x_n'$ of convex combinations of the $x_n$ such that $x_n' \restriction E$ converges uniformly to 0. Clearly $x_n' \xrightarrow{w^*} x^{**}$ as well and since $Y \cap B_1(X^*) \subseteq E$, it follows that $\|x_n'\|_Y = \sup_{y^* \in Y \cap B_1(X^*)} |<x_n', y^*>| \to 0$ as $n \to \infty$. Let then by 2.9 $x_n^{**} \in X^{**}$ be such that $x_n^{**} \in Y^{\perp}$ and $\|x_n' - x_n^{**}\|_{**} \to 0$. Then $x_n^{**} \in Y^{\perp}$, $x_n^{**} \xrightarrow{w^*} x^{**}$ and $R(x_n^{**}) \le \|x_n' - x_n^{**}\|_{**} \to 0$, hence $x^{**} \in (Y^{\perp})_{(1)}$, and we are done. □

## §4. The Piatetski-Shapiro rank on U

We specialize now the preceding ideas to the case of the closed sets of uniqueness.

For $E \in K(\mathbb{T})$, we define the subspace $J(E)$ of $A$ by

$$J(E) = \{f \in A : \mathrm{supp}(f) \cap E = \emptyset\},$$

i.e. $J(E)$ is the space of functions in $A$ which vanish on some nbhd of the closed set $E$. Note that $J(E)$ is an ideal in $A$, i.e. a subspace closed also under multiplication by any $f \in A$.

The description of closed $\mathcal{U}$-sets in terms of pseudofunctions (II.4.1) gives by duality immediately the following

Theorem 1 (Piatetski-Shapiro [2]). Let $E \subseteq \mathbb{T}$ be closed. Then $E$ is a set of uniqueness iff the ideal $J(E)$ is $w^*$-dense in $A$.

Proof.. This is equivalent to saying that $E$ is a set of multiplicity iff $\overline{J(E)}^{w^*} \neq A$, which by Hahn-Banach is equivalent to the existence of $S \in PF$ with $S \neq 0$ but $\langle f, S\rangle = 0$ for all $f \in J(E)$, i.e. $\mathrm{supp}(S) \subseteq E$. □

Definition 2. The Piatetski-Shapiro rank $[E]_{PS}$ of $E \in U$ is the ordinal

$$[E]_{PS} = \mathrm{ord}(J(E)).$$

It follows from 2.6 that $[E]_{PS}$ is a rank on U. We have now the following result conjectured by Kechris and proved by Solovay in early 1984, whose proof however remains still unpublished.

<u>Theorem 3</u> (Solovay [2]). The Piatetski-Shapiro rank $[E]_{PS}$ is a $\underset{\sim}{\Pi}^1_1$-rank on U.

<u>Proof</u>. Solovay's original proof of this result used relatively little analysis, but a good deal of logic. We give here an alternative proof, based on the work in §3. Fix a dense in the open unit ball of PF countable set D, closed under rational multiplication, and consider the tree rank associated to J(E). By Theorem 3.5, we know that if $E \neq \emptyset$ $[E]_{PS} = rk_T(J(E))$, so for the computation of the complexity of $[E]_{PS}$, we can work with $rk_T(J(E))$. Now the tree-rank is defined by the equality

$$\omega \cdot rk_T(J(E)) = \sup_{\epsilon \in \mathbb{Q} \cap (0,1)} ht(T^{\epsilon}_{J(E)})$$

so that by the computations we have done in Lemma 1.2 concerning heights of well founded trees, it is enough to prove that for each $\epsilon \in \mathbb{Q} \cap (0, 1)$, the function $E \mapsto T^{\epsilon}_{J(E)}$ from K(T) into $\{0, 1\}^{\mathrm{Seq}\ D}$ (with the product topology) is a Borel function. And by the definition of the trees $T^{\epsilon}_{J(E)}$, this reduces to checking that for each x $\in$ PF, $\delta > 0$, $C = \{E \in K(T) : \|x\|_{J(E)} \leq \delta\}$ is a Borel subset of K(T). Now

$$C = \bigcap_{\substack{f \in A, \|f\|_A \leq 1 \\ |<x,f>| > \delta}} \{E \in K(T) : \mathrm{supp}(f) \cap E \neq \emptyset\}$$

so C is closed in K(T) and we are done. □

The following result was proved earlier directly by McGehee—see Chapter VI.2.4.

Corollary 4 (McGehee [2]). The Piatetski-Shapiro rank is unbounded on U, i.e. for each $\alpha < \omega_1$ there is $E \in U$ with $[E]_{PS} \geq \alpha$.

Proof. Obvious from Theorem 3, 1.7 and IV.2.2. □

We consider now the R-rank on $J(E)^{\perp}$. Since J(E) is an ideal, so is each

$$J^{\alpha}(E) = (J(E))^{(\alpha)}$$

as we can easily prove by induction. In particular

$$J^{\alpha}(E) = A \text{ iff } 1 \in J^{\alpha}(E).$$

Another consequence is that we can apply to the $J^{\alpha}(E)$'s the following simple fact.

Lemma 5. Let $Y \subseteq A$ be an ideal and define

$$\text{hull}(Y) = \{x \in \mathbb{T} : \forall f \in Y(f(x) = 0)\}.$$

Then

$$Y = A \Leftrightarrow \text{hull}(Y) = \emptyset.$$

In particular, since $\text{hull}(Y) = \text{hull}(\bar{Y})$,

$$Y = A \Leftrightarrow \overline{Y} = A$$

Proof. The direction $\Rightarrow$ is obvious. If now hull(Y) $= \emptyset$, one can find for each $x \in \mathbf{T}$ a function $f_x \in Y$ which is not 0 in some nbhd of x. Multiplying $f_x$ by $\overline{f}_x$ if necessary we can assume that actually $f_x > 0$ in a nbhd of x and $f_x \geq 0$ everywhere. By compactness then there is $f \in Y$ which is non-0 everywhere. By Wiener's Theorem II.2.3, $1/f \in A$ hence $1 \in Y$, so that $Y = A$.

Note that the conclusion $\overline{Y} = A \Rightarrow Y = A$ can be easily established directly, since if $\overline{Y} = A$ there is $f \in Y$ with $\|1 - f\|_A < 1$, so that by standard Banach algebra theory f is invertible, and so $1 \in Y$. □

For $E \in K(\mathbf{T})$, let

$$PM(E) = J(E)^{\perp},$$

be the set of pseudomeasures with support contained in E and define

$$PM_{\alpha}(E) = PM(E)_{(\alpha)}$$

using the derivation of §3 (related to the R-rank). Note that for $X = PF$, $X^{**} = PM$ one has for $S \in PM$,

$$R(S) =^{\text{def}} \text{dist}(S, PF) = \overline{\lim}\ |S(n)|.$$

To see this note first that for all $x \in PF$, $\|S - x\|_{PM} \geq \overline{\lim}\ |S(n)|$, so R(S)

$\geq \overline{\lim} |S(n)|$. On the other hand if we define for $n \in \mathbb{N}$, $(S)_n$ by

$$(S)_n(j) = \begin{cases} S(j), \text{ if } |j| \leq n, \\ 0, \text{ if } |j| > n, \end{cases}$$

clearly $(S)_n \in PF$ and

$$\lim_n \|S - (S)_n\|_{PM} = \overline{\lim} |S(n)|$$

so $R(S) \leq \overline{\lim} |S(n)|$.

It follows that the derivative $Z_{(1)}$ in PM is defined by

$$Z_{(1)} = \{S \in Z : \text{There is a sequence } S_n \text{ in } Z \text{ with } S_n \xrightarrow{w^*} S \text{ and } \overline{\lim_k} |S_n(k)| \to 0\}.$$

By 3.11 we have

(i) $PM_\alpha(E) = (J^\alpha(E))^\perp$

(ii) $E \in U \Leftrightarrow \exists\alpha < \omega_1 \, (PM_\alpha(E) = \{0\})$

(iii) For $E \in U$,

$$\begin{aligned} rk_R(PM(E)) &= \text{least } \alpha(PM_\alpha(E) = \{0\}) \\ &= \text{least } \alpha \, (\overline{J^\alpha(E)} = A). \end{aligned}$$

Using the preceding lemma we have then

Theorem 6. If $E \in K(\mathbf{T})$ is a set of uniqueness, then

$$[E]_{PS} = rk_R(PM(E)).$$

So we have three equivalent ranks on U, the Piatetski-Shapiro rank, the tree-rank on J(E) and the R-rank on PM(E). (The only exception is the case $E = \emptyset$, where $[E]_{PS} = rk_R(PM(E)) = 0$, but $rk_T(J(E)) = 1$).

For each countable ordinal $\alpha$, let

$$U^{(\alpha)} = \{E \in K(\mathbf{T}) : [E]_{PS} \leq \alpha\}$$

so that the $U^{(\alpha)}$ form an increasing sequence of Borel sets with union U. Moreover from Corollary 4 $U^{(\alpha)} \subsetneq U$, for each $\alpha < \omega_1$.

Proposition 7. For each $\alpha$, $U^{(\alpha)}$ is an ideal of closed sets.

Proof. If $E_1 \subseteq E_2$, $J(E_2) \subseteq J(E_1)$, hence by the definition of $[E]_{PS}$, $[E_1]_{PS} \leq [E_2]_{PS}$, so $U^{(\alpha)}$ is downward closed under inclusion. It remains to show that the union of two sets in $U^{(\alpha)}$ is again in $U^{(\alpha)}$. It is clearly enough to prove, by induction on $\beta$, that for E, F in $K(\mathbf{T})$, $f \in J^{\beta}(E)$ and $g \in J^{\beta}(F)$ imply $f \cdot g \in J^{\beta}(E \cup F)$. This is immediate for $\beta = 0$, and clearly goes through at limit ordinals. So suppose we know it for $\beta$, and let $f \in J^{\beta+1}(E)$, $g \in J^{\beta+1}(F)$. Let $f_n \xrightarrow{w^*} f$, $f_n \in J^{\beta}(E)$ and $g_n \xrightarrow{w^*} g$, $g_n \in J^{\beta}(F)$. Let M be such that all $f_n$, $g_m$ and $f_n \cdot g_m$ are in $B_M(A)$. As $B_M(A)$ is metrizable in the weak*-topology, since $f_n \cdot g_m \xrightarrow{w^*} f_n \cdot g$ as $m \to \infty$ and $f_n \cdot g \xrightarrow{w^*} f \cdot g$ as $n \to \infty$, there is a sequence $h_k$ of functions of the form $f_n \cdot g_m$ with $h_k \xrightarrow{w^*} f \cdot g$. By the induction hypothesis each $h_k$ is in $J^\beta(E \cup F)$, hence $f \cdot g$ is in $J^{\beta+1}(E \cup F)$ and we are done. □

Proposition 8. The rank $[E]_{PS}$ is invariant under translations, so in particular $U^{(\alpha)}$ is closed under translations, for each $\alpha$.

Proof. For f in A and $a \in \mathbb{R}$, let $f_a$ be defined by $f_a(x) = f(x + a)$. Then clearly $\hat{f_a}(n) = e^{ina}\hat{f}(n)$, hence for each a, $f_a \in A$ (and $\|f_a\|_A = \|f\|_A$). Moreover

$$f_a \in J(E) \Leftrightarrow f \in J(E + a)$$

where $E + a = \{(x + a) \bmod 2\pi : x \in E\}$. Using this and the immediate fact that $f^n \xrightarrow{w^*} f \Leftrightarrow f_a^n \xrightarrow{w^*} f_a$, one easily checks by induction that for all $\alpha$,

$$f_a \in J^\alpha(E) \Leftrightarrow f \in J^\alpha(E + a),$$

hence for all a, $[E]_{PS} = [E + a]_{PS}$. □

Theorem 9 (Piatetski-Shapiro [2] for $\alpha = 1$). For all $\alpha$, $U^{(\alpha)}$ is closed under dilations and contractions, i.e. if E, F are closed subsets of $[0, 2\pi]$, $t > 0$ and $F = tE$ ($= \{tx : x \in E\}$), then $E \in U^{(\alpha)} \Leftrightarrow F \in U^{(\alpha)}$.

Proof. Using the same remarks as in the proof of the Marcinkiewicz-Zygmund Theorem II.4.2, and using Propositions 7 and 8 above, the proof reduces to showing that if $t > 0$, and E, tE are subsets of $(0, 2\pi)$, then $[tE]_{PS} \geq [E]_{PS}$. Recall that for each $S \in PM(E)$ we defined in the proof of II.4.2 a pseudomeasure $S_t$ by

$$S_t(n) = \langle e^{-itnx} \cdot \varphi, S\rangle$$

where $\varphi \in C^\infty(\mathbb{T})$ is 1 in a nbhd of E and 0 in a nbhd of 0, and proved there that $S_t \in PM(tE)$. It is enough to show that actually for every $\alpha$,

$$(*) \qquad S \in PM_\alpha(E) \Rightarrow S_t \in PM_\alpha(tE).$$

Because then if $PM_\alpha(E) \neq \{0\}$ let $S \neq 0$ be in $PM_\alpha(E)$, say with $S(n_0) \neq 0$. If $S' = e^{-in_0x} \cdot S$, then $S'(0) = S(n_0) \neq 0$, so $S'_t(0) = S'(0) \neq 0$. But it is easy to check by induction on $\alpha$ that for every $g \in A$ and $T \in PM_\alpha(E)$, $g \cdot T \in PM_\alpha(E)$. Thus $S' \in PM_\alpha(E)$ and $0 \neq S'_t \in PM_\alpha(tE)$. This shows that $rk_R(PM(tE)) = [tE]_{PS} \geq [E]_{PS} = rk_R(PM(E))$.

Now (*) follows easily from the following:

(†) If $S^n \in PM(E)$ and $S^n \xrightarrow{w^*} S$, then $S^n_t \xrightarrow{w^*} S_t$. If $S^n \in PM(E)$ and $R(S^n) \to 0$, then $R(S^n_t) \to 0$.

To prove (†) notice first that for all $p \in \mathbb{Z}$, $S^n_t(p) \to S_t(p)$ since $S^n \xrightarrow{w^*} S$. Moreover as in the proof of II.4.2 $\|S^n_t\|_{PM}$ is bounded by $\|S^n\|_{PM} \cdot \sup_{0\leq\epsilon\leq1} \|e^{-i\epsilon x}\varphi\|_A$, which is bounded independently of n. This shows that $S^n_t \xrightarrow{w^*} S_t$.

It remains to show that if $R(S^n) \to 0$, then $R(S^n_t) \to 0$. If not, towards a contradiction, we can assume that for all n, $R(S^n_t) > \epsilon > 0$, hence we can choose an increasing sequence $\{|k^n_j|\}_{j\in\mathbb{N}}$ for each n with $|S^n_t(k^n_j)| > \epsilon$. As in the proof of II.4.2, by going to a subsequence, there is $\epsilon_n \in [0, 1]$ with $T_n = (e^{-i\epsilon_n x}\varphi) \cdot S^n$

satisfying $R(T_n) = \overline{\lim_k} |T_n(k)| > \epsilon$. But since $R(S^n) = \text{dist}(S^n, PF) \to 0$, there is $x_n \in PF$ with $\|S^n - x_n\|_{PM} \to 0$, and therefore

$$\|T_n - (e^{-i\epsilon_n x}\varphi) \cdot x_n\|_{PM} \leq \|e^{-i\epsilon_n x}\varphi\|_A \cdot \|S^n - x_n\|_{PM} \to 0$$

so, as $(e^{-i\epsilon_n x}\varphi) \cdot x_n \in PF$, we have $R(T_n) \to 0$, a contradiction. □

The only closed set of Piatetski-Shapiro rank 0 is ∅ for which $J(E) = A$. The class $U^{(1)}$ of uniqueness sets of rank $\leq 1$ is usually denoted by

$$U' = U^{(1)}$$

and we will study it in the next section. To avoid pedantry we will refer to the sets in $U'$ as <u>sets of rank</u> 1.

## §5. The class U′ of uniqueness sets of rank 1

We defined

$$U' = \{E \in U : [E]_{PS} \leq 1\}$$
$$= \{\emptyset\} \cup \{E \in U : [E]_{PS} = 1\}$$

to be the class of all closed subsets E of $\mathbb{T}$ for which J(E) is $w^*$-sequentially dense in A. As J(E) is an ideal, this amounts to saying that the function 1 is a $w^*$-limit of a sequence of elements of J(E), and we get at once that the sets in $U'$ are exactly those satisfying the Piatetski-Shapiro criterion in III.1.3, i.e.

Proposition 1. For $E \in K(T)$, $E \in U'$ iff there exists a sequence $f_n$ of functions in $A$ with $\mathrm{supp}(f_n) \cap E = \emptyset$, $\|f_n\|_A \leq M$ for some $M$ independent of $n$, $\hat{f_n}(0) \to 1$ and $\hat{f_n}(m) \to 0$ for $m \neq 0$, as $n \to \infty$.

Going back to the proof of III.1.4 that $H^{(n)}$ sets are U-sets, recall that we constructed sequences satisfying the hypotheses of the preceding criterion, so that we get

Proposition 2. For all n, the $H^{(n)}$-sets are in $U'$.

Although most of the usual examples of U-sets are in fact $U'$-sets, it is often not so immediate to prove it. We will do this next for the countable sets. But first let us recall some of the notions introduced in §2 and see how they apply in the present context to give another very useful criterion for being in $U'$.

Let $Z \subseteq PM$ be a set, $Z \neq \{0\}$. Consider as in §2 the quantity

$$\eta(Z) = \inf\left\{\frac{R(S)}{\|S\|_{PM}} : S \in Z, S \neq 0\right\}.$$

Recall also that we defined in §2, for $Y$ a subspace of $A$,

$$s(Y) = \inf\left\{\frac{\|x\|_Y}{\|x\|_{PM}} : x \in PF, x \neq 0\right\}.$$

It might be suspected that when $Z = Y^{\perp}$, $\eta(Z)$ is the "dual" notion of $s(Y)$. In fact one has the following precise relationship

<u>Theorem 3</u> (McGehee [2]). Let Y be a subspace of A with $\overline{Y} \neq A$. Then

$$s(Y) = \frac{\eta(Y^{\perp})}{1 + \eta(Y^{\perp})}.$$

In particular, Y is w*-sequentially dense in A iff $\eta(Y^{\perp}) > 0$.

For $E \in K(\mathbb{T})$, $E \neq \emptyset$, we let

$$\eta(E) =^{\text{def}} \eta(PM(E)) = \inf\left\{\frac{R(S)}{\|S\|_{PM}} : S \in PM,\ S \neq 0,\ \text{supp}\,(S) \subseteq E\right\}.$$

<u>Corollary 4</u>. A nonempty set $E \in K(\mathbb{T})$ is in U′ iff $\eta(E) > 0$.

<u>Proof of Theorem 3</u>. The second statement follows immediately from the first and 2.8. Let $s = s(Y)$ and $\eta = \eta(Y^{\perp})$. We have already seen at the end of § 2 that $s \geq \frac{\eta}{1 + \eta}$.

For the converse, let $S \in Y^{\perp}$, $S \neq 0$ and for $n \in \mathbb{N}$ define as before

$$(S)_n\,(j) = \begin{cases} S(j), \text{ for } |j| \leq n, \\ 0, \text{ for } |j| > n, \end{cases}$$

and

$$(S)^n = S - (S)_n.$$

Clearly $(S)_n \in PF$, $\|(S)_n\|_{PM} \to \|S\|_{PM}$ and $\|(S)^n\|_{PM} \to R(S)$. Let $\gamma_n = \frac{\|(S)^n\|_{PM}}{\|(S)_n\|_{PM}}$ (for n big enough so that $\|(S)_n\|_{PM} \neq 0$), and let $x_n = (1 + \gamma_n)\,(S)_n$. Then $x_n \in PF$, $x_n \neq 0$ and hence

$$s \le \frac{\|x_n\|_Y}{\|x_n\|_{PM}} = \frac{\mathrm{dist}(x_n, Y^\perp)}{\|x_n\|_{PM}} \le \frac{\|x_n - S\|_{PM}}{\|x_n\|_{PM}}.$$

Now $\|x_n\|_{PM} = (1 + \gamma_n)\,\|(S)_n\|_{PM}$ and

$$\|x_n - S\|_{PM} = \|\gamma_n(S)_n - (S)^n\|_{PM}$$

$$= \max(\|(S)^n\|_{PM}, \gamma_n\|(S)_n\|_{PM}) = \|(S)^n\|_{PM}$$

so that $s \le \dfrac{\|(S)^n\|_{PM}}{(1 + \gamma_n)\|(S)_n\|_{PM}} = \dfrac{\gamma_n}{1 + \gamma_n}$. Letting $n \to \infty$ and taking the infimum over $S \in Y^\perp$ we get finally $s \le \dfrac{\eta}{1 + \eta}$ and we are done. □

The remainder of this section is devoted to the proof of the following theorem, essentially due to Loomis.

<u>Theorem 5</u>. Every countable closed set is in U′.

The key fact, due to Loomis [1], is that every pseudomeasure S with countable support is in fact almost periodic, which implies in particular that $R(S) = \|S\|_{PM}$. So for each nonempty countable closed set E, $\eta(E) = 1$, and Corollary 4 gives Theorem 5.

For the rest of this section, define for $S \in PM$ and $n \in \mathbb{Z}$ the n-<u>translate</u> $S_n$ of S by $S_n(k) = S(n + k)$.

<u>Definition 6</u>. An element S of PM is <u>almost periodic</u> (a.p.) if for every $\epsilon > 0$ there are finitely many translates $(S_{n_j})_{j=1\ldots,p}$ of S such that the balls $B(PM; S_{n_j}, \epsilon) =$

$B(S_{n_j}, \epsilon)$ cover the set $\{S_n : n \in \mathbb{Z}\}$ of all translates of S (i.e. the set $\{S_n : n \in \mathbb{Z}\}$ is precompact in the norm topology of PM).

The term "almost-periodic" comes from the fact that this property can be rephrased as follows: Say that $p \in \mathbb{N}$ is an $\epsilon$-almost period for S if $\|S - S_p\|_{PM} \leq \epsilon$. Then S is a.p. iff for each $\epsilon > 0$ there is an $n \in \mathbb{N}$ such that each interval of length n of $\mathbb{Z}$ contains an $\epsilon$-almost period of S.

Example. Let $x \in [0, 2\pi]$, and $S^x(k) = e^{-ikx} = \hat{\delta}_x(k)$, where $\delta_x$ is the Dirac measure at x. The pseudomeasure $S^x$ is a.p. In fact one has $S^x_n - S^x_m = (e^{-inx} - e^{-imx}) \cdot S^x$. So for each $\epsilon$, if we choose a finite $F \subseteq \mathbb{Z}$ such that for all n in $\mathbb{Z}$ there is $m \in F$ with $|e^{-inx} - e^{-imx}| \leq \epsilon$, F clearly witnesses that $S^x$ is a.p.

We will give now Loomis' proof through a series of propositions.

Proposition 7. Let $AP = \{S \in PM : S \text{ is a.p.}\}$. Then AP is a (norm) closed subspace of PM.

(By the example above, it also contains the (Fourier transforms of the) Dirac measures. In fact it is the least (norm) closed subspace of PM containing the Dirac measures (see Katznelson [1], Loomis [2], Benedetto [2]), but we will not use this result here. These references contain also much further information on the theory of almost periodic functions originated by H. Bohr).

Proof. (i) Let S, T be a.p. We want to show that aS + bT is a.p., where a, b $\neq$ 0. Let $\epsilon > 0$, $F_1$ witness that S is a.p. for $\frac{\epsilon}{4|a|}$ and $F_2$ witness that T is a.p. for

$\frac{\epsilon}{4|b|}$. For each pair $(n_1, n_2) \in F_1 \times F_2$ choose m such that $\|S_m - S_{n_1}\|_{PM} \leq \frac{\epsilon}{4|a|}$ and $\|T_m - T_{n_2}\|_{PM} \leq \frac{\epsilon}{4|b|}$ if there is such an m, and let F be the finite subset of $\mathbb{Z}$ consisting of these m's. We claim F works: Given any $p \in \mathbb{Z}$, there is $n_1 \in F_1$ and $n_2 \in F_2$ with $\|S_p - S_{n_1}\|_{PM} \leq \frac{\epsilon}{4|a|}$ and $\|T_p - T_{n_2}\|_{PM} \leq \frac{\epsilon}{4|b|}$, so some m in F satisfies the same inequalities, hence $\|S_p - S_m\|_{PM} \leq \frac{\epsilon}{2|a|}$ and $\|T_p - T_m\|_{PM} \leq \frac{\epsilon}{2|b|}$ and $\|(aS + bT)_p - (aS + bT)_m\|_{PM} \leq \epsilon$.

(ii) If $(S^n)$ are a.p. and $S^n \to S$ in PM, S is a.p.: Note that for all p, $S^n_p \to S_p$ in PM. For $\epsilon > 0$ given, choose $n_0$ so that $\|S^{n_0} - S\|_{PM} < \frac{\epsilon}{3}$ and let F witness that $S^{n_0}$ is a.p. with $\frac{\epsilon}{3}$. Then for each $m \in \mathbb{Z}$ there is a $p \in F$ with $\|S^{n_0}_m - S^{n_0}_p\|_{PM} \leq \frac{\epsilon}{3}$, and $\|S_m - S_p\|_{PM} \leq \|S_m - S^{n_0}_m\|_{PM} + \|S^{n_0}_m - S^{n_0}_p\|_{PM} + \|S^{n_0}_p - S_p\|_{PM} \leq \epsilon$, as $\|S_m - S^{n_0}_m\|_{PM} = \|S - S^{n_0}\|_{PM}$ for all m. □

<u>Proposition 8</u>. AP is closed under multiplication by functions in A.

<u>Proof</u>. Let $S \in AP$, and $f \in A$, so that

$$(f \cdot S)(n) = \sum_{k+m=n} \hat{f}(k) \cdot S(m).$$

This gives

$$\begin{aligned}(f \cdot S)_p(n) &= \sum_{k+m=n+p} \hat{f}(k) \cdot S(m) \\ &= \sum_{k+m=n} \hat{f}(k) \cdot S(m + p) \\ &= (f \cdot S_p)(n)\end{aligned}$$

So $(f \cdot S)_p = f \cdot S_p$ and $\|(f \cdot S)_p - (f \cdot S)_n\|_{PM} \le \|f\|_A \cdot \|S_p - S_n\|_{PM}$, which gives immediately the proposition. □

Let $x \in \mathbf{T}$. We say that $S \in PM$ is <u>almost-periodic</u> <u>at</u> <u>x</u> (a.p. at x) if for some $f \in A$ with $f(x) \neq 0$, $f \cdot S$ is a.p. It follows from Proposition 8 that if S is a.p., S is a.p. everywhere. The converse also holds:

<u>Proposition</u> <u>9</u>. If for all $x \in \mathbf{T}$, $S \in PM$ is a.p. at x, then $S \in AP$.

<u>Proof</u>. Notice first that in the definition of a.p. at x, we can require the $f \in A$ to be $\ge 0$, by replacing it by $f \cdot \overline{f}$, using Proposition 8. By compactness, if S is a.p. everywhere we can find functions $f_j \in A$, $j = 1,\dots, k$, with $f_j \cdot S \in AP$ for all j and $\Sigma f_j > 0$ on $\mathbf{T}$. By Wiener's Theorem (II.2.3) $f = 1/\sum_j f_j$ is in A, and by Propositions 7 and 8 $S = f \cdot \sum_j (f_j \cdot S)$ is AP. □

The next goal is to strengthen the preceding result by getting the same conclusion for an $S \in PM$ which is a.p. at all points except possibly one.

We start with an intermediate step:

<u>Proposition</u> <u>10</u>. Let $S \in PM$ be a.p. at every $x \neq 0$. Then for each $n \in \mathbf{Z}$, $(S - S_n)$ is a.p.

<u>Proof</u>. By definition, $S - S_n = (1 - e^{-inx}) \cdot S$, and the function $\varphi(x) = 1 - e^{-inx}$ is in A, and $\varphi(0) = 0$. Using Lemma II.2.2 (and the remarks following it), there is for each $\epsilon > 0$ a function $f_\epsilon \in A$ vanishing in a neighborhood of 0, with

$\|\varphi \cdot S - f_\epsilon \cdot S\|_{PM} \leq \epsilon \cdot \|S\|_{PM}$. By Proposition 7, it is enough to check that $f_\epsilon \cdot S \in AP$, and by Proposition 9, that $f_\epsilon \cdot S$ is a.p. everywhere. This is clear for $x \neq 0$, as S itself is a.p. at x. For $x = 0$, let $g \in A$ be such that $g(0) \neq 0$ but $f_\epsilon \cdot g = 0$, which is possible as $f_\epsilon$ is 0 on a neighborhood of 0. Then $g \cdot (f_\epsilon \cdot S) = (g \cdot f_\epsilon) \cdot S = 0$ is a.p., and we are done. □

Lemma 11. Let $S \in PM$. If for all $n \in \mathbf{Z}$ $S - S_n$ is a.p., S is a.p.

Proof. First notice that if $S^1, \ldots, S^N$ are a.p. pseudomeasures then $\forall \epsilon > 0$ $\exists F$ finite such that F witnesses $S^j$ is a.p. for $\epsilon$, for all $j = 1, \ldots, N$ (i.e. $S^1, \ldots, S^N$ are uniformly a.p.). To see this, we argue as in Proposition 7: Choose $F^j$ finite witnessing that $S^j$ is a.p. for $\epsilon/2$, and for each sequence $\vec{m} = (m_j)$ in $\prod_{j=1}^{N} F^j$, choose $k(\vec{m}) = k$ such that for all j, $\|S^j_{m_j} - S^j_k\|_{PM} \leq \epsilon/2$ if there is such a k. Let $F = \{k : \exists \vec{m} \in \prod_{j=1}^N F^j (k = k(\vec{m}))\}$. Then F is finite, and for any $p \in \mathbf{Z}$ there is $\vec{m} \in \prod_{j=1}^N F^j$ with $\|S^j_p - S^j_{m_j}\|_{PM} \leq \epsilon/2$, so that $k = k(\vec{m})$ is defined, belongs to F, and $\|S^j_p - S^j_k\|_{PM} \leq \epsilon$, for all $j = 1, \ldots, N$.

Next we may clearly assume that S is real valued, i.e. $S(k) \in \mathbf{R}$ for all $k \in \mathbf{Z}$. Let $M = \sup S$ and $m = \inf S$. We prove now that for $\epsilon > 0$ given, we can find finite $F \subseteq \mathbf{Z}$ such that for all $n \in \mathbf{Z}$ there are $k, \ell \in F$ with $S_n(k) > M - \epsilon$ and $S_n(\ell) < m + \epsilon$.

To see this, choose $a, b \in \mathbf{Z}$ with $S(a) > M - \frac{\epsilon}{4}$ and $S(b) < m + \frac{\epsilon}{4}$, and F witnessing that $S_a - S_b$ is a.p. for $\frac{\epsilon}{2}$. So for $n \in \mathbf{Z}$, there is $k \in F$ such that

$$\|(S_a - S_b)_{a-n} - (S_a - S_b)_k\|_{PM} \leq \frac{\epsilon}{2}.$$

Considering the value at $n - a$, one gets

$$|S(a) - S(b) - (S(k + n) - S(b + k + n - a))| \leq \epsilon/2$$

so

$$S(k + n) \geq S(a) - S(b) + S(b + k + n - a) - \frac{\epsilon}{2}$$

hence

$$S(k + n) > M - \frac{\epsilon}{4} - (m + \frac{\epsilon}{4}) + m - \frac{\epsilon}{2} = M - \epsilon.$$

The proof for the infimum m is similar.

To finish the proof, consider the finite F found above for $\epsilon$, and the finite family $(S - S_k)_{k \in F}$. By our first remark, there is a finite $F^*$ such that for all n $\in \mathbb{Z}$ there is a $p \in F^*$ with $\|(S - S_k)_n - (S - S_k)_p\|_{PM} \leq \epsilon$, for all $k \in F$. This gives for all q, $|S_n(q) - S_k(n + q) - S_p(q) + S_k(p + q)| \leq \epsilon$, hence $S_n(q) - S_p(q) \leq S_k(n + q) - S_k(p + q) + \epsilon$ for all $k \in F$, and by choosing $k \in F$ so that $S_{p+q}(k) = S_k(p + q) \geq M - \epsilon$,

$$S_n(q) - S_p(q) \leq M - (M - \epsilon) + \epsilon = 2\epsilon.$$

Working with the infimum m gives similarly $S_n(q) - S_p(q) \geq - 2\epsilon$. Thus $F^*$ witnesses that S is a.p. for $2\epsilon$, and we are done. □

<u>Proposition 12</u>. If $S \in PM$ is a.p. at all points except possibly one, S is a.p.

<u>Proof</u>. For each $T \in PM$ and each $x \in \mathbb{R}$, define $T^x \in PM$ by $T^x(k) = e^{-ikx}$ $T(k)$. We check that if T is a.p., so is $T^x$. To see this, note that $(T^x)_n\ (k) =$

$e^{-i(n+k)x} \cdot T_n(k)$, hence $\|T^x_n - T^x_m\|_{PM} \leq \|T_n - T_m\|_{PM} + \|T\|_{PM} \cdot |e^{-inx} - e^{-imx}|$. Given $\epsilon > 0$, choose F witnessing that T is a.p. for $\frac{\epsilon}{3}$, so that for each n $\in \mathbf{Z}$ $\exists k = k(n) \in F$ with $\|T_n - T_{k(n)}\|_{PM} \leq \frac{\epsilon}{3}$. Choose then, by the compactness of $\mathbf{T}$, for each $k \in F$ a finite set $F_k \subseteq \{n \in \mathbf{Z} : k(n) = k\}$ such that for all $n \in \mathbf{Z}$ with $k(n) = k$ there is some $m \in F_k$ with $\|T\|_{PM} \cdot |e^{-inx} - e^{-imx}| \leq \frac{\epsilon}{3}$. One checks immediately that $\cup_{k \in F} F_k$ witness that $T^x$ is a.p. for $\epsilon$.

Let now $x_0$ be the exceptional point for S. Then by a simple translation argument and the preceding fact $S^{-x_0}$ is a.p. except possibly at 0, hence by Proposition 10 and Lemma 11 $S^{-x_0}$ is a.p. and so is $S = (S^{-x_0})^{x_0}$. □

<u>Theorem</u> <u>13</u> (Loomis [1]). Let $S \in PM$ have countable support. Then S is almost periodic.

<u>Proof</u>. Consider $E = \{x \in \mathbf{T} : S \text{ is not a.p. at } x\}$. We want to show that $E = \emptyset$. Note that E is closed and $E \subseteq \text{supp}(S)$. So if $E \neq \emptyset$, towards a contradiction, E has an isolated point $x_0$. Choose $f \in A$ which is 1 in a nbhd of $x_0$ and 0 in a nbhd of $E - \{x_0\}$. Then $f \cdot S$ is a.p. at every point except possibly $x_0$, hence is a.p. by Proposition 12. So S is a.p. at $x_0$, a contradiction. □

<u>Proof</u> <u>of</u> <u>Theorem</u> <u>5</u>. We will show that if $E \neq \emptyset$ is closed and countable then $\eta(E) = 1$. For that it is enough to show that if S is a.p. then $R(S) = \|S\|_{PM}$. If this fails, towards a contradiction, there is $m_0 \in \mathbf{Z}$ with $\|S\|_{PM} = |S(m_0)| > R(S)$, so for some N and $\epsilon > 0$,

$$(*) \qquad |n| \geq N \Rightarrow |S(n) - S(m_0)| > \epsilon.$$

**Choose F finite witnessing that S is a.p. for $\epsilon$. Then for any $n \in \mathbb{Z}$ there is $k \in F$ with $|S_n(m_0 - n) - S_k(m_0 - n)| \leq \epsilon$, hence $|S(m_0) - S(k + m_0 - n)| \leq \epsilon$. Since F is finite, this contradicts (*), choosing n large enough.** □

# Chapter VI. Decomposing U-sets into Simpler Sets

In this chapter, we consider the following problem: Are there nice families B of simply characterizable U-sets such that every U-set is a countable union of sets in B? Typical such candidates could be for instance $B = \cup_n H^{(n)}$ or $B = U'$, the classes of $H^{(n)}$-sets or U-sets of rank 1, respectively.

In view of the Characterization Problem, the best would be to find at least a Borel such family B. (The above examples are clearly so). This is the Borel Basis problem for U. In the first section, we study in a general setting Borel bases for $\underset{\sim}{\Pi}^1_1$ $\sigma$-ideals of closed sets. In the second section, we discuss a decomposition theorem for U due to Piatetski-Shapiro. Finally in the third section, we state the Borel Basis Problem for U and discuss some related issues. The problem itself will be solved—negatively—in Chapter VII, a result of Debs and Saint Raymond.

## §1. Borel bases for $\sigma$-ideals of closed sets

In this section, we go back to the framework of IV.3, i.e. E is a compact, metrizable space and K(E) is the space of closed subsets of E with the Hausdorff topology.

Let B be a hereditary subset of K(E), i.e. downward closed under inclusion. We defined $B_\sigma$, the $\sigma$-ideal generated by B, as $\{F \in K(E) : \exists(F_n)\ (F_n \in B$ for all n and $F = \bigcup_n F_n)\}$.

<u>Definition</u> <u>1</u>. Let I be a $\sigma$-ideal of closed sets in a compact metrizable space E. We say that $B \subseteq I$ is a <u>basis</u> for I if B is hereditary and $I = B_\sigma$.

Unless otherwise stated, most of the results about $\sigma$-ideals with Borel bases which follow are taken from Kechris-Louveau-Woodin [1]. (Note however that in that paper a basis is not required to be hereditary).

<u>Proposition</u> <u>2</u>. Let E be a compact, metrizable space. If B is a Borel hereditary subset of K(E), then the $\sigma$-ideal $B_\sigma$ is $\underset{\sim}{\Pi}^1_1$.

<u>Proof</u>. Let $\{V_n\}$ be a basis for E. Letting for $F \subseteq E$, $K(F) = \{K \in K(E) : K \subseteq F\}$, we claim that

$$(*)\quad F \in B_\sigma \Leftrightarrow \forall K \in K(F)\,(K \neq \emptyset \Rightarrow \exists n\,(K \cap V_n \neq \emptyset \;\&\; \overline{K \cap V_n} \in B))$$

which immediately gives that $B_\sigma$ is $\underset{\sim}{\Pi}^1_1$.

To prove (*), suppose first $F \in B_\sigma$. Then if $K \in K(F)$, $K \neq \emptyset$ we have that $K \in B_\sigma$ as well, i.e. $K = \bigcup_p K_p$ with $K_p \in B$. By the Baire Category Theorem there is some n with $K \cap V_n \neq \emptyset$ and $K \cap V_n \subseteq K_p$ for some p, hence $\overline{K \cap V_n} \in B$ and one direction of (*) is proved. Conversely, if $F \notin B_\sigma$, consider the family of open sets V in E such that $F \cap V$ is covered by countably many sets in B. There is clearly a largest such open set, say W, and $K = F - W \neq \emptyset$. Now if $V_n \cap K \neq \emptyset$, then $\overline{V_n \cap K} \notin B$ (otherwise $V_n \cap F$ is covered by $\overline{V_n \cap K}$ and W, so $V_n \subseteq$

W, thus $V_n \cap K = \emptyset$). So K is a counterexample to the right hand side of (*) and we are done. □

For example the $\sigma$-ideals $H_\sigma$, $H_\sigma^{(n)}$, $(\bigcup_n H^{(n)})_\sigma$, $U'_\sigma$ are all $\underset{\sim}{\Pi}^1_1$, so by IV.2.6 they are $\underset{\sim}{\Pi}^1_1$-complete.

So every $\sigma$-ideal generated by a Borel basis is $\underset{\sim}{\Pi}^1_1$. The converse is false, since if $P \subseteq E$ is $\underset{\sim}{\Pi}^1_1$ but not Borel, and

$$I = K(P) = \{K \in K(E) : K \subseteq P\}$$

then I is $\underset{\sim}{\Pi}^1_1$ but cannot have a Borel basis B, since then P would be Borel, as

$$x \in P \Leftrightarrow \{x\} \in B.$$

In fact we will see in this section that true $\underset{\sim}{\Pi}^1_1$ $\sigma$-ideals with Borel bases enjoy very particular properties. A good typical example of a true $\underset{\sim}{\Pi}^1_1$ $\sigma$-ideal with a Borel basis is $K_\omega(E)$ (E uncountable), the set of countable closed subsets of E, which admits the closed set $\{\emptyset\} \cup \{\{x\} : x \in E\}$ as a basis.

<u>Proposition</u> <u>3</u>. Let E be compact, metrizable. Let I be a $\underset{\sim}{\Pi}^1_1$ $\sigma$-ideal in K(E). The following are equivalent:

(i) I admits a Borel basis,

(ii) There is a $\underset{\sim}{\Sigma}^1_1$ subset P of I such that I is the $\sigma$-ideal generated by P, i.e.

$$I = \{F : \exists (F_n)\ (F_n \in P \text{ for all } n \text{ and } F \subseteq \cup_n F_n)\},$$

(iii) If

$$I^{loc} = \{F \in K(E) : \exists \text{ open } V\ (V \cap F \neq \emptyset \text{ and } \overline{V \cap F} \in I)\}$$

the set $I^{loc}$ is Borel in K(E).

Proof. Clearly (i) $\Rightarrow$ (ii). For (ii) $\Rightarrow$ (i), note that if $R \in \underset{\sim}{\Sigma}^1_1$ is a subset of I, then $her(R) = \{K \in K(E) : \exists F \in R(K \subseteq F)\}$ is $\underset{\sim}{\Sigma}^1_1$ as well. Starting with P we construct Borel sets $B_n$ as follows: By the Lusin Separation Theorem (see IV.1) let $B_o$ be Borel with $P \subseteq B_o \subseteq I$. Suppose $B_n$ has been constructed with $B_n \subseteq I$. Then $her(B_n)$ is $\underset{\sim}{\Sigma}^1_1$ and contained in I, so we can find again Borel $B_{n+1}$ with $her(B_n) \subseteq B_{n+1} \subseteq I$. Then $B = \cup_n B_n$ is Borel, contained in I, and hereditary. Since $P \subseteq B$ obviously $B_\sigma = I$. (This proof used essentially that I is $\underset{\sim}{\Pi}^1_1$. In fact the analog of Proposition 2 fails if one drops the assumption that B is hereditary, even for $G_\delta$ B's).

(i) $\Rightarrow$ (iii). As in the proof of Proposition 1, one gets, if B is a Borel basis for I,

$$F \in I^{loc} \leftrightarrow \exists n\ (F \cap V_n \neq \emptyset \text{ and } \overline{F \cap V_n} \in B)$$

hence $I^{loc}$ is Borel. (Here $\{V_n\}$ is an open basis for E).

(iii) ⇒ (ii). Define a set $C \subseteq K(E) \times \mathbb{N}$ by

$$(F, n) \in C \Leftrightarrow F \in I^{loc} \text{ and } F \cap V_n \neq \emptyset \text{ and } \overline{F \cap V_n} \in I.$$

This set is clearly $\underset{\sim}{\Pi}^1_1$ in $K(E) \times \mathbb{N}$. Moreover for each $F \in I^{loc}$, there is an $n$ with $(F, n) \in C$. So we are in a position to apply Novikov's Theorem V.1.8: There is a Borel function $f : K(E) \to \mathbb{N}$ such that for all $F \in I^{loc}$, $F \cap V_{f(F)} \neq \emptyset$ and $\overline{F \cap V_{f(F)}} \in I$. Let $P = \{\overline{F \cap V_{f(F)}} : F \in I^{loc}\}$. Then $P$ is a $\underset{\sim}{\Sigma}^1_1$ subset of $I$. We claim now that every set in $I$ is covered by countably many sets in $P$. If not, then as in the proof of Proposition 2, there is $K \in I$, $K \neq \emptyset$ such that for every open $V$ with $K \cap V \neq \emptyset$ we have that $K \cap V$ is not contained in a set in $P$. But then $K \in I^{loc}$, so $K \cap V_{f(K)} \neq \emptyset$ and $K \cap V_{f(K)} \subseteq \overline{K \cap V_{f(K)}} \in P$, a contradiction. □

Our next goal is to define a natural $\underset{\sim}{\Pi}^1_1$-rank on the $\underset{\sim}{\Pi}^1_1$ $\sigma$-ideal generated by a Borel hereditary set $B \subseteq K(E)$. In case $B = \{\emptyset\} \cup \{\text{singletons}\}$, this is done by considering the classical Cantor-Bendixson derivative on closed sets. We generalize this in a straightforward fashion as follows:

Let $B \subseteq K(E)$ be hereditary. For each $F \in K(E)$, define the B-<u>derivative</u> $F^{(1)}_B$ by

$$F^{(1)}_B = \{x \in F : \forall \text{ open } V(x \in V \Rightarrow \overline{F \cap V} \notin B)\}$$

and then by induction, let

$$F^{(0)}_B = F,$$

$$F_B^{(\alpha+1)} = \left(F_B^{(\alpha)}\right)_B^{(1)},$$

$$F_B^{(\lambda)} = \bigcap_{\alpha<\lambda} F_B^{(\alpha)}, \text{ if } \lambda \text{ is limit.}$$

The sequence $F_B^{(\alpha)}$ is a decreasing sequence of closed sets in E, hence stabilizes at some countable ordinal. It is easy to verify that $F_B^{(\alpha)}$ stabilizes at $\emptyset$ iff $F \in B_\sigma$. Hence we define the <u>Cantor-Bendixson rank associated with the B-derivative,</u> in symbols $rk_B$ by,

$$rk_B(F) = \begin{cases} \text{the least } \alpha \text{ such that } F_B^{(\alpha)} = \emptyset, \text{ if such exists;} \\ \omega_1, \text{ otherwise.} \end{cases}$$

Thus $rk_B$ is a rank on $I = B_\sigma$, i.e.

$$rk_B(F) < \omega_1 \Leftrightarrow F \in B_\sigma.$$

Note that if B is an ideal (i.e. closed under finite unions),

$$B = \{F \in B_\sigma : rk_B(F) \leq 1\}.$$

<u>Theorem 4</u>. Let E be a compact, metrizable space. Let $B \subseteq K(E)$ be Borel and hereditary. Then the Cantor-Bendixson rank $rk_B$ associated with the B-derivative is a $\underset{\sim}{\Pi}_1^1$-rank on the $\underset{\sim}{\Pi}_1^1$ $\sigma$-ideal $B_\sigma$ generated by B.

<u>Proof</u>. Let LO denote the set of $R \subseteq \mathbb{Q}$ such that $0 \in R$ and $\forall x \in R(x \geq 0)$. Viewing subsets of $\mathbb{Q}$ as members of $2^{\mathbb{Q}} = \{0, 1\}^{\mathbb{Q}}$ we have that LO is a closed subset of the Polish space $2^{\mathbb{Q}}$. Let also WO be the set of all $R \in LO$ such that

R is wellordered (under the ordering of the rationals). Then WO is a $\underset{\sim}{\Pi}^1_1$ subset of $2^Q$. If $R \in WO$ we denote by $\ell h(R)$ the countable ordinal corresponding to R. Then as R varies over WO, $\ell h(R)$ gives all the non-0 countable ordinals. Moreover the map $R \mapsto \ell h(R)$ is a $\underset{\sim}{\Pi}^1_1$-rank on WO. (The proof of this is similar to that of V.1.2).

In order to show that $rk_B$ is a $\underset{\sim}{\Pi}^1_1$-rank on $I = B_\sigma$ it is enough to find relations $S_1, S_2 \subseteq K(E) \times K(E)$ with $S_1 \in \underset{\sim}{\Sigma}^1_1$ and $S_2 \in \underset{\sim}{\Pi}^1_1$ such that if $K, L \in K(E)$ and $L \in I$ we have

$$rk_B(K) \leq rk_B(L) \Leftrightarrow (K, L) \in S_1 \Leftrightarrow (K, L) \in S_2.$$

Because if $\leq^*$, $<^*$ are the two relations associated with $rk_B$ (as in the beginning of V.1), then for any $K, L \in K(E)$,

$$K \leq^* L \Leftrightarrow K \in I \ \& \ [(K, L) \in S_2 \vee (L, K) \notin S_1]$$

and

$$K <^* L \Leftrightarrow K \in I \ \& \ (L, K) \notin S_1$$

so that $\leq^*$, $<^*$ are both $\underset{\sim}{\Pi}^1_1$ and we are done.

To find $S_1, S_2$ consider first the relation $T \subseteq 2^Q \times K(E)^Q \times K(E)$ given by

$$\begin{aligned}(R, h, K) \in T \Leftrightarrow R \in LO \ \& \ h(0) = K \ \& \\ \forall q \in R\Big[h(q) \neq \emptyset \ \& \\ h(q) \subseteq \big(\bigcap_{\substack{p<q \\ p\in R}} h(p)^{(1)}_B\big)\Big].\end{aligned}$$

Since B is Borel it is easy to check that $F \mapsto F_B^{(1)}$ is a Borel function from K(E) into K(E) and thus T is a Borel set in $2^{\mathbb{Q}} \times K(E)^{\mathbb{Q}} \times K(E)$. We claim now that for $K \in I$, $K \neq \emptyset$

$$(*) \qquad R \in WO \;\&\; \ell h(R) \leq rk_B(K) \Leftrightarrow \exists h\ (R, h, K) \in T.$$

The direction $\Rightarrow$ is obvious. For the direction $\Leftarrow$ let h be such that $(R, h, K) \in T$. Note first that if $q \in R$ there is $\alpha < rk_B(K)$ such that $\bigcap_{\substack{p<q\\ p\in R}} h(p)_B^{(1)} \subsetneqq K_B^{(\alpha+1)}$. This is true since otherwise for all $\alpha < rk_B(K)$, $\emptyset \neq h(q) \subseteq \bigcap_{\substack{p<q\\ p\in R}} h(p)_B^{(1)} \subseteq K_B^{(\alpha+1)}$, which clearly violates the fact that $rk_B(K)$ is a successor ordinal. Let then for $q \in R$, $f(q) =$ least $\alpha < rk_B(K)$ such that $\bigcap_{\substack{p<q\\ p\in R}} h(p)_B^{(1)} \subsetneqq K_B^{(\alpha+1)}$. We claim that if $q, r \in R$ then $q < r \Rightarrow f(q) < f(r)$, which of course implies that $R \in WO$ and $\ell h(R) \leq rk_B(K)$. Indeed note that for $q \in R$, $\bigcap_{\substack{p<q\\ p\in R}} h(p)_B^{(1)} \subseteq \bigcap_{\alpha<f(q)} K_B^{(\alpha+1)} = K_B^{(f(q))}$, so $h(q) \subseteq \bigcap_{\substack{p<q\\ p\in R}} h(p)_B^{(1)} \subseteq K_B^{(f(q))}$ and $h(q)_B^{(1)} \subseteq K_B^{(f(q)+1)}$. So if $q < r$, $\bigcap_{\substack{p<r\\ p\in R}} h(p)_B^{(1)} \subseteq h(q)_B^{(1)} \subseteq K_B^{(f(q)+1)}$ thus $f(q) < f(r)$, and we are done.

Similarly define $T^*$ by

$$(R, h, K) \in T^* \Leftrightarrow R \in LO \;\&\; h(0) = K \;\&\; \forall q \in R(h(q) \neq \emptyset \;\&\; h(q) = \bigcap_{\substack{p<q\\ p\in R}} h(p)_B^{(1)}) \;\&\; \bigcap_{p\in R} h(p)_B^{(1)} = \emptyset.$$

Then again $T^*$ is Borel and we can check as before that for $K \in I$, $K \neq \emptyset$

$$(**) \qquad R \in WO \;\&\; \ell h(R) = rk_B(K) \Leftrightarrow \exists h\ (R, h, K) \in T^*$$

while for $R \in WO$, $K \neq \emptyset$

$$(***) \qquad K \in I \ \& \ \ell h(R) = rk_B(K) \Leftrightarrow \exists h\,(R, h, K) \in T^*.$$

We can complete now the proof as follows: First note that for nonempty K, $L \in K(E)$ with $L \in I$ we have

$$rk_B(K) \leq rk_B(L) \Leftrightarrow \exists R[(R \in WO \ \& \ \ell h(R) \leq rk_B(L)) \ \& \ (K \in I \ \& \ \ell h(R) = rk_B(K))]$$

so that by (*), (***) the expression on the right hand side is $\underset{\sim}{\Sigma}{}^1_1$, while also

$$rk_B(K) \leq rk_B(L) \Leftrightarrow K \in I \ \& \ \forall R_1 \ \forall R_2 \ \{[(R_1 \in WO \ \& \ \ell h(R_1) = rk_B(K)) \ \& \\ (R_2 \in WO \ \& \ \ell h(R_2) = rk_B(L))] \Rightarrow \ell h(R_1) \leq \ell h(R_2)\},$$

so that by (**) and the fact that $\ell h$ is a $\underset{\sim}{\Pi}{}^1_1$-rank on WO, we have that the right hand side expression is $\underset{\sim}{\Pi}{}^1_1$ and we are done. □

In particular, by taking $B = \{\emptyset\} \cup \{\text{singletons}\}$ the preceding result shows that the classical Cantor-Bendixson rank is a $\underset{\sim}{\Pi}{}^1_1$-rank of $K_\omega(E)$.

We show now how to construct—under certain assumptions—complicated sets in $B_\sigma$, and also sets with special properties which are outside $B_\sigma$. The first construction is very simple, and is just an elaboration on how to get complicated countable closed sets. The second construction is more involved (the first place where a similar construction was used is probably in Hurewicz [1], for proving his Theorem IV.3.2), and will be the basis of many later constructions.

We first need an easy lemma:

Lemma 5. Let $X$ be a Polish space, $E \subseteq X$ a nowhere dense closed set, $D \subseteq X$ dense and $V$ open with $E \subseteq V$. Then there exists a countable discrete set $D_E \subseteq D \cap V$ with $D_E \cap E = \emptyset$, such that $\overline{D_E} = D_E \cup E$.

Proof. Let $\{V_n\}$ be a basis for $X$, and pick for each $n$ such that $V_n \cap E \neq \emptyset$, $V_n \subseteq V$, a point $x_n \in (V_n \cap D) - E$, with $\mathrm{dist}(x_n, E) \leq \frac{1}{n}$. The set $D_E = \{x_n\}$ clearly works. □

Theorem 6. Let $E$ be compact, metrizable, $B$ and $B'$ hereditary subsets of $K(E)$ with $B \subseteq B'$, and $B$ consisting of nowhere dense sets. If every non-empty open set in $E$ contains a $B_\sigma$-set $F$ with $\mathrm{rk}_{B'}(F) > 1$, then $\mathrm{rk}_{B'}$ is unbounded on $B_\sigma$. In particular, if $B = B'$ is Borel and the previous hypotheses hold, the $\sigma$-ideal $B_\sigma$ is $\underset{\sim}{\Pi}^1_1$-complete.

Proof. The second assertion follows from the first, Theorem 4, Theorem V.1.7 and Theorem IV.3.1. For the first assertion, we prove by induction on $\alpha$ that for every open $V \neq \emptyset$, $V$ contains an $F \in B_\sigma$ with $\mathrm{rk}_{B'}(F) \geq \alpha + 1$. This is the hypothesis for $\alpha = 1$. Suppose it is true for $\alpha$, and choose $F_0 \subseteq V$, $F_0 \in B_\sigma$ with $\mathrm{rk}_{B'}(F_0) \geq 2$. By the hypothesis, $F_0$ is nowhere dense so we can find $D$ discrete countable $\subseteq V$ with $D \cap F_0 = \emptyset$ and $\bar{D} = F_0 \cup D$, say $D = \{x_1, \ldots, x_n, \ldots\}$. Choose for each $n$ $W_n$ open, with $\bar{W}_n \subseteq V - F_0$, $\mathrm{diam}(W_n) \leq \frac{1}{n}$, and $\bar{W}_n \cap D = \{x_n\}$. By the induction hypothesis, let $F_n \in B_\sigma$, $F_n \subseteq W_n$ and $\mathrm{rk}_{B'}(F_n) \geq \alpha + 1$. The set $F = F_0 \cup \bigcup_n F_n$ is in $B_\sigma$ and contained in $V$. Moreover $F_n \subseteq F \cap \bar{W}_n$, so $(F \cap \bar{W}_n)^{(\alpha)}_{B'} \neq \emptyset$. Then $F_0 \subseteq \overline{\bigcup_n (F \cap \bar{W}_n)^{(\alpha)}_{B'}} \subseteq F^{(\alpha)}_{B'}$, hence $\emptyset \neq (F_0)^{(1)}_{B'} \subseteq F^{(\alpha+1)}_{B'}$ and $\mathrm{rk}_{B'}(F) \geq \alpha + 2$. The limit case is similar, by choosing increasing $\alpha_n \to \lambda$ and $F_n \subseteq W_n$

with $\mathrm{rk}_{B'}(F_n) \geq \alpha_n + 1$. □

<u>Remark</u>. Applying this result to a Borel (hence $G_\delta$) $\sigma$-ideal I and a Borel basis B for it which is an ideal, one gets that there is an open set $V \neq \emptyset$ such that $B \cap K(V) = I \cap K(V)$. Then by an immediate Baire Category argument, one gets that there exists a sequence $\{E_n\}$ of closed subsets of E with $E = \cup_n E_n$ and such that $\cup_n(I \cap K(E_n)) \subseteq B$. In other words, Borel bases for Borel $\sigma$-ideals are essentially trivial.

We come now to the key construction lemma, which may look a bit technical, but embodies many future constructions of complicated sets.

<u>Lemma</u> <u>7</u> (Kechris-Louveau-Woodin [1]). Let E be compact, metrizable, $B \subseteq K(E)$ hereditary. Assume:

$$(*) \text{ If } V \neq \emptyset \text{ is open in E, there is } F \subseteq V \text{ with } F \in B_\sigma - B.$$

Let D be a dense set in E and let $\{J_n\}$ be a sequence of nonempty hereditary open sets in K(E) such that:

$$(**) \text{ If } F \in J_n \text{ and } x \in D, \text{ then } F \cup \{x\} \in J_n.$$

Then there is a closed set $K \notin B_\sigma$ and a sequence $\{K_n\}$ of elements of $B_\sigma$, such that for any <u>closed</u> set F,

$$F \subseteq K - \cup_n K_n \Rightarrow F \in \cap_n J_n.$$

<u>Proof</u>. We will construct inductively for each $s \in \mathrm{Seq}\ \mathbb{N}$ a closed set $K_s$ and an open set $V_s$ satisfying:

(1) $V_s \neq \emptyset$, $K_s \subseteq V_s$, $K_s \in B_\sigma - B$,

(2) $n \neq m \Rightarrow \overline{V_{s^\frown(n)}} \cap \overline{V_{s^\frown(m)}} = \emptyset$,

(3) $\overline{V_{s^\frown(n)}} \subseteq V_s$, $\overline{V_{s^\frown(n)}} \cap K_s = \emptyset$

(4) $\mathrm{diam}(V_{s^\frown(n)}) \leq 2^{-\ell h(s)}$,

(5) $\overline{\bigcup_n V_{s^\frown(n)}} = \bigcup_n \overline{V_{s^\frown(n)}} \cup K_s$,

(6) $K_s \subseteq \overline{\bigcup_n K_{s^\frown(n)}}$,

(7) for every closed set $F \subseteq \bigcup_{\ell h(s)=n+1} V_s$, $F \in J_n$.

The construction is by induction on $k = \ell h(s)$. First note that the hypothesis (*) implies that any set in $B$, hence in $B_\sigma$, must be nowhere dense, for if some $V \neq \emptyset$ open satisfies $\bar{V} \in B$, then clearly $B_\sigma \cap K(V) = B \cap K(V)$ $(= K(V))$, contradicting (*). We start with $V_\emptyset = E$, and $K_\emptyset \in B_\sigma - B$ given by (*). Suppose we have constructed $V_s$, $K_s$ for $\ell h(s) \leq k$ satisfying (1)–(7). For each $s$ of length $k$, $K_s$ is nowhere dense and $K_s \subseteq V_s$ so we can find $D_s$ discrete countable, $D_s \subseteq V_s$, $D_s \cap K_s = \emptyset$, such that $\bar{D}_s = D_s \cup K_s$ by Lemma 5, and we can arrange so that $D_s \subseteq D$, our fixed dense set in $E$.

Enumerate the countable set $\bigcup_{\ell h(s)=k} D_s$ as $\{x_1,\ldots, x_n,\ldots\}$. Consider first $x_1$. By (**), $\{x_1\}$ is in $J_k$, and $\{x_1\} = \bigcap_N \bar{B}(x_1, \frac{1}{N})$, where $B(x_1, \frac{1}{N}) = \{x \in X : \mathrm{dist}(x, x_1) < \frac{1}{N}\}$. As $J_k$ is open, $\bar{B}(x_1, \frac{1}{N_1}) \in J_k$ for $N_1$ large enough. By choosing $N_1$ large, we can also insure that $\bar{B}(x_1, \frac{1}{N_1})$ has diameter $\leq 2^{-k}$, and if $s = s(x_1)$ is the (unique) $s$ with $x_1 \in D_s$, that $\bar{B}(x_1, \frac{1}{N_1}) \subseteq V_s - K_s$ and $\bar{B}(x_1, \frac{1}{N_1}) \cap D_s = \{x_1\}$. If $x_1$ is the $n^{th}$ point in some fixed enumeration of $D_s$, we let $V_{s^\frown(n)} = B(x_1, \frac{1}{N_1})$, and choose

$K_{s^\frown(n)} \subseteq V_{s^\frown(n)}$, $K_{s^\frown(n)} \in B_\sigma - B$ by (*). We now look at $x_2$, which is (say) the $m^{th}$ point in $D_t$. Applying (**), we know that $\bar{B}(x_1, \frac{1}{N_1}) \cup \{x_2\} \in J_k$, hence as $J_k$ is open, the sets $\bar{B}(x_1, \frac{1}{N_1}) \cup \bar{B}(x_2, \frac{1}{N})$ belong to $J_k$ as well for large enough N. And we can choose $N_2$ large enough to also insure that $\mathrm{diam}(\bar{B}(x_2, \frac{1}{N_2})) \leq 2^{-k}$, $\bar{B}(x_2, \frac{1}{N_2}) \subseteq V_t - K_t$, $\bar{B}(x_2, \frac{1}{N_2}) \cap D_t = \{x_2\}$ and $\bar{B}(x_2, \frac{1}{N_2}) \cap \bar{B}(x_1, \frac{1}{N_1}) = \emptyset$. We then set $V_{t^\frown(m)} = B(x_2, \frac{1}{N_2})$ and choose by (*) $K_{t^\frown(m)} \subseteq V_{t^\frown(m)}$, $K_{t^\frown(m)} \in B_\sigma - B$. Continuing this way, we construct inductively for all s of length k and all $n \in \mathbb{N}$ sets $V_{s^\frown(n)}$ and $K_{s^\frown(n)}$, and it is routine to check that (1)–(6) hold for them. For (7), note that if F closed satisfies $F \subseteq \bigcup_{\ell h(s)=k+1} V_s$, then by compactness F is covered by a finite sequence of such $V_s$, hence $F \in J_k$ by our construction.

Let $H = \bigcap_{n\in\mathbb{N}} \bigcup_{\ell h(s)=n} V_s$ and $K = H \cup \left(\bigcup_{s\in \mathrm{Seq}\mathbb{N}} K_s\right)$. We claim that K and the sequence $\{K_s\}$ satisfy the conclusions of the lemma.

First K is closed, in fact $K = \bigcap_n \overline{\bigcup_{\ell h(s)=n} V_s}$. Indeed if $x \in K$, then either $x \in H$ in which case clearly $x \in \bigcup_{\ell h(s)=n} V_s$ for all n, or else $x \in K_t$, for some t. Then by (6) and (1), $x \in \overline{\bigcup_{\ell h(s)=n} V_s}$, if $n \geq \ell h(t)$, while obviously $x \in \overline{\bigcup_{\ell h(s)=n} V_s}$, if $n < \ell h(t)$. So $x \in \bigcap_n \overline{\bigcup_{\ell h(s)=n} V_s}$. Conversely if for all n, $x \in \overline{\bigcup_{\ell h(s)=n} V_s}$, but $x \notin \bigcup_s K_s$, then by (2) and (5) for each n there is an s with $\ell h(s) = n$ and $x \in \bar{V}_s$, so that by (3) $x \in H$.

By (7) if F is closed and $F \subseteq K - \bigcup_s K_s = H$, then $F \in J_n$ for all n. So it remains only to show that $K \notin B_\sigma$.

Suppose $K \in B_\sigma$, towards a contradiction. Then by the Baire Category Theorem, there is V open in E such that $V \cap K \neq \emptyset$ and $\overline{V \cap K} \in B$. Let $x \in V$

$\cap$ K. If $x \in H$, then by (2), (3) and (4) there is unique $\epsilon \in \mathbb{N}^{\mathbb{N}}$ such that for all n, $x \in V_{\epsilon \restriction n}$. Since $\operatorname{diam}(V_{\epsilon \restriction n}) \to 0$ there is an n with $V_{\epsilon \restriction n} \subseteq V$. But then $K_{\epsilon \restriction n} \subseteq V \cap K$, so $K_{\epsilon \restriction n} \in B$ contradicting (1). If on the other hand $x \in K_s$, for some s, then by (5) there is $p \in \mathbb{N}$ with $V_{s^\frown(p)} \subseteq V$ and so $K_{s^\frown(p)} \subseteq V \cap K$, thus $K_{s^\frown(p)} \in B$, a contradiction again. This finishes the proof of the lemma. □

In the preceding lemma, no particular definability conditions were imposed on B. However in the applications, one way to insure condition (*) is via definability properties. Typically, we will consider cases in which B is Borel, but such that for any non-empty open V, $B_\sigma \cap K(V)$ is a true $\underset{\sim}{\Pi}_1^1$ set, so that (*) is automatically satisfied.

We have then the following immediate

<u>Corollary 8</u>. Let E be compact, metrizable, I a $\underset{\sim}{\Pi}_1^1$ $\sigma$-ideal of closed subsets of E, $D \subseteq E$ dense in E and $J_n \subseteq K(E)$ nonempty open, hereditary such that ($F \in J_n$ and $x \in D \Rightarrow F \cup \{x\} \in J_n$). If I has a Borel basis, but $I \cap K(V)$ is true $\underset{\sim}{\Pi}_1^1$ for all non-empty open $V \subseteq E$, then there is $K \notin I$ and $\{K_n\}$ with $K_n \in I$ such that if F is closed, $F \subseteq K - \bigcup_n K_n$ then $F \in \bigcap_n J_n$.

Note also the interesting feature of the proof of Lemma 7, that it constructs closed sets by appropriately constructing $G_\delta$ and $F_\sigma$ sets and taking their union.

Before applying Lemma 7, let us introduce a definition

<u>Definition 9</u>. Let E be compact, metrizable, I a $\sigma$-ideal of closed subsets of E. We

define $I^{int}$ as the class of subsets P of E which are in I from the interior, i.e. those for which for every closed $F \subseteq P$, $F \in I$.

We say that I is <u>calibrated</u> if for every closed $F \subseteq E$ and every sequence $\{F_n\}$ of sets in I, if $F - \bigcup_n F_n$ is in $I^{int}$ then $F \in I$.

For example the $\sigma$-ideal of null sets for a positive measure on E is calibrated. On the other hand, the $\sigma$-ideal of nowhere dense sets in a perfect E is not calibrated.

The following is an interesting closure property of $I^{int}$ for calibrated I.

<u>Proposition</u> <u>10</u>. Let E be compact, metrizable, and I a $\sigma$-ideal in K(E). If I is calibrated, the $\underset{\sim}{\Sigma}^0_3$ (i.e $G_{\delta\sigma}$) sets in $I^{int}$ form a $\sigma$-ideal, i.e. are closed under countable unions.

<u>Proof</u>. We want to show that if $H_n$ is $\underset{\sim}{\Sigma}^0_3$ and in $I^{int}$ and $H = \bigcup_n H_n$, then $H \in I^{int}$. We can clearly assume that each $H_n$ is $G_\delta$, and that H is closed. Write $H - H_n$ as $\bigcup_p E^n_p$, with $E^n_p$ closed. We argue by contradiction: Suppose $H \notin I$. As $H = \bigcup_p E^1_p \cup H_1$ and $H_1 \in I^{int}$, by the calibration of I there is a $p_1$ such that $E^1_{p_1} \notin I$. Now

$$E^1_{p_1} = \bigcup_p (E^1_{p_1} \cap E^2_p) \cup (H_2 \cap E^1_{p_1}) \text{ and } H_2 \cap E^1_{p_1} \in I^{int}$$

hence by the calibration of I there must be $p_2$ with $E^1_{p_1} \cap E^2_{p_2} \notin I$. Continuing this way, we can define a sequence $\{p_n\}$ with $E^1_{p_1} \cap E^2_{p_2} \cap ... \cap E^n_{p_n} \notin I$ for all n. But then in particular $\forall n(E^1_{p_1} \cap \ldots \cap E^n_{p_n} \neq \emptyset)$, hence by compactness $\bigcap_n E^n_{p_n} \neq \emptyset$. But this contradicts that $H = \bigcup_n H_n$ and we are done. □

Note that if I is a $\sigma$-ideal in K(E), there is always a natural extension of I to a $\sigma$-ideal on all of power(E), given by

$$I^{ext} = \{P \subseteq E : \exists\{F_n\} \, (F_n \in I \text{ for all } n \text{ and } P \subseteq \bigcup_n F_n)\}.$$

Thus $I^{ext}$ consists of all subsets of E which are in I from the exterior. Clearly $I^{ext} \subseteq I^{int}$ and $I^{ext} \cap K(E) = I^{int} \cap K(E) = I$.

We turn now to one of the most useful applications of Lemma 7, due to Debs and Saint Raymond:

Theorem 11 (Debs-Saint Raymond [1]). Let I be a $\sigma$-ideal in K(E) for some compact metrizable space E. Assume the following:

(i) For every closed $F \notin I$, $I \cap K(F)$ is true $\underset{\sim}{\Pi}^1_1$,

(ii) I is calibrated,

(iii) I admits a Borel basis.

Then for P a $\underset{\sim}{\Sigma}^1_1$ subset of E,

$$P \in I^{int} \text{ iff } P \in I^{ext}.$$

Proof. As we noticed before, the inclusion $I^{ext} \subseteq I^{int}$ is always true. So let P be

$\underset{\sim}{\Sigma}^1_1$ in E, and assume $P \notin I^{ext}$, i.e. cannot be covered by a sequence of sets in I. We will construct a closed subset of P which is not in I. Let $G \subseteq E \times 2^{\mathbb{N}}$ be $G_\delta$ with $\text{proj}(G) = P$. Consider all open sets in $E \times 2^{\mathbb{N}}$ such that $\text{proj}(G \cap V) \in I^{ext}$. Their union $V_0$ is clearly an open set with the same property. Let $G^* = G - V_0$ and $P^* = \text{proj}(G^*)$. Then $G^*$ is $G_\delta$, $P^*$ is $\underset{\sim}{\Sigma}^1_1$, and $P^* \notin I^{ext}$, else P would be in $I^{ext}$ too. Moreover by the maximality of $V_0$, if $V \subseteq E \times 2^{\mathbb{N}}$ is open and $G^* \cap V \neq \emptyset$, then $\text{proj}(G^* \cap V) \notin I^{ext}$.

Let $E^* = \overline{G^*}$ (= the closure of $G^*$ in $E \times 2^{\mathbb{N}}$). Let

$$I^* = \{F \in K(E^*) : \text{proj}(F) \in I\}.$$

Clearly $I^*$ is a $\underset{\sim}{\Pi}^1_1$ $\sigma$-ideal in $K(E^*)$. Moreover if B is a Borel basis for I, $B^* = \{F \in K(E^*) : \text{proj}(F) \in B\}$ is hereditary, Borel and a basis for $I^*$. (If $F \in I^*$ and $K = \text{proj}(F) \in I$, then $K = \cup_n K_n$ with $K_n \in B$, so that $F = \cup_n F_n$, where $F_n = (K_n \times 2^{\mathbb{N}}) \cap F \in B^*$).

Write now $G^* = \cap_n V^*_n$, where $V^*_n$ is open in $E^*$, and let $J^*_n = K(V^*_n)$. We claim that the hypotheses of Corollary 8 hold for $E^*$, $I^*$, $D^* = G^*$ and $J^*_n$.

First $B^*$ is Borel hereditary in $K(E^*)$ and $I^* = B^*_\sigma$, so $I^*$ has a Borel basis. Also $G^*$ is dense in $E^*$ and $J^*_n = K(V^*_n)$ is open hereditary in $K(E^*)$. Moreover as $G^* \subseteq V^*_n$, if $F \in J^*_n$ and $x \in G^*$ then $F \cup \{x\} \in J^*_n$. Finally, let $\emptyset \neq V^* \subseteq E^*$ be open in $E^*$ in order to show that $I^* \cap K(V^*)$ is not Borel. If it is, towards a contradiction, choose first V nonempty open in $E^*$ with $\overline{V} \subseteq V^*$ (closures are taken in $E^*$). Then $I^* \cap K(\overline{V})$ is Borel as well. Since $V \cap G^* \neq \emptyset$, it follows that F =

$\overline{\text{proj}\,(V \cap G^*)} \notin I$ by the above property of $G^*$. For each $K \in K(F)$, $K^* = (K \times 2^{\mathbb{N}}) \cap \overline{V}$ satisfies $\text{proj}(K^*) = K$, hence $K \in I \cap K(F) \Leftrightarrow K^* \in I^* \cap K(\overline{V})$. But $K \mapsto K^*$ is a Borel map from $K(F)$ into $K(\overline{V})$, hence $I \cap K(F)$ is Borel as well, contradicting (i).

So we can apply Corollary 8 to get a closed set $K \subseteq E^*$ with $K \notin I^*$ and a sequence $\{K_n\}$ with $K_n \in I^*$ such that if $H \in K(E^*)$ and $H \subseteq K - \bigcup_n K_n$ then $H \in \bigcap_n J_n^*$. Let $F = \text{proj}(K)$, $F_n = \text{proj}(K_n)$. Then $F \notin I$ and for all n, $F_n \in I$. We claim that $F - \bigcup_n F_n \subseteq P^* \subseteq P$ which, since I is calibrated, implies that there is some closed subset of $F - \bigcup_n F_n$, and hence P, not in I and completes the proof. To prove our claim let $x \in F - \bigcup_n F_n$. Then there is $\epsilon \in 2^{\mathbb{N}}$ with $(x, \epsilon) \in K$ and $(x, \epsilon) \notin K_n$, for all n. Then $\{(x, \epsilon)\} \in J_n^*$ for all n, i.e. $(x, \epsilon) \in \bigcap_n V_n^* = G^*$, so that $x \in P^*$. □

As for Lemma 7, there are two ways of using Theorem 11: If one works with a $\sigma$-ideal I satisfying the hypotheses (i), (ii) and (iii), Theorem 11 gives a complete description of the behavior of Borel (even $\underset{\sim}{\Sigma}_1^1$) subsets of E with respect to I. This approach will be used in Chapter VIII to solve the Category Problem.

On the other hand, if one wants to prove that a $\sigma$-ideal I has no Borel basis, it is enough to prove (i) (that I is not Borel, on any closed set $F \notin I$) and (ii) (that I is calibrated), and to construct a $G_\delta$ set P in $I^{int}$ which cannot be covered by countably many sets in I. This is the way the Borel Basis Problem for U will be handled in the next chapter.

## §2. The class $U_1$ and the decomposition theorem of Piatetski-Shapiro

We associate now with each closed set $E \subseteq \mathbf{T}$ another ideal of functions in A, denoted by I(E), defined by

$$I(E) = \{f \in A : \forall x \in E(f(x) = 0)\}$$

i.e. f is in I(E) if f vanishes on E—but not necessarily in a nbhd of E.

Clearly for all E, $J(E) \subseteq I(E)$, and I(E) is a (norm-) closed linear subspace of A which is also an ideal.

Definition 1. (i) A set $E \in K(\mathbf{T})$ is a $U_1$-set if I(E) is $w^*$-dense in A. We let $U_1 = \{E \in K(\mathbf{T}) : E \text{ is a } U_1\text{-set}\}$. Let also $M_1 = K(\mathbf{T}) - U_1$.

(ii) We define the Piatetski-Shapiro rank $[E]^1_{PS}$ on $U_1$ as in V.4 by $[E]^1_{PS} = \text{ord}(I(E))$, so that if we define the sequence $I^\alpha(E)$ of iterates of I(E) by closing under $w^*$-limits of sequences in A, then

$$[E]^1_{PS} = \begin{cases} \text{least } \alpha\ (I^\alpha(E) = A), \text{ if } \overline{I(E)}^{w^*} = A, \\ \omega_1, \text{ otherwise.} \end{cases}$$

(iii) We define $U_1'$ as the set

$$U_1' = \{E \in U_1 : [E]^1_{PS} \le 1\}.$$

Most of the remarks we made for J(E) and U go through for I(E) and $U_1$. E.g., $I^{\alpha}(E)$ is an ideal for each countable $\alpha$, and hence $I^{\alpha}(E) = A$ iff $1 \in I^{\alpha}(E)$. In particular

$$E \in U_1' \Leftrightarrow \exists\{f_n\} (f_n \in A \ \&\ f_n = 0 \text{ on } E \ \&\ f_n \xrightarrow{w^*} 1).$$

One can also define a tree-rank and an R-rank for $U_1$ and prove that they coincide with $[E]^1_{PS}$ (if $E \neq \emptyset$).

Let also

$$N(E) = I(E)^{\perp}$$

so that $N(E) \subseteq PM(E)$ ($= J(E)^{\perp}$), and define

$$\eta_1(E) = \eta(N(E))$$

as in V.5 (for $E \neq \emptyset$), i.e.

$$\eta_1(E) = \inf\left\{\frac{R(S)}{\|S\|_{PM}} : S \neq 0 \ \&\ S \in N(E)\right\}.$$

Then by V.5.3, if $E \neq \emptyset$,

$$E \in U_1' \Leftrightarrow \eta_1(E) > 0.$$

There is however an important definability difference between I(E) and J(E).

In a sense to be described precisely in Chapter X.2, the map $E \mapsto J(E)$ is Borel but $E \mapsto I(E)$ is not. Because of that we do not know how to classify the complexity of $U_1$ and $U_1'$. The only known results are

(i) $U_1'$ is $\underset{\sim}{\Sigma}_1^1$ (by its definition),

(ii) $U_1$ is $\underset{\sim}{\Pi}_1^1$-hard (follows from IV.2.6 and the easy fact that $U \subseteq U_1 \subseteq U_0$), hence it is not $\underset{\sim}{\Sigma}_1^1$ (but it is $\underset{\sim}{\Delta}_2^1$—we omit the proof of that here).

In particular it is open if $U_1'$ is Borel or whether $U_1$ is $\underset{\sim}{\Pi}_1^1$.

From the inclusion $J(E) \subseteq I(E)$, we immediately get

<u>Proposition 2</u>. For every $E \in K(\mathbb{T})$, if $E$ is in $U$, then $E$ is in $U_1$ and $[E]_{PS}^1 \leq [E]_{PS}$; hence in particular $U' \subseteq U_1'$.

We will see in the next chapter that the inclusion is proper, and that the inequality between the ranks cannot be reversed, in a very strong sense.

We consider now the $\sigma$-ideal $(U_1')_\sigma$ generated by the (clearly hereditary) family $U_1'$, and the corresponding Cantor-Bendixson rank, that we will denote by $rk_1(E)$. The relation between $rk_1(E)$ and $[E]_{PS}^1$ in given by

<u>Theorem 3</u>. (i) For every set $E \in K(\mathbb{T})$, $rk_1(E) \leq [E]_{PS}^1$.

(ii) (Piatetski-Shapiro [2]) In particular, $U_1$ (and hence $U$) is contained in $(U_1')_\sigma$, i.e. every set in $U_1$ is a countable union of sets in $U_1'$.

Proof. The second assertion follows from the first, as if $E \in U_1$, $[E]^1_{PS} < \omega_1$ thus $rk_1(E) < \omega_1$, and so $E \in (U'_1)_\sigma$.

Let for $E \in K(T)$, $E_1^{(\alpha)} = E_{U'_1}^{(\alpha)}$ be the sequence of derivatives of E for the Cantor-Bendixson derivation associated with $U'_1$. We will prove by induction on $\alpha$ that

$$I(E_1^{(\alpha)}) \supseteq I^\alpha(E)$$

This clearly implies that if $I^\alpha(E) = A$, then $E_1^{(\alpha)} = \emptyset$, hence $rk_1(E) \leq [E]^1_{PS}$.

Suppose we know, for all E, that $I(E_1^{(1)}) \supseteq I^1(E)$. If the result has been proved for $\alpha$, we get

$$\begin{aligned} I(E_1^{(\alpha+1)}) &= I(E_1^{(\alpha)}{}_1^{(1)}) \\ &\supseteq I^1(E_1^{(\alpha)}) \\ &= (I(E_1^{(\alpha)}))^{(1)} \\ &\supseteq (I^\alpha(E))^{(1)} = I^{\alpha+1}(E). \end{aligned}$$

Similarly if the result is known for all $\alpha < \lambda$ limit,

$$I(E_1^{(\lambda)}) \supseteq \cup_{\alpha<\lambda} I(E_1^{(\alpha)})$$

$$\supseteq \bigcup_{\alpha<\lambda} I^{\alpha}(E)$$

$$= I^{\lambda}(E).$$

It remains to show that $I(E_1^{(1)}) \supseteq I^1(E)$. Let $x \in E_1^{(1)}$ and $f \in I^1(E)$. We want to show that $f(x) = 0$. Let $\{V_n\}$ be a decreasing sequence of nbhds of x with $\mathrm{diam}(V_n) \to 0$. Then for each n, $E_n = \overline{E \cap V_n} \notin U_1'$, as $x \in E_1^{(1)}$, hence $I^1(E_n) \neq A$, and by V.4.5, $\mathrm{hull}(I^1(E_n)) \neq \emptyset$. Pick $x_n \in \mathrm{hull}(I^1(E_n)) \subseteq E_n$. Then $x_n \to x$. Now as $E_n \subseteq E$, $I(E) \subseteq I(E_n)$ and hence $I^1(E) \subseteq I^1(E_n)$. So for all n, $f \in I^1(E_n)$, hence $f(x_n) = 0$, so $f(x) = 0$. □

The preceding decomposition theorem was the original reason for the introduction of the Piatetski-Shapiro rank. It gives in particular a decomposition theorem for U, with basis $U \cap U_1'$. We will argue later that in some sense this result is best possible (see the second remark in VII.4).

We note also that one can give a direct proof of the decomposition theorem which avoids the use of ordinals and transfinite induction as follows: Let $E \in U_1$. Put $W = \bigcup\{V : V$ open in $\mathbf{T}$ and $E \cap V$ can be covered by countably many sets in $U_1'\}$. If $E \subseteq W$ we are done. Else let $E' = E - W \neq \emptyset$. Then $\overline{E' \cap V} \notin U_1'$ for all open V with $E' \cap V \neq \emptyset$. Since $\emptyset \neq E' \in U_1$ we have $I(E') \neq \overline{I(E')}^{w^*} = A$, so $I^1(E') \neq I(E')$ (by V.2.2). Choose $f \in I^1(E') - I(E')$. Then there is open V with $V \cap E' \neq \emptyset$ and $f \neq 0$ on $\overline{V}$. Put $E^* = \overline{E' \cap V}$. Then $\mathrm{hull}(I^1(E^*)) \subseteq E^* \cap \mathrm{hull}(I^1(E')) = \emptyset$, since $f \neq 0$ on $E^*$. Thus $I^1(E^*) = A$, i.e. $E^* \in U_1'$, a contradiction.

Another proof of the decomposition theorem will be given in Chapter VIII.4 (at the end).

We turn now to an application of this result to the unboundedness of the Piatetski-Shapiro rank on U, that we proved in V.4.4. We will present here the direct proof of unboundedness due to McGehee, which is "softer", as it does not rely on the Salem-Zygmund Theorem. In the next chapter (VII.2) we will prove the unboundedness of the Piatetski-Shapiro rank on any K(E), where E is a closed set of multiplicity. This will be important for applying the methods of Section 1.

Note also that conversely the unboundedness of the Piatetski-Shapiro rank on U, together with V.4.3 and V.1.7, gives that U is not Borel, hence is $\underset{\sim}{\Pi}^1_1$-complete by IV.3.1, so gives an alternative proof of the Theorem of Solovay and Kaufman. In particular it implies the existence of many M-sets—for example M-sets of measure 0 (Menshov's Theorem). And this by only constructing U-sets! (Recall our remarks on overspill after V.1.7).

Theorem 4 (McGehee [2]). The Piatetski-Shapiro rank $[E]_{PS}$ is unbounded on U. In fact, the Cantor-Bendixson rank $rk_1(E)$ on $(U_1')_\sigma$ is unbounded on $H_\sigma$, the $\sigma$-ideal generated by the H-sets.

To prove this result, we can use Theorem 1.6 with $B = H$, $B' = U_1'$. Since $U_1'$ is an ideal (this can be proved exactly as in V.4.7), $E \in U_1' \Leftrightarrow rk_1(E) \leq 1$, so by 1.6 it is enough to prove

Theorem 5 (Piatetski-Shapiro [2]). For every non-∅ open set V in $\mathbf{T}$, there is an

$H_\sigma$-set contained in V, which is not in $U'_1$.

This in turn follows easily from

Lemma 6. For every non-empty open set V in $\mathbb{T}$ and every $\epsilon > 0$, there is an H-set E, $E \subseteq V$ with $\eta_1(E) \leq \epsilon$.

Proof of Theorem 5 (from the Lemma). Let V be open nonempty in $\mathbb{T}$ and choose nonempty disjoint open intervals $V_n$ in V converging to some point $x \in V$. By Lemma 6, let $E_n \in H$, $E_n \subseteq V_n$ be such that $\eta_1(E_n) \leq \frac{1}{n}$, and let $E = \bigcup_n E_n \cup \{x\}$. Then E is a closed subset of V, and E is clearly in $H_\sigma$. Now $I(E) \subseteq I(E_n)$, hence $N(E_n) \subseteq N(E)$. So $\eta_1(E) \leq \eta_1(E_n)$, hence $\eta_1(E) = 0$ and $E \notin U'_1$. □

To prove Lemma 6, we will use

Lemma 7 (Piatetski-Shapiro [2]). Let $0 < a \leq 1$ and $N \geq 3$, and let $E_N^a = \{2\pi a \left(\sum_{j=1}^{\infty} \epsilon_j N^{-j}\right) : \epsilon_j \in \mathbb{N}, 0 \leq \epsilon_j < N, \epsilon_j \neq 1\}$. Then $\eta_1(E_N^a) \leq C \cdot N^{-1}$, where C is a constant independent of a and N.

Lemma 6 follows easily from Lemma 7, by noting that in case a is rational, $E_N^a$ is an H-set, since

$$N^k \cdot E_N^a \subseteq 2\pi a\mathbb{Z} + E_N^a \pmod{2\pi}$$

which is a finite union of translates of $E_N^a$ and is therefore not dense in $\mathbb{T}$. If V is any non-$\emptyset$ open set in $\mathbb{T}$, we can choose small enough rational a so that some rational (relative to $2\pi$) translate of $E_N^a$ is contained in V. But H-sets are clearly preserved under rational translations, as is $\eta_1$. Finally choose N large enough so that $C \cdot N^{-1} \leq \epsilon$.

Proof of Lemma 7. It is enough to construct a probability measure $\mu$ supported by $E_N^a$ with $R(\hat{\mu}) = \overline{\lim}\, |\hat{\mu}(n)| \leq C \cdot N^{-1}$, for clearly $\|\hat{\mu}\|_{PM} = \mu(\mathbf{T}) = 1$, and $\hat{\mu} \in I(E_N^a)^{\perp}$. To do this, let $\nu_S$ be the canonical uniform measure on $S = \{0, 2, \ldots, N-1\}$, i.e. $\nu_S(\{k\}) = \frac{1}{N-1}$ for $k \in S$, and let $\nu$ be the product measure of $\nu_S$ on $S^{\mathbb{N}-\{0\}}$. Let $\varphi : S^{\mathbb{N}-\{0\}} \to [0, 2\pi]$ be defined by $\varphi(\epsilon) = \sum_{j=1}^{\infty} 2\pi a\epsilon_j N^{-j}$. The function $\varphi$ is clearly continuous, with $\varphi[S^{\mathbb{N}-\{0\}}] = E_N^a$. Let $\mu$ be the image of $\nu$ by $\varphi$, i.e. for f continuous on $\mathbf{T}$, $\int f d\mu = \int (f \circ \varphi) d\nu$. The measure $\mu$ is clearly a probability measure on $E_N^a$. For each real number t, let $\hat{\mu}(t) = \int e^{-itx} d\mu(x)$, so that

$$\hat{\mu}(t) = \prod_{k=1}^{\infty} \Big(\frac{1}{N-1} \sum_{j \in S} e^{-2\pi i a t N^{-k} j}\Big).$$

An immediate computation yields, letting $\varphi_k(t) = \frac{1}{N-1} \sum_{j \in S} e^{-2\pi i a t N^{-k} j}$,

$$\hat{\mu}(t) = \varphi_1(t) \cdot \hat{\mu}(tN^{-1})$$

and

$$\hat{\mu}(t) = \varphi_1(t) \cdot \varphi_2(t) \cdot \hat{\mu}(tN^{-2}).$$

Clearly for all t, $|\hat{\mu}(t)| \leq 1$ and $|\varphi_k(t)| \leq 1$, hence

$$|\hat{\mu}(t)| \leq |\hat{\mu}(tN^{-1})|$$

and

$$|\hat{\mu}(t)| \leq |\varphi_1(t)| \cdot |\varphi_2(t)|$$

for all $t \in \mathbb{R}$. Therefore it is enough to find a $b > 0$ such that for $t \in [b, bN]$, $|\varphi_1(t)| \cdot |\varphi_2(t)| \leq C \cdot N^{-1}$. Since then $|\hat{\mu}(n)| \leq C \cdot N^{-1}$ for $n \geq b$, and also

$$|\hat{\mu}(-n)| = |\hat{\mu}(n)| \leq C \cdot N^{-1} \text{ for } n \leq -b.$$

Let $A_1 = \sum_{j=0}^{N-1} e^{-it2\pi ajN^{-1}}$, $B_1 = e^{-it2\pi aN^{-1}}$, and $A_2 = \sum_{j=0}^{N-1} e^{-it2\pi ajN^{-2}}$, $B_2 = e^{-it2\pi aN^{-2}}$, so that $\varphi_1(t) \cdot \varphi_2(t) = \frac{1}{(N-1)^2}(A_1 - B_1) \cdot (A_2 - B_2)$. Clearly $|A_1|$ and $|A_2|$ are $\leq N$, and $|B_1|, |B_2| \leq 1$, thus $|A_1 B_2 + A_2 B_1 - B_1 B_2| \leq 2N + 1$. So if we can bound $|A_1 A_2|$ by a linear function in N, we will be done. If $atN^{-1}$ is not an integer,

$$A_1 = \frac{1 - e^{-it2\pi a}}{1 - e^{-it2\pi aN^{-1}}} \text{ and } A_2 = \frac{1 - e^{-it2\pi aN^{-1}}}{1 - e^{-it2\pi aN^{-2}}}$$

so that $A_1 A_2 = \frac{1 - e^{-it2\pi a}}{1 - e^{-it2\pi aN^{-2}}}$, and $|A_1 A_2| = \frac{|\sin(\pi at)|}{|\sin(\pi at\, N^{-2})|}$. Take $b = \frac{N}{2a}$. If $t \in [b, bN]$, then $\pi atN^{-2} \in [\frac{\pi}{2N}, \frac{\pi}{2}]$. On this interval $\sin x \geq \sin \frac{\pi}{2N} \geq \frac{1}{N}$. So $|A_1 A_2| \leq N$ for $t \in [b, bN]$ with $atN^{-1}$ not an integer, hence for all $t \in [b, bN]$ by continuity. This proves Lemma 7. □

The preceding construction gives very specific $H_\sigma$-sets of $rk_1 \geq 2$. We will see in VII.2 another proof of this, much less explicit but of wider applicability.

Let

$$U_1^{(\alpha)} = \{E \in U_1 : [E]_{PS}^1 \leq \alpha\}.$$

We end this section with the following proposition

Proposition 8 (Piatetski-Shapiro [2] for $\alpha = 1$). The class $U_1^{(\alpha)}$ is an ideal, closed under translations, dilations and contractions, and thus so is $U_1$.

Proof. Closure under translations is obvious and closure under unions is as in V.4.7. For the last assertion, it is enough to show that if $E, F \subseteq (0, 2\pi)$ are closed, with $F = tE = \{tx : x \in E\}$, $t > 0$, then $[tE]^1_{PS} \geq [E]^1_{PS}$. As in Theorem II.4.2, associate with any $S \in PM(E)$ the corresponding $S_t \in PM(F)$. We have seen in the proof of V.4.9 that if $S^n \in PM(E)$, $S^n \xrightarrow{w^*} S$ then $S^n_t \xrightarrow{w^*} S_t$ and if $S^n \in PM(E)$, $R(S^n) \to 0$ then $R(S^n_t) \to 0$. So it is enough to show that if $S \in N(E)$, then $S_t \in N(F)$.

Notice that since $f \in I(E)$ iff $\langle f, \delta_x\rangle = \langle f, \hat{\delta}_x\rangle = 0$ for all Dirac measures $\delta_x$ with $x \in E$, we have by Hahn-Banach that $N(E)$ is the $w^*$-closure (in PM) of the linear subspace of PM generated by the $\hat{\delta}_x$, $x \in E$. So by V.2.2 $N(E)$ is the iterated $w^*$-sequential closure of the linear subspace generated by the $\hat{\delta}_x$, $x \in E$. Now we have proved in V.4.9 that if $S^n \xrightarrow{w^*} S$ then $S^n_t \xrightarrow{w^*} S_t$. Moreover a direct calculation gives that $(\hat{\delta}_x)_t = \hat{\delta}_{tx}$ for $x \in E$, so that if $S \in N(E)$, then $S_t \in N(F)$ and we are done. □

## §3. The Borel Basis Problem for U and relations between U, $U_1$ and $U_0$

The theorem of Solovay and Kaufman IV.2.2 states that there is no definably simple characterization of the closed sets of uniqueness. So the next question is whether every U-set can be decomposed into a countable union of sets admitting a simple characterization. So we have the following

Problem. Does U admit a Borel basis?

In view of the Piatetski-Shapiro Theorem 2.3(ii) and recalling that U has a Borel basis if it has a $\underset{\sim}{\Sigma}_1^1$ one by 1.3, the above problem would be positively solved if one could prove the inclusion $U_1' \subseteq U$ (and hence $U = U_1 = (U_1')_\sigma$). Unfortunately this inclusion is false, by a theorem of Körner that we will discuss in the next chapter. And in fact Körner's result will be one of the ingredients in the negative solution to the Borel Basis Problem for U. In this section, we establish some results relating U, $U_1$ and $U_0$ (the class of closed sets of extended uniqueness, see II.5 and IV.2), that stress the differences between these notions.

Recall that we defined in II.5 the closed sets of extended uniqueness or $U_0$-sets as those $E \in K(T)$ for which no measure supported by E is a pseudofunction (except 0 of course). As the set M(E) of measures supported by E is clearly a subset of $N(E) = I(E)^\perp$ (identifying here $\mu$ with $\hat{\mu}$), we have the inclusion

$$M(E) \subseteq N(E) \subseteq PM(E)$$

as well as the relation

$$(*) \quad U \subseteq U_1 \subseteq U_0.$$

One of the results we want to get is that the second inclusion in (*) is proper if the first one is. Let us start with some preliminary results concerning the relation between I(E) and J(E).

Proposition 1. Let $E \in K(\mathbf{T})$. Then

(i) J(E) is the smallest ideal J in A with hull(J) = E, hence $\overline{J(E)}$ is the smallest closed ideal with this property.

(ii) I(E) is the largest ideal J in A with hull(J) = E. Similarly, if "ideal" is replaced by "closed ideal".

Proof. The proof of (ii) is obvious as $J \subseteq I(\text{hull}(J))$ for any ideal J, and I(E) is closed in A.

For (i), one clearly has hull(J(E)) = E. So it is enough to show that if J is any ideal in A with hull(J) $\subseteq$ E and $f \in J(E)$, then $f \in J$. Let F = supp(f). Then $F \cap E = \emptyset$, hence by compactness we can find in J a $g \geq 0$ with $g > 0$ on F. But we can easily extend $g \restriction F$ to some $h \in A$ with $h > 0$ everywhere, hence $1/h \in A$ by Wiener's Theorem, so that $g/h \in J$ and $g/h = 1$ on F. But then $f = f \cdot g/h \in J$ and we are done. □

To get that $U_1 \subseteq U$, it would have been enough to know that for all E, I(E) is contained in the $w^*$-closure of J(E) in A. Even more, one can ask the following question: Is I(E) the strong closure of J(E), i.e. is it true that there is a unique closed ideal J with hull(J) = E? This is also equivalent, in view of Proposition 1, to the question: Are all closed ideals of A of form I(E)? This problem is known as the problem of harmonic synthesis, and has been solved negatively by Malliavin in 1959. This leads to the following definition.

Definition 2. A set $E \in K(\mathbf{T})$ is a set of synthesis if $\overline{J(E)} = I(E)$. We denote by $\mathcal{S}$ the family of sets of synthesis.

We will study the class $\mathcal{S}$ in Chapter X. Note that any set in $U_1 - U$ must be of non-synthesis, hence the result of Körner of the next chapter will in particular prove Malliavin's Theorem.

We give here two positive results about sets of synthesis which give some information about the relation between $U$ and $U_1$.

Theorem 3 (Ditkin). Let $E \in K(\mathbf{T})$ have countable boundary. Then $E$ is a set of synthesis.

Proof. Let $E$ have countable boundary. We want to show that $\overline{J(E)} = I(E)$. By duality, this is the same as $N(E) = PM(E)$, so it is enough to show that for $S \in PM(E)$ and $f \in I(E)$, $\langle f, S\rangle = 0$. As $\langle f, S\rangle = \langle 1, f \cdot S\rangle$, it is enough to show that $f \cdot S = 0$ or equivalently $\text{supp}(f \cdot S) = \emptyset$. Now $\text{supp}(f \cdot S) \subseteq \text{supp}(f) \cap \text{supp}(S) \subseteq E \cap \overline{(\mathbf{T} - E)} = \partial E$, the boundary of $E$. So $\text{supp}(f \cdot S)$ is countable, and if nonempty, towards a contradiction, it has an isolated point $x_0$, which by translation we may assume is the point 0. Consider the trapezoidal functions $\tau_\epsilon = \tau_{0,\epsilon}$ of II.2. We saw (II.2.2) that for $f \in A$ with $f(0) = 0$, $\|\tau_\epsilon f\|_A \to 0$ as $\epsilon \to 0$. Since 0 is isolated in $\text{supp}(f \cdot S)$, we have $\tau_\epsilon \cdot (f \cdot S) = \tau_{\epsilon'} \cdot (f \cdot S)$ for $\epsilon, \epsilon'$ small enough. But $\|(\tau_\epsilon f) \cdot S\|_{PM} \le \|\tau_\epsilon f\|_A \cdot \|S\|_{PM} \to 0$ as $\epsilon \to 0$, thus $(\tau_\epsilon\, f) \cdot S = \tau_\epsilon \cdot (f \cdot S) = 0$ for all small enough $\epsilon$. But then if $g \in A$ is supported in $[-\epsilon, \epsilon]$, we have $g = \tau_\epsilon g$ hence $(gf) \cdot S = (g\, \tau_\epsilon f) \cdot S = g \cdot ((\tau_\epsilon f) \cdot S) = 0$, so that $0 \notin \text{supp}(f \cdot S)$ a contradiction. □

<u>Theorem 4</u> (Herz's Criterion). Let $E \in K(T)$ be such that for infinitely many $N \in \mathbb{N}$ the points $0 \cdot \frac{2\pi}{N}, 1 \cdot \frac{2\pi}{N}, \dots, (N-1) \cdot \frac{2\pi}{N}$ are either in E or else at distance $\geq \frac{2\pi}{N}$ from E. Then E is a set of synthesis.

To prove this we need first a lemma.

<u>Lemma 5</u> (Herz's Lemma). Let $f \in A$. For each $N = 1, 2, \dots$, let $f_N$ be the function which is equal to f at $0 \cdot \frac{2\pi}{N}, 1 \cdot \frac{2\pi}{N}, \dots, (N-1) \cdot \frac{2\pi}{N}$ and linear in between. Then

(i) $\|f_N\|_A \leq \|f\|_A$,

(ii) $\|f_N - f\|_A \to 0$ as $N \to \infty$.

<u>Proof</u>. Assuming (i) let us prove (ii). First suppose $g \in C^1(T)$. Then the derivative of $g - g_N$ converges uniformly to 0 at all points except at the rational multiples of $2\pi$. Then by II.2.1 $\|g_N - g\|_A \to 0$. If now $f \in A$, let $\epsilon > 0$ be given and consider $g(x) = \sum_{j=-p}^{p} \hat{f}(j)\, e^{ijx}$, where p is chosen so that $\|g - f\|_A < \epsilon$. Then by (i), $\|(g - f)_N\|_A \leq \|g - f\|_A < \epsilon$, hence as $g \in C^1(T)$, $\|f_N - f\|_A < 3\epsilon$, for N big enough and we are done.

We prove now (i). Since $(\Sigma\, a_n e^{inx})_N = \Sigma\, a_n (e^{inx})_N$ it is enough to prove (i) for $e^{inx}$. Let $\psi_h = \psi_{0,h}$ be the usual triangular function (see II.2), so that $\psi_h(x) = \sum_{m=-\infty}^{\infty} \left(\frac{\sin(mh/2)}{mh/2}\right)^2 e^{imx}$. An immediate computation gives for $f(x) = e^{inx}$,

$$f_N(x) = \sum_{k=0}^{N-1} e^{\frac{2\pi k}{N} ni} \cdot \psi_{2\pi/N}\left(x - \frac{2\pi k}{N}\right) \cdot \frac{1}{N}$$

$$= \sum_{k=0}^{N-1} e^{\frac{2\pi k}{N} ni} \cdot \frac{1}{N} \cdot \sum_{m} \left(\frac{\sin(mh/2)}{mh/2}\right)^2 e^{im(x-\frac{2\pi k}{N})}$$

where $h = \frac{2\pi}{N}$,

$$= \sum_m \left(\frac{1}{N} \cdot \sum_{k=0}^{N-1} e^{\frac{2\pi k}{N}(n-m)i}\right) \cdot e^{imx} \cdot \left(\frac{\sin(mh/2)}{mh/2}\right)^2.$$

Now $N^{-1} \sum_{k=0}^{N-1} e^{\frac{2\pi k}{N}(n-m)i}$ is equal to 1 if $n \equiv m (\text{mod } N)$, and 0 if $n \not\equiv m \ (\text{mod } N)$, because if $\theta = e^{2\pi i(\frac{n-m}{N})}$, then $\theta \neq 1$ and so this sum is equal to $N^{-1} \cdot \sum_{k=0}^{N-1} \theta^k = \frac{(\theta^N - 1)}{N(\theta - 1)} = 0$. So $f_N(x) = \sum_{m \equiv n(\text{mod } N)} \left[\frac{\sin(mh/2)}{mh/2}\right]^2 e^{imx}$ and $\|f_N\|_A = \sum_{m \equiv n(\text{mod } N)} \left[\frac{\sin(mh/2)}{mh/2}\right]^2 = f_N(0) = f(0) = 1$. □

<u>Proof of Theorem 4</u>. Let $E \in K(\mathbb{T})$ and $N_1, N_2, \ldots, N_i, \ldots$ be such that any point $\frac{2\pi k}{N_i}$, $k = 0, \ldots, N_{i-1}$ is at distance $\geq \frac{2\pi}{N_i}$ from E or else is in E. Let $f \in I(E)$ and $\epsilon > 0$. By Lemma 5, the sequence $f_N$ converges to f in A, so we can choose an N in the sequence $N_i$ such that $\|f_N - f\|_A < \frac{\epsilon}{2}$. Now note that each time $E \cap \left[\frac{2\pi k}{N}, \frac{2\pi(k+1)}{N}\right] \neq \emptyset$, $k = 0, \ldots, N-1$, the endpoints are in E hence $f_N$ is 0 on this interval. So $f_N$ is 0 in a nbhd of each point in E, except possibly for some (finitely many) points of the form $\frac{2\pi k}{N}$ which are in E. Fix $\delta > 0$. For each such point $\frac{2\pi k}{N}$ it must be that at one of $\frac{2\pi(k-1)}{N}, \frac{2\pi(k+1)}{N}$ f is not 0. Define now $g_\delta$ to be 0 in $[\frac{2\pi k}{N} - \delta, \frac{2\pi k}{N} + \delta]$ and linear in $[\frac{2\pi(k-1)}{N}, \frac{2\pi(k-1)}{N} - \delta]$, $[\frac{2\pi k}{N} + \delta, \frac{2\pi(k+1)}{N}]$ with the same value as f at $\frac{2\pi(k-1)}{N}, \frac{2\pi(k+1)}{N}$. Then $g_\delta \in J(E)$, and $g_\delta \to f_N$ in A, when $\delta \to 0$ (as we can easily check using II.2.2), so that for small enough $\delta$, $\|g_\delta - f_N\|_A < \epsilon/2$ and so finally $\|f - g_\delta\|_A < \epsilon$. Thus $f \in \overline{J(E)}$ and we are done. □

Herz's criterion can be used to show that specific sets are of synthesis: e.g.

the Cantor set, since Herz's criterion is satisfied with the sequence $N_j = 3^j$. Moreover it gives some way of "regularizing sets" to get sets of synthesis. Define the <u>Herz Transform</u> $E^h$ of a closed set $E \subseteq \mathbb{T}$ by

$$E^h = E \cup \left\{\frac{2\pi k}{2^n} : n \in \mathbb{N},\, k < 2^n,\, \operatorname{dist}\left|\frac{2\pi k}{2^n}, E\right| < \frac{2\pi}{2^n}\right\}.$$

Clearly $E^h$ is closed, and is the union of E with a countable discrete set. Moreover by Theorem 4, $E^h$ is always a set of synthesis.

<u>Proposition 6</u>. The following are equivalent:

(i) $U_1' \not\subseteq U$,

(ii) $U_1 \neq U$,

(iii) $U_1$ is not a $\sigma$-ideal,

and they all imply $U \subsetneq U_1 \subsetneq U_0$ and $S \neq K(\mathbb{T})$.

<u>Proof</u>. Clearly (i) $\Rightarrow$ (ii). If $U_1' \subseteq U$, $(U_1')_\sigma \subseteq U$, as U as a $\sigma$-ideal. So $U \subseteq U_1 \subseteq (U_1')_\sigma \subseteq U$ and (ii) $\Rightarrow$ (i). (iii) $\Rightarrow$ (ii) is trivial, again since U is a $\sigma$-ideal. Conversely suppose $U_1$ is a $\sigma$-ideal. Let $E \in U_1$. As singletons are in $U_1$ the Herz Transform $E^h$ is in $U_1$ too, i.e. $I(E^h)$ is w*-dense in A. But as $E^h \in S$, $\overline{J(E^h)} = I(E^h)$, hence $J(E^h)$ is w*-dense in A and $E^h \in U$. A fortiori $E \in U$, so $U = U_1$. Clearly (iii) implies $U_1 \subsetneq U_0$ as $U_0$ is a $\sigma$-ideal. Since for $E \in S$, $E \in U$ iff $E \in U_1$, (ii) implies $S \neq K(\mathbb{T})$. □

So from Körner's Theorem in the next chapter that $U_1' \subsetneq U$, we will have that $U \subsetneqq U_1 \subsetneqq U_0$ and $U_1$ is not a $\sigma$-ideal.

We conclude with some facts on sets which are locally of uniqueness.

Definition 7. Define, following the general definition for $\sigma$-ideals (1.3), for $E \in K(\mathbf{T})$,

$$E \in U^{loc} \Leftrightarrow \exists \text{ open } V(E \cap V \neq \emptyset \text{ and } \overline{E \cap V} \in U).$$

We call these sets of local uniqueness. The closed sets not in $U^{loc}$ are called sets of pure multiplicity, or $M^p$-sets.

Clearly $U^{loc}$ is $\underset{\sim}{\Pi}^1_1$ and we proved in 1.3 that U admits a Borel basis iff $U^{loc}$ is Borel. The following result gives some information on $U^{loc}$ and $M^p$.

Proposition 8. Let $E \in K(\mathbf{T})$. Then

(i) (Piatetski-Shapiro [1]) $E \in M^p \Leftrightarrow \overline{J(E)}^{w^*} \subseteq I(E)$,

(ii) If $E \in U^{loc}$, then

$$\text{least } \alpha(J^{\alpha}(E) \nsubseteq I(E)) = \text{least } \alpha\,[\exists \text{ open } V\ (V \cap E \neq \emptyset \text{ and } [\overline{V \cap E}]_{PS} = \alpha)],$$

so that the two "natural" ranks on $U^{loc}$ coincide. (It is easy to check that this rank is also a $\underset{\sim}{\Pi}^1_1$-rank).

(iii) A closed set E is in $M^p$ iff $E = \text{supp}(S)$ for a pseudofunction S. In particular every M-set contains a nonempty $M^p$-set. (Note that by definition the empty set is in $M^p$).

We prove first a lemma.

Lemma 9. Let $E \in K(\mathbb{T})$ and $x_0 \in E$. Let $V$ be open with $x_0 \in V$ and let $E' = \overline{E \cap V}$. Then

$$x_0 \in \text{hull}(J^\alpha(E)) \Leftrightarrow x_0 \in \text{hull}(J^\alpha(E')),$$

and

$$x_0 \in \text{hull}(I^\alpha(E)) \Leftrightarrow x_0 \in \text{hull}(I^\alpha(E')).$$

Proof. As $E' \subseteq E$, $J^\alpha(E) \subseteq J^\alpha(E')$ and $I^\alpha(E) \subseteq I^\alpha(E')$, hence $\text{hull}(J^\alpha(E')) \subseteq \text{hull}(J^\alpha(E))$ and $\text{hull}(I^\alpha(E')) \subseteq \text{hull}(I^\alpha(E))$.

To see the converse, assume that $x_0 \notin \text{hull}(J^\alpha(E'))$, and let $g \in A$ be such that $g(x_0) = 1$, $\text{supp}(g) \subseteq V$. By induction on $\alpha$, one easily checks that if $f \in J^\alpha(E')$, $f \cdot g \in J^\alpha(E)$, as this is true for $\alpha = 0$. By the assumption, there is an $f \in J^\alpha(E')$ with $f(x_0) \neq 0$. But then $f \cdot g \in J^\alpha(E)$ and $f(x_0) \cdot g(x_0) = f(x_0) \neq 0$, i.e. $x_0 \notin \text{hull}(J^\alpha(E))$. The proof for $I(E)$ is similar. □

Proof of Proposition 8. Suppose first $E \in U^{loc}$, and let $V$ be open with $V \cap E \neq \emptyset$ and such that $[\overline{E \cap V}]_{PS} = \alpha$ is minimal. Choose $x_0 \in E \cap V$. Set $E' = \overline{E \cap V}$, so that $J^\alpha(E') = A$. Then $x_0 \notin \text{hull}(J^\alpha(E'))$ $( = \emptyset)$, and by Lemma 9, $x_0 \notin \text{hull}(J^\alpha(E))$, while $x_0 \in E = \text{hull}(I(E))$. This gives $J^\alpha(E) \not\subseteq I(E)$, and proves $\Leftarrow$ of part (i) and half of (ii). Conversely, let $\alpha$ be least with $J^\alpha(E) \not\subseteq I(E)$, and let $f \in J^\alpha(E) - I(E)$. Let $V$ be open with $E \cap V \neq \emptyset$ and $f$ non-zero on $\overline{V}$ (this is possible since $f \notin I(E)$). Then if $E' = \overline{E \cap V}$, $\mathrm{hull}(J^{\alpha}(E')) \subseteq E' \cap \mathrm{hull}(J^{\alpha}(E)) = \emptyset$, so $J^{\alpha}(E') = A$, i.e. $E' \in U$ and $[E']_{PS} \leq \alpha$. This proves $\Rightarrow$ of (i) and the second half of (ii).

For (iii), let first $S \in PF$ and $V$ be open with $V \cap \mathrm{supp}(S) \neq \emptyset$. Then there is $f \in A$ with support contained in $V$ and $T = f \cdot S \neq 0$. But clearly $T \in PF$ and $\mathrm{supp}(T) \subseteq V \cap \mathrm{supp}(S)$, so that $\overline{V \cap \mathrm{supp}(S)}$ is an M-set.

Conversely let $\emptyset \neq E \in M^{p}$. Let $\{V_n\}$ enumerate all basic open sets such that $V_n \cap E \neq \emptyset$. Then for each n there are nonzero pseudofunctions with support contained in $V_n \cap E$. We will define inductively on n a pseudofunction $S_n \in PM(E)$ with $\|S_n\|_{PM} = 1$, a function $f_n \in A$ with support contained in $V_n$ and a sequence of positive numbers $\epsilon^{(n)}_{n+1}, \epsilon^{(n)}_{n+2}, \epsilon^{(n)}_{n+3}, \ldots$ as follows: For $n = 0$, let $S_0 \in PF$ be such that $\|S_0\|_{PM} = 1$ and $\mathrm{supp}(S_0) \subseteq V_0 \cap E$. Let $f_0 \in A$ be such that $\mathrm{supp}(f_0) \subseteq V_0$ and $\langle f_0, S_0\rangle \neq 0$. Choose positive $\epsilon^{(0)}_1, \epsilon^{(0)}_2, \ldots$ so that $\|f_0\|_A \cdot (\sum_{m\geq 1} \epsilon^{(0)}_m) < |\langle f_0, S_0\rangle|$. Then for every $S'_1, S'_2, \ldots$ with $\|S'_m\|_{PM} = 1$ and every $0 < \delta_m \leq \epsilon^{(0)}_m$ we have $\langle f_0, S_0 + \sum_{m\geq 1} \delta_m S'_m\rangle \neq 0$. Let now $\delta_1 = \epsilon^{(0)}_1$ and find $S_1 \in PF$, $f_1 \in A$ such that $\|S_1\|_{PM} = 1$, $\mathrm{supp}(S_1) \subseteq V_1 \cap E$, $\mathrm{supp}(f_1) \subseteq V_1$ and $\langle f_1, S_0 + \delta_1 S_1\rangle \neq 0$. Choose positive $\epsilon^{(1)}_2, \epsilon^{(1)}_3, \ldots$ so that $\|f_1\|_A \cdot (\sum_{m\geq 2} \epsilon^{(1)}_m) < |\langle f_1, S_0 + \delta_1 S_1\rangle|$. Then for every $S'_2, S'_3, \ldots$ with $\|S'_m\|_{PM} = 1$ and every $0 < \delta_m \leq \epsilon^{(1)}_m$ we have $\langle f_1, S_0 + \delta_1 S_1 + \sum_{m\geq 2} \delta_m S'_m\rangle \neq 0$. Let now $\delta_2 = \min\{\epsilon^{(0)}_2, \epsilon^{(1)}_2\}$ and proceed as before with $V_2$, etc. Letting now in general $\delta_n = \min\{\epsilon^{(0)}_n, \epsilon^{(1)}_n, \ldots, \epsilon^{(n-1)}_n\}$, we have that $\Sigma\, \delta_n \leq \Sigma\, \epsilon^{(0)}_n < \infty$ and thus if $S = S_0 + \sum_{n\geq 1} \delta_n S_n$, then $S \in PF$ and $\mathrm{supp}(S) \subseteq E$. Finally $\mathrm{supp}(S) = E$ since $\langle f_n, S\rangle \neq 0$

for each n. □

<u>Remark</u>. The preceding proof shows also that if $M_1^p$ is defined in the obvious way, then $E \in M_1^p \Leftrightarrow I(E)$ is w*-closed.

We will come back to the relationship between the various Piatetski-Shapiro and Cantor-Bendixson ranks on U and $U_1$ in the next chapter, after we prove Körner's Theorem that $U_1' \subsetneqq U$. The structure of $U_0$ will be studied in Chapter VIII. This chapter contains also further results on $U_1$ and its relationship to U.

# Chapter VII. The Shrinking Method, the Theorem of Körner and Kaufman, and the Solution to the Borel Basis Problem for U

A standard technique in the subject under study is the use of multiplication of some given pseudofunction S by an appropriate $f \in A$ in order to "shrink" the support of S to a set with certain desirable properties. In this chapter we discuss various ways of doing that, and present a number of applications to the structure of sets of uniqueness based on this technique and the results of the earlier chapters. These include Kaufman's proof of the Körner theorem on the existence of Helson (thus $U_1'$-)sets of multiplicity and the Debs-Saint Raymond result on the non-existence of a Borel basis for U.

## §1. Sets of interior uniqueness

Recall that a set $P \subseteq \mathbb{T}$ is a set of interior uniqueness if every closed subset of P is a set of uniqueness. Trivially every set of uniqueness is of interior uniqueness, and the two notions coincide on $F_\sigma$ sets, by Bary's Theorem I.5.1. The question whether these notions coincide for $G_{\delta\sigma}$ sets (The Interior Problem—see I.6), or even for $G_\delta$ sets is to our knowledge open, although it appears more likely to be false. To some extent, the family $U^{int}$ of sets of interior uniqueness seems much more manageable then the family of sets of uniqueness, because it is directly related to closed U-sets. A typical example is the Union Problem. As we discussed in I.6, the Union Problem for $G_\delta$ sets of uniqueness is open. However we will see in this section that it can be solved affirmatively for $G_\delta$ (and thus $G_{\delta\sigma}$) sets of interior uniqueness. The proof is a simple application of the shrinking method.

<u>Lemma 1</u>. Let $Z \subseteq A$ be a convex set. For $S \in PF$, define $Z \cdot S = \{g \cdot S : g \in Z\}$. Then

$$\overline{Z}^{w^*} \cdot S \subseteq \overline{Z \cdot S}.$$

<u>Proof</u>. We know that $f \in A$ and $S \in PF$ imply $f \cdot S \in PF$, so that for $Z \subseteq A$, $Z \cdot S \subseteq PF$. Moreover if $Z$ is convex, clearly $Z \cdot S$ is convex in $PF$. Let $T \in PF$, $T \notin \overline{Z \cdot S}$. By Hahn-Banach, there is an $f$ in $A$ with

$$\mathrm{Re}\,(\langle f, T\rangle) > a = \sup\{\mathrm{Re}(\langle f, g \cdot S\rangle) : g \in Z\}.$$

Now $\langle f, g \cdot S\rangle = \langle g, f \cdot S\rangle$, so $\{g \in A : \mathrm{Re}(\langle f, g \cdot S\rangle)\ (= \mathrm{Re}(\langle g, f \cdot S\rangle)) \leq a\}$ is a $w^*$-closed subset of $A$ containing $Z$, hence containing $\overline{Z}^{w^*}$. So $T \notin \overline{Z}^{w^*} \cdot S$ and we are done. □

<u>Lemma 2</u>. Let $S \in PF$, $E \in U$ and $\epsilon > 0$. There is an $f \in J(E)$ so that letting $T = f \cdot S$, $\|S - T\|_{PM} < \epsilon$, $\mathrm{supp}(T) \subseteq \mathrm{supp}(S)$ and $\mathrm{supp}(T) \cap E = \emptyset$ (i.e. without losing much of $S$, one can shrink its support to avoid a given U-set).

<u>Proof</u>. We apply Lemma 1 with $Z = J(E)$. As $E \in U$ we have $1 \in \overline{J(E)}^{w^*}$, and Lemma 1 gives

$$1 \cdot S = S \in \overline{J(E) \cdot S}$$

which is exactly the content of the lemma. □

<u>Theorem 3</u> (Kechris-Louveau, Debs-Saint Raymond [1]). The $\sigma$-ideal U of closed sets of uniqueness is calibrated, i.e. if $\{E_n\}$ are in U, E is closed and $E - \bigcup_n E_n$ is in $U^{int}$, then $E \in U$ (cf. definition VI.1.9).

<u>Proof</u>. Suppose towards a contradiction that $E \notin U$, and let $S \in PF$, with say $\|S\|_{PM} = 1$, be supported by E. One defines by induction a sequence $S_n$ of pseudofunctions with $S_0 = S$, $\|S_n - S_{n+1}\|_{PM} \leq 2^{-(n+2)}$ and $\text{supp}(S_{n+1}) \subseteq \text{supp}(S_n) - E_n$, by applying Lemma 2 to $S_n$, $E_n$ and $2^{-(n+2)}$ to get $S_{n+1} = f_n \cdot S_n$. Then $T = \lim S_n$ is in PF, and $\|T\|_{PM} \geq \|S\|_{PM} - \|S - T\|_{PM} \geq \frac{1}{2}$. Moreover we have $\text{supp}(T) \subseteq E - \bigcup_n E_n$. But supp(T) is in M and this contradicts that $E - \bigcup_n E_n \in U^{int}$. □

Using Proposition VI.1.10, the preceding result immediately yields

<u>Theorem 4</u> (Kechris-Louveau, Debs-Saint Raymond [1]). The union of countably many $\underset{\sim}{\Sigma}^0_3$ ($\equiv G_{\delta\sigma}$) sets of interior uniqueness is also of interior uniqueness.

<u>Problem 5</u>. Is the union of countably many Borel sets of interior uniqueness also of interior uniqueness?

We do not know that even for $\underset{\sim}{\Pi}^0_3$ (i.e., $F_{\sigma\delta}$) sets.

Note also that from Theorem 4 a negative solution of the Union Problem for $G_{\delta\sigma}$ sets (which appears more likely than a positive one) implies a negative solution to the Interior Problem as well.

The next standard lemma (occurring essentially already in Kahane-Katznelson [1]) is extremely useful in many shrinking arguments.

Lemma 6. Let $S \in PF$ and $\varphi \in A$. For $n = 1, 2, 3, \ldots$, define

$$\varphi_n(x) = \varphi(nx).$$

Then

(i) $(\varphi_n \cdot S)(k) \to \hat{\varphi}(0) \cdot S(k)$ as $n \to \infty$, $\forall k \in \mathbb{Z}$,

(ii) $\|\varphi_n \cdot S\|_{PM} \to \|\hat{\varphi}\|_{PM} \cdot \|S\|_{PM}$ as $n \to \infty$.

Proof. (i) Note that if $n$ does not divide $k$, $\hat{\varphi}_n(k) = 0$, and if $k = n \cdot \ell$, $\hat{\varphi}_n(k) = \hat{\varphi}(\ell)$. So

$$(\varphi_n \cdot S)(k) = \sum_{\ell} \hat{\varphi}(\ell) \cdot S(k - n \cdot \ell).$$

Fix $\epsilon > 0$ and choose $N$ such that $\sum_{|\ell|>N} |\hat{\varphi}(\ell)| \leq \epsilon$. Then

$$\left|\sum_{|\ell|>N} \hat{\varphi}(\ell) \cdot S(k - n \cdot \ell)\right| \leq \epsilon \|S\|_{PM}.$$

Now for fixed $k$, $\ell \neq 0$, $|k - n \cdot \ell| \to \infty$ with $n$, so we can find $n_0$ such that for $n > n_0$, $|S(k - \ell \cdot n)| \leq \frac{\epsilon}{2N}$ for all $|\ell| \leq N$, except $\ell = 0$ (as $S \in PF$). This gives, for $n > n_0$, $|(\varphi_n \cdot S)(k) - \hat{\varphi}(0) \cdot S(k)| \leq \epsilon \|\hat{\varphi}\|_{PM} + \epsilon \|S\|_{PM}$ and we are done.

(ii) Fix again $\epsilon$, let $N$ be as before, and choose $M$ such that for $|j| \geq M$,

$|S(j)| \leq \frac{\epsilon}{2N}$. If $n \geq 2M$, then for any k, at most one value of $(k - n \cdot \ell)_{|\ell| \leq N}$ is in $[-M, M]$, say for $\ell = \ell(k)$. One then gets as before

$$|(\varphi_n \cdot S)(k) - \hat{\varphi}(\ell(k)) \cdot S(k - n \cdot \ell(k))| \leq \epsilon \cdot \|\hat{\varphi}\|_{PM} + \epsilon \cdot \|S\|_{PM}$$

for $n \geq 2M$, so that $\|\varphi_n \cdot S\|_{PM} \leq \|\hat{\varphi}\|_{PM} \cdot \|S\|_{PM} + \epsilon(\|\hat{\varphi}\|_{PM} + \|S\|_{PM})$. On the other hand, if $m_0$ and $n_0$ are chosen so that $|\hat{\varphi}(m_0) \cdot S(n_0)| \geq \|\hat{\varphi}\|_{PM} \cdot \|S\|_{PM} - \epsilon$, let $\psi = e^{-im_0x}\varphi$, so that $\hat{\psi}(0) = \hat{\varphi}(m_0)$, and apply (i) to $\psi$. We get $(\psi_n \cdot S)(n_0) \to \hat{\psi}(0) \cdot S(n_0)$ as $n \to \infty$, so we can find $k_0 \geq 2M$ such that for $n \geq k_0$ $|(\psi_n \cdot S)(n_0) - \hat{\psi}(0) \cdot S(n_0)| < \epsilon$, hence $\|\psi_n \cdot S\|_{PM} \geq \|\hat{\varphi}\|_{PM} \cdot \|S\|_{PM} - 2\epsilon$. Now $\psi_n \cdot S = (e^{-im_0nx}\varphi_n) \cdot S$ hence $\|\psi_n \cdot S\|_{PM} \leq \|\varphi_n \cdot S\|_{PM}$. We finally get for $n \geq k_0$, $\|\hat{\varphi}\|_{PM} \cdot \|S\|_{PM} - 2\epsilon \leq \|\varphi_n \cdot S\|_{PM} \leq \|\hat{\varphi}\|_{PM} \cdot \|S\|_{PM} + \epsilon(\|\hat{\varphi}\|_{PM} + \|S\|_{PM})$, and we are done. □

Using the preceding lemma, the following interesting splitting property of M- and $M_0$-sets can be established.

<u>Theorem 7</u> (Kaufman). Let $E \in K(\mathbf{T})$ be an M-set. Then we can find a family $\{E_x\}_{x \in 2^{\mathbb{N}}}$ of pairwise disjoint (i.e $x \neq y \Rightarrow E_x \cap E_y = \emptyset$) M-sets contained in E. Similarly replacing M by $M_0$.

<u>Proof</u>. We will need the following lemma

<u>Lemma 8</u>. Given $\epsilon > 0$ there are functions $\varphi^0, \varphi^1 \in C^\infty(\mathbf{T})$, $\varphi^0 \geq 0$, $\varphi^1 \geq 0$ with $\text{supp}(\varphi^0) \cap \text{supp}(\varphi^1) = \emptyset$ and

$$\widehat{\varphi^i}(0) = 1, |\widehat{\varphi^i}(m)| \leq \epsilon, \forall m \neq 0; i = 0, 1.$$

Granting this lemma we proceed as follows. Let S be in PF with $\mathrm{supp}(S) \subseteq E$ and $\|S\|_{PM} = 1$. We will define for each $u \in \mathrm{Seq}\{0, 1\}$ a pseudofunction $S_u$ such that $S_\emptyset = S$ and

(i) $\|S_{u^\frown(i)} - S_u\|_{PM} \leq \dfrac{1}{2^{\ell h(u)+2}}$ ; $i = 0, 1,$

(ii) $\mathrm{supp}(S_{u^\frown(0)}) \cap \mathrm{supp}(S_{u^\frown(1)}) = \emptyset,$

(iii) $\mathrm{supp}(S_{u^\frown(i)}) \subseteq \mathrm{supp}(S_u)$ ; $i = 0, 1.$

Then for $x \in 2^{\mathbb{N}}$ let $S_x = \lim_{n\to\infty} S_{x \restriction n}$, which exists in PF and is non-0 by (i). By (ii), (iii) if $E_x = \mathrm{supp}(S_x)$, $\{E_x\}$ is the required family.

We construct $S_u$ by induction on $\ell h(u)$. Given $S_u$, we define $S_{u^\frown(0)}, S_{u^\frown(1)}$ as follows: By Lemma 6 applied to $\varphi^i - 1$ ($\varphi^i$ as in Lemma 8) we have $\|(\varphi^i)_n \cdot S_u - S_u\|_{PM} \to \|\widehat{(\varphi^i - 1)}\|_{PM} \cdot \|S_u\|_{PM} \leq \epsilon \cdot \|S_u\|_{PM}$. Choose now $\epsilon = 2^{-(\ell h(u)+3)} \cdot \|S_u\|_{PM}^{-1}$ and then choose n big enough so that if

$$S_{u^\frown(i)} = (\varphi^i)_n \cdot S_u \; ; i = 0, 1$$

then $\|S_{u^\frown(i)} - S_u\|_{PM} \leq 2^{-(\ell h(u)+2)}$. Obviously (i) is satisfied and (ii), (iii) are automatic from the definition of $S_{u^\frown(i)}$ and the properties of $\varphi^i$.

For case of $M_0$ notice that if $S = \hat{\mu}$, where $\mu$ is a probability measure, then each $S_u$ is also of the form $\hat{\nu}$ for some positive measure $\nu$, since $\varphi^i \geq 0$. (Notice here the easy fact that if $f \in A$ and $\rho, \sigma \in M(T)$ are such that $d\rho = fd\sigma$, then

$\hat{\rho} = f \cdot \hat{\sigma}$). It is easy to check now that if $\{\mu_i\}$ are positive measures and $\hat{\mu}_i \xrightarrow{w^*} S$, then $\{\mu_i\}$ $w^*$-converges to some positive measure $\mu$ in the weak$^*$-topology of $M(\mathbb{T})$ and $S = \hat{\mu}$. Thus each $S_x$ is also of the form $\hat{\nu}$ for some positive measure $\nu$ and we are done.

It remains to prove Lemma 8. Let for each $N = 1, 2, \ldots$ $\sigma_N$ be the measure concentrating on $\{\frac{2\pi k}{N} : k = 0, 1, \ldots, N-1\}$ with $\sigma_N\left(\{\frac{2\pi k}{N}\}\right) = \frac{1}{N}$. Let also $\mu$ be a probability measure on $\mathbb{T}$ with $\hat{\mu}(n) \to 0$ as $|n| \to \infty$, and $\text{supp}(\mu)$ having (Lebesgue) measure 0. (This exists by Menshov's Theorem). Since $\hat{\sigma}_N(m) = 0$ except for the multiples of N it follows that for sufficiently large N, $|\hat{\mu}(m) \cdot \hat{\sigma}_N(m)| \le \epsilon/4$ for all $m \neq 0$. Let $\nu = \mu * \sigma_N$ be the <u>convolution</u> of $\mu$, $\sigma_N$ defined as usual by $\int f d(\mu * \sigma_N) = \iint f(x + y)\, d\mu(x)\, d\sigma_N(y)$ or equivalently $(\mu * \sigma_N)(P) = \int \mu(P - x)\, d\sigma_N(x)$, where $P - x = \{y : y + x \in P\}$ (addition $x + y$ is of course here modulo $2\pi$ if we view $\mu$, $\sigma_N$ as measures on $[0, 2\pi)$). Then $\nu$ concentrates in the union of finitely many translates of the support of $\mu$, so its support has measure 0 as well. Thus we can find $\varphi^0 \in C^\infty(\mathbb{T})$, $\varphi^0 \ge 0$ which is 0 in a nbhd of $\text{supp}(\nu)$ and $\widehat{\varphi^0}(0) = 1$, $|\widehat{\varphi^0}(m)| \le \epsilon$, $\forall m \neq 0$. (For that just choose $0 \le \varphi \le 1$ in $C^\infty(\mathbb{T})$ with $\lambda(\{x : \varphi(x) \neq 1\})$ very small and then normalize $\varphi$ to produce $\varphi^0$ with $\widehat{\varphi^0}(0) = 1$). Also $\hat{\nu}(m) = \hat{\mu}(m) \cdot \hat{\sigma}_N(m)$, so $\hat{\nu}(0) = 1$, $|\hat{\nu}(m)| \le \epsilon/4$ if $m \neq 0$. Finally choose $\varphi^1$ with support disjoint from that of $\varphi^0$, $\varphi^1 \in C^\infty(\mathbb{T})$, $\varphi^1 \ge 0$, to be a close enough approximation to $\nu$.

(This can be done as follows: Let V be a nbhd of $\text{supp}(\nu)$ with $\bar{V} \cap \text{supp}(\varphi^0) = \emptyset$. Let $\psi \in A$ be such that $\psi \ge 0$, $\psi = 1$ off V and $\psi = 0$ on $\text{supp}(\nu)$. We claim that we can approximate $\hat{\nu}$ in PM by positive trigonometric polynomials as close as we want. Granting that let $P \ge 0$ be a trigonometric polynomial with

$\|\hat{P} - \hat{\nu}\|_{PM} \le \frac{\epsilon}{8} \cdot \max(1, \|\psi\|_A^{-1})$. Let then $\varphi^* \in C^\infty(\mathbf{T})$, $P \ge \varphi^* \ge 0$ be such that $\varphi^* = P$ on V, and $\mathrm{supp}(\varphi^*) \cap \mathrm{supp}(\varphi^0) = \emptyset$. Then $\|\hat{P} - \widehat{\varphi^*}\|_{PM} = \int(P - \varphi^*)d\lambda = \int_{\mathbf{T}-V} (P - \varphi^*)d\lambda \le \int(P - \varphi^*)\,\psi d\lambda \le \int P\psi d\lambda = <\psi, \hat{P}>$. But as $\psi = 0$ on $\mathrm{supp}(\nu)$, $<\psi, \hat{P}> = <\psi, \hat{P} - \hat{\nu}> \le \frac{\epsilon}{8}\|\psi\|_A^{-1} \cdot \|\psi\|_A = \frac{\epsilon}{8}$. So $\|\hat{\nu} - \widehat{\varphi^*}\|_{PM} \le \frac{\epsilon}{4}$. We can normalize now $\varphi^*$ to get $\varphi^1$ with all the required properties.

To prove our claim about approximating $\hat{\nu}$ by trigonometric polynomials let $K_n(x) = \sum_{m=-n}^{n} (1 - \frac{|m|}{n+1})e^{imx}$, $n = 0, 1, 2, \ldots$, be the Fejér kernel, so that $K_n \ge 0$. Given any $\delta > 0$ choose $n_0$ so that $|\hat{\nu}(m)| < \delta$ if $|m| > n_0$ (since $\hat{\nu} \in PF$). Then choose $n_1 > n_0$ such that $\frac{n_0}{n_1+1} \cdot \|\hat{\nu}\|_{PM} < \delta$. Let $d\rho_{n_1} = K_{n_1}\,d\lambda$ and consider $\rho = \rho_{n_1} * \nu$. Then $\hat{\rho}(m) = \hat{\rho}_{n_1}(m) \cdot \hat{\nu}(m) = \hat{K}_{n_1}(m) \cdot \hat{\nu}(m)$, so $\rho$ is a trigonometric polynomial, i.e. $\hat{\rho} = \hat{P}$ with P a trigonometric polynomial. As $\rho$ is a positive measure we have $P \ge 0$. Finally $|\hat{P}(m) - \hat{\nu}(m)|$ is equal to $|\hat{\nu}(m)| < \delta$, if $|m| > n_1$, and is equal to $\frac{|m|}{n_1+1} \cdot |\hat{\nu}(m)|$, if $|m| < n_1$. If $|m| > n_0$ this is at most $|\hat{\nu}(m)| < \delta$ again, while if $|m| \le n_0$ it is $< \delta$ by our choice of $n_1$. So $\|\hat{P} - \hat{\nu}\|_{PM} \le \delta$ and we are done). □

According to the terminology of Kechris-Louveau-Woodin [1] a closed set $E \subseteq \mathbf{T}$ is U-thin if it contains no uncountable family of pairwise disjoint M-sets. Similarly we define $U_0$-thin sets. By the preceding result the U-thin sets are exactly the U-sets and similarly for $U_0$.

Again adopting a definition of Kechris-Louveau-Woodin [1] we say that U is strongly calibrated if for every $E \in M$ and $P \subseteq \mathbf{T} \times 2^{\mathbb{N}}$, P a $\underset{\sim}{\Sigma}_1^1$ set with proj(P) $= E$, there is $F \in K(\mathbf{T} \times 2^{\mathbb{N}})$, $F \subseteq P$ with proj(F) $\in M$. It is easy to check that this property implies calibration. However it is not known if U is strongly calibrated. (An equivalent form of this question has been also asked by Kaufman). It follows from a result of Kechris-Louveau-Woodin [1] (see 3.2.5 in that paper) and Theorem 7 that if U is strongly calibrated then the class of Borel (in fact even $\underset{\sim}{\Sigma}^1_1$) sets in $U^{int}$ is closed under countable unions. (By standard capacitability arguments, see e.g. IX.1.5, $U_0$ is strongly calibrated).

## §2. Approximating M-sets by $H_\sigma$-sets

In the previous section, we saw how to shrink the support of a pseudofunction in order to avoid a given U-set. Here, we show how to approximate pseudofunctions by pseudomeasures with support in U, in fact in $H_\sigma$, by a shrinking method due to Kaufman. We will apply it to compute the complexity of U restricted to any M-set.

Theorem 1 (Kaufman [7]). Let $S \in PF$. Then for every $\epsilon > 0$ there is $T \in PM$ with $\|S - T\|_{PM} \leq \epsilon$, $\text{supp}(T) \subseteq \text{supp}(S)$, and supp(T) a finite union of H-sets. If moreover S is a probability measure (i.e. $S = \hat{\mu}$ for some probability measure $\mu$ on $\mathbf{T}$), one can choose T to be a probability measure as well.

Proof. We start with $S \in PF$, $S \neq 0$ and $\epsilon > 0$. Let $N \in \mathbb{N}$ be such that $(4\|S\|_{PM} + 1)/N \leq \epsilon$. We construct by induction a sequence of pseudofunctions $T_k$, and numbers $m_k$ and $n_k$ satisfying

(i) $n_0 = n_1 = .. = n_{N-1} = 0 < n_N < n_{N+1} < \dots$; $m_0 = m_1 = .. = m_{N-1} = 0 < m_N < m_{N+1} < \dots$,

(ii) $T_0 = T_1 = .... = T_{N-1} = S$, and $\|T_k\|_{PM} < 2\,\|S\|_{PM}$,

(iii) For $|j| \leq n_{k+N-1}$, $|T_{k+N}(j) - T_k(j)| \leq 2^{-(k+1)}$,

(iv) $\mathrm{supp}(T_{k+N}) \subseteq \mathrm{supp}(T_k)$,

(v) $m_k \cdot \mathrm{supp}(T_k) \subseteq [0, 2^{-k}]$,

(vi) For $|j| \geq n_{k+N}$, $|T_{k+N}(j) - T_k(j)| \leq 2^{-(k+1)}$.

Suppose the construction has been carried out up to $k + N - 1$. To define $T_{k+N}$, choose $\varphi^{k+N} = \varphi \in A$ with $\hat{\varphi}(0) = 1$, $\mathrm{supp}(\varphi) \subseteq [0, 2^{-(k+N)}]$ and $\varphi \geq 0$, so that $\|\hat{\varphi}\|_{PM} = \hat{\varphi}(0) = 1$. Apply Lemma 1.6 to $T_k$ and $\varphi$: We can find $m_{k+N} > m_{k+N-1}$ such that $T_{k+N} = \varphi_{m_{k+N}} \cdot T_k$ satisfies $\left|T_{k+N}(j) - T_k(j)\right| \leq 2^{-(k+1)}$ for $|j| \leq n_{k+N-1}$ (as $(\varphi_p \cdot T_k)(j) \to T_k(j)$ with $p \to \infty$) and such that $\|T_{k+N}\|_{PM} < 2\|S\|_{PM}$ (as $\|\varphi_p \cdot T_k\|_{PM} \to \|T_k\|_{PM}$ with $p \to \infty$). By construction, $\mathrm{supp}(T_{k+N}) \subseteq \mathrm{supp}(T_k)$, and $m_{k+N} \cdot \mathrm{supp}(T_{k+N}) \subseteq m_{k+N} \cdot \mathrm{supp}(\varphi_{m_{k+N}}) \subseteq \mathrm{supp}(\varphi)$. So this takes care of (i)–(v). Finally as $T_k$, $T_{k+N}$ are pseudofunctions we can choose $n_{k+N}$ large enough to insure (vi).

Fix n with $0 \leq n \leq N - 1$. By (iii), the sequence $T_{n+kN}$ constructed above converges with $k \to \infty$ on each exponential, and is uniformly bounded by (ii), hence $w^*$-converges in PM to some $T^n$. And as $\mathrm{supp}(T^n) \subseteq \bigcap_k \mathrm{supp}(T_{n+kN})$, $\mathrm{supp}(T^n) \subseteq \mathrm{supp}(S)$ by (ii), (iv).

By (v), the sequence $m_{n+kN}$ witnesses that $\mathrm{supp}(T^n)$ is an H-set—even of a very special form, for not only the sequence $\{m_{n+kN} \cdot \mathrm{supp}(T^n)\}_{k \in \mathbb{N}}$ avoids some interval, but in fact converges to $\{0\}$. (Such sets are called <u>Dirichlet</u> sets).

Define $T = \frac{1}{N} \cdot \sum_{n=0}^{N-1} T^n$. By the preceding discussion, T satisfies the requirements in Theorem 1, with the possible exception of the inequality

$\|S - T\|_{PM} \leq \epsilon$. To check this, fix $j \in \mathbf{Z}$. Then

$$|S(j) - T(j)| \leq \frac{1}{N} \sum_{n=0}^{N-1} |S(j) - T^n(j)|$$

$$\leq \frac{1}{N} \sum_{n=0}^{N-1} \sum_{\ell=0}^{\infty} |T_{n+\ell N}(j) - T_{n+(\ell+1)N}(j)|$$

$$\leq \frac{1}{N} \sum_{k=0}^{\infty} |T_{k+N}(j) - T_k(j)|.$$

Now notice that there is at most one value of k for which $|T_{k+N}(j) - T_k(j)| > 2^{-(k+1)}$, namely the k for which $n_{k+N-1} < |j| < n_{k+N}$, by (iii) and (vi). For this k we obviously have $\left|T_{k+N}(j) - T_k(j)\right| \leq 4\ \|S\|_{PM}$, so that finally we obtain

$$|S(j) - T(j)| \leq \frac{1}{N}(1 + 4 \cdot \|S\|_{PM}) \leq \epsilon$$

and we are done.

Finally assume that $S = \hat{\mu}$, $\mu$ a probability measure. We claim then that we can choose the $T_k$ to be probability measures as well, i.e. $T_k = \hat{\nu}_k$ for some probability measure $\nu_k$. As in the proof of 1.7 this will show that $T^n$ and thus T are probability measures too.

To define $\nu_{k+N}$ we proceed as before: Since for each $f \in A$ and $\nu$, $\mu \in M(\mathbf{T})$ with $d\nu = fd\mu$, we have $\hat{\nu} = f \cdot \hat{\mu}$, it follows (recalling that $\varphi \geq 0$) that $\|\varphi_p \cdot \hat{\nu}_k\|_{PM} = \varphi_p \cdot \hat{\nu}_k(0) = \int \varphi_p \, d\nu_k \to \hat{\nu}_k(0) = 1$ (by 1.6), so we can choose $m_{k+N}$ large enough to insure that $\nu_{k+N}$ given by

$$d\nu_{k+N} = \left(\int \varphi_{m_{k+N}} \, d\nu_k\right)^{-1} \cdot \varphi_{m_{k+N}} \, d\nu_k$$

is a probability measure, and $\hat{\nu}_{k+N}$ still satisfies (i) — (vi) (with the appropriately chosen $n_{k+N}$). □

The next result, a local version of the theorem of Solovay and Kaufman was proved independently by Kaufman [8], Kechris-Louveau (by the argument that follows) and Debs-Saint Raymond [1] (by adopting a method of Kaufman [7]).

<u>Theorem</u> <u>2</u> (Kaufman [8], Kechris-Louveau, Debs-Saint Raymond [1]). Let E be a closed set of multiplicity. Then the Piatetski-Shapiro rank is unbounded on the $H_\sigma$ subsets of E. In particular, U ∩ K(E) is $\underset{\sim}{\Pi}^1_1$-complete.

<u>Proof</u>. The second assertion follows from the first using the Boundedness Theorem V.1.7, the fact that the Piatetski-Shapiro rank is a $\underset{\sim}{\Pi}^1_1$-rank on U (V.4.3) and Theorem IV.3.1.

To prove the first assertion, we work with the R-rank of U, which by V.4.6 we know is the same as the Piatetski-Shapiro rank. What we will prove is that for each countable ordinal $\alpha$, if $E \in M$, $\epsilon > 0$ and $S \in PM(E) \cap PF$, then there is $F \subseteq E$, $F \in H_\sigma$ and $T \in PM$ with $T \in PM_\alpha(F)$ and $\|T - S\|_{PM} \leq \epsilon$ (recall here the notation of V.4).

We proceed by induction on $\alpha$. For $\alpha = 0$, this is given by Theorem 1. Let $\alpha_n = \alpha - 1$ if $\alpha$ is successor, and $\alpha_n \to \alpha$, $\alpha_n$ increasing, if $\alpha$ is limit. Given $0 \neq S \in PF \cap PM(E)$ and $\epsilon > 0$, we can find by (the proof of) Theorem 1 a sequence $S_n \in PF$, $S_n \xrightarrow{w^*} T \in PM$, such that if $\mathrm{supp}(S_n) \subseteq E_n \subseteq E$, $\{E_n\}$ is a decreasing sequence with $\mathrm{supp}(T) \subseteq \bigcap_n E_n$, $\bigcap_n E_n$ is a finite union of H-sets, and $\|T - S\|_{PM} \leq \epsilon$. By the induction hypothesis, find $T_n \in PM$ with $\|S_n - T_n\|_{PM} \leq \frac{1}{n}$, $\mathrm{supp}(T_n) \subseteq F_n \subseteq E_n$, $F_n \in H_\sigma$, and $T_n \in PM_{\alpha_n}(F_n)$. Let $F = (\bigcap_n E_n) \cup \bigcup_n F_n$. Clearly F is a closed $H_\sigma$-set contained in E. Also as $T_n \in PM_{\alpha_n}(F_n)$, $T_n \in PM_{\alpha_n}(F)$. Moreover $R(T_n) \leq \|S_n - T_n\|_{PM} \leq \frac{1}{n}$, and $T_n \xrightarrow{w^*} T$ as $S_n \xrightarrow{w^*} T$. By the definition of $PM_\alpha(F)$, this gives $T \in PM_\alpha(F)$, and we are done. □

Note that the results of Section 1 (that U is calibrated) and Section 2 (that U is $\underset{\sim}{\Pi}^1_1$-complete on each M-set) are just what is needed to apply the constructions in VI.1. (See the discussion after Theorem VI.1.11).

## §3. Helson sets of multiplicity

Definition 1. A set $E \in K(\mathbf{T})$ is a Helson set if every continuous function on E can be extended to a function in A.

Let us denote for $E \in K(\mathbf{T})$, by A(E) the set $\{f \restriction E : f \in A\} \subseteq C(E)$ (= the set of continuous functions on E). This forms a Banach space, with norm $\|f\|_{A(E)} = \inf\{\|g\|_A : g \in A, g \restriction E = f\}$, i.e.

$$A(E) = A/I(E)$$

and its dual is

$$A(E)^* = N(E) = I(E)^\perp.$$

Note that for $f \in A(E)$ and $x \in E$, $|f(x)| = |g(x)| \le \|g\|_A$ for any $g$ with $g \restriction E = f$. Hence

$$\|f\|_{C(E)} \le \|f\|_{A(E)}$$

i.e. the identity $Id : A(E) \to C(E)$ is continuous.

We will make use of the following classical result:

<u>Lemma 2</u>. Let X, Y be Banach spaces, $T : X \to Y$ a linear continuous map. Then the following are equivalent,

(i) T is onto,

(ii) $T^* : Y^* \to X^*$ defined by $\langle x, T^*y^*\rangle = \langle Tx, y^*\rangle$ is such that for some $c > 0$, $\|y^*\|_* \le c\ \|T^*y^*\|_*$.

Moreover, if T is $1 - 1$ and (i) or (ii) holds, then $T^{-1}$ is continuous, with norm the least c for which (ii) holds.

<u>Proof</u>. (i) $\Rightarrow$ (ii). Suppose T is onto, so that $Y = \bigcup_n \overline{T(B_n(X))}$. By the Baire Category Theorem, and using translation, there is n such that $B_1(Y) \subseteq \overline{T(B_n(X))}$. But then by a direct computation

$$\|T^*y^*\|_* = \sup_{x \in B_1(X)} |\langle x, T^*y^*\rangle|$$

$$= \sup_{x \in B_1(X)} |\langle Tx, y^*\rangle|$$

$$= \frac{1}{n} \sup_{x \in B_n(X)} |\langle Tx, y^* \rangle|$$

$$= \frac{1}{n} \sup\{|\langle y, y^* \rangle| : y \in \overline{T(B_n(X))}\}$$

$$\geq \frac{1}{n} \|y^*\|_*.$$

(ii) $\Rightarrow$ (i). Suppose $\|y^*\|_* \leq c\ \|T^*y^*\|_*$. We show first that $B_{1/c}(Y) \subseteq \overline{T(B_1(X))}$ : For if $y \notin \overline{T(B_1(X))}$, then by Hahn-Banach there is a $y^* \in Y^*$ with $|\langle y, y^* \rangle| > 1$ but $|\langle z, y^* \rangle| \leq 1$ on $\overline{T(B_1(X))}$, hence $\|T^*y^*\|_* \leq 1$. This implies $\|y^*\|_* \leq c$, and as $|\langle y, y^* \rangle| > 1$, $\|y\| > 1/c$.

Fix now $\epsilon > 0$. By the preceding fact, for each $y \in Y$ there is a $z \in X$ with $\|z\| \leq c \cdot \|y\|$ and $\|y - Tz\| \leq \epsilon \cdot c \cdot \|y\|$. We can apply this to get a sequence $y_0 = y$, with corresponding $z_0$, $y_1 = y_0 - Tz_0$, with corresponding $z_1$, etc. So we get $y_n$, $z_n$ with $y_0 = y$, $\|z_n\| \leq c \cdot \|y_n\|$, $\|y_n - Tz_n\| \leq \epsilon \cdot c \cdot \|y_n\|$ and $y_{n+1} = y_n - Tz_n$. We have then $\|z_n\| \leq \epsilon^n \cdot c^{n+1} \cdot \|y\|$, so for $\epsilon$ small enough $z = \sum_n z_n$ exists, and has norm $\|z\| \leq \frac{c}{1-\epsilon c} \|y\|$. Moreover $\|y - T(\sum_{n \leq N} z_n)\| = \|y_{N+1}\| \to 0$, hence $y = Tz$. This shows that T is onto. Moreover if T is bijection, the preceding shows (as $z = T^{-1}y$) that $\|T^{-1}y\| \leq \frac{c}{1-\epsilon c} \|y\|$ for all $\epsilon$ small enough, i.e. $\|T^{-1}\| \leq c$, for any c satisfying (ii). Conversely, $\|y^*\|_* = \sup_{y \in B_1(Y)} |\langle y, y^* \rangle| \leq \sup_{x \in B_{\|T^{-1}\|}(X)} |\langle Tx, y^* \rangle| \leq \|T^{-1}\| \cdot \|T^*y^*\|_*$, so $\|T^{-1}\|$ is the least such c. □

Recall that M(E) is the space of all measures supported by E. It is clearly a closed (in fact $w^*$-closed) subspace of $M(\mathbb{T})$. Moreover $M(E) = C(E)^*$, i.e. M(E) is also the dual of the space of continuous functions on E. The map $\mu \mapsto \hat{\mu}$ is 1–1 from M(E) into N(E), so it will be convenient to write M(E) as well for its range.

Thus $M(E) \subseteq N(E)$. Also we will write

$$\|\mu\|_{PM} = \|\hat{\mu}\|_{PM}.$$

Corollary 3. Let $E \in K(\mathbb{T})$. Then the following are equivalent for $E \neq \emptyset$:

(i) E is a Helson set,

(ii) $M(E) = N(E)$,

(iii) $\alpha(E) =^{\text{def}} \sup\left\{\frac{\|\mu\|_M}{\|\mu\|_{PM}} : \mu \in M(E),\ \mu \neq 0\right\} < \infty.$

Proof. (i) $\Leftrightarrow$ (iii) follows from the first assertion of the preceding lemma applied to $X = A(E)$, $Y = C(E)$, $T = Id : X \to Y$. Also (ii) $\Rightarrow$ (iii) follows from the last conclusion of the preceding lemma applied to $X = M(E)$, $Y = N(E)$, $T(\mu) = \hat{\mu}$. Finally, if $T : X \to Y$ is an isomorphism between Banach spaces, so is $T^* : Y^* \to X^*$. Applying this to $X = A(E)$, $Y = C(E)$ and $T = Id$ we notice that $T^*(\mu) = \hat{\mu}$ and this shows that (i) $\Rightarrow$ (ii). □

The constant $\alpha(E)$ is called the Helson constant of the set E. Note that (iii) can be expressed also by saying that

$$\inf\left\{\frac{\|\mu\|_{PM}}{\|\mu\|_M} : \mu \in M(E),\ \mu \neq 0\right\} > 0.$$

One also associates with closed sets another constant, which is sometimes denoted by s(E), but for consistency with our notation and to avoid conflict with the quantity s(Y) defined in V.2.7 we will denote as $\eta_M(E)$, defined by

$$\eta_M(E) = \inf\left\{\frac{R(\mu)}{\|\mu\|_M} : \mu \in M(E),\ \mu \neq 0\right\}$$

where as usual

$$R(\mu) = R(\hat{\mu}) = \overline{\lim}\ |\hat{\mu}(n)|.$$

Note that if E is Helson, then $M(E) = N(E)$, so $\eta_M(E) \leq \eta_1(E)$ (as defined in Section VI.2). (This is because for $\mu \in M(E)$, $\|\mu\|_{PM} \leq \|\mu\|_M$).

Theorem 4 (see Kahane-Salem [1]). Let $E \in K(\mathbf{T})$. Then the following are equivalent,

(i) E is Helson,

(ii) $\eta_M(E) > 0$.

So in particular if E is Helson, $\eta_1(E) > 0$ and $E \in U_1'$.

Proof. One direction is easy: if $\eta_M(E) > 0$, then for $\mu \in M(E)$ $\|\mu\|_M \leq \frac{1}{\eta_M(E)} \cdot R(\mu) \leq \frac{1}{\eta_M(E)} \|\mu\|_{PM}$, hence E is Helson with constant $\alpha(E) \leq \frac{1}{\eta_M(E)}$. To prove the converse, we need first two lemmas.

Let $\mu_0$ be the uniform distribution on the finite set $\{-1, +1\}^k$, and for $\vec{c} \in \mathbb{C}^k$, $\vec{a} \in \{-1, +1\}^k$ let $\vec{c} \cdot \vec{a} = \sum_{i=1}^{k} c_i a_i$. Let also $\|\vec{c}\| = (\sum_{j=1}^{k} |c_j|^2)^{1/2}$.

Lemma 5. $\int |\vec{c} \cdot \vec{a}|\, d\mu_0(\vec{a}) \geq \frac{\sqrt{3}}{3} \|\vec{c}\|$.

Proof. Let $f(\vec{a}) = |\vec{c} \cdot \vec{a}|$. By Hölder's inequality

$$\int |f|^2 d\mu_0 = \int |f|^{2/3} \cdot |f|^{4/3}\, d\mu_0$$

$$\leq \left(\int |f| d\mu_0\right)^{2/3} \cdot \left(\int |f|^4 d\mu_0\right)^{1/3}.$$

One easily computes

$$\int |f|^2 d\mu_0 = \frac{1}{2^k} \sum_{\vec{a}} \left(\sum_{j=1}^{k} |a_j|^2\, |c_j|^2 + \sum_{j\neq i} a_i\, \bar{a}_j\, c_i\, \bar{c}_j\right)$$

$$= \|\vec{c}\,\|^2, \text{ as the } a_j\text{'s are independent}$$

and

$$\int |f|^4 d\mu_0 = \sum_j |c_j|^4 + 3 \sum_{i\neq j} |c_i|^2\, |c_j|^2 \leq 3\, \|\vec{c}\,\|^4$$

which by the preceding inequality gives the result. □

<u>Lemma</u> <u>6</u>. Let $\mu \in M(\mathbb{T})$, $k \geq 2$ and $\epsilon > 0$ be given. There exists integers $n_1, \ldots, n_k$ such that for all $\vec{a} \in \{-1, +1\}^k$, $d\nu = d\nu(\vec{a}) = \left(\sum_{j=1}^{k} a_j\, e^{in_j t}\right) d\mu$ satisfies

$$\|\nu\|_{PM} \leq \|\mu\|_{PM} + (k-1)\, R(\mu) + \epsilon.$$

<u>Proof</u>. Choose first N such that for $|j| \geq N$, $|\hat{\mu}(j)| \leq R(\mu) + \frac{\epsilon}{k-1}$ and then $n_1, \ldots, n_k$ such that $n_{j+1} - n_j > 2N$. For each $p \in \mathbb{Z}$, $\hat{\nu}(p) = \sum_{j=1}^{k} a_j\, \hat{\mu}(p - n_j)$, and by the choice of the $n_j$'s at most one of the $p - n_j$ is in $[-N, N]$, hence for all p $|\hat{\nu}(p)| \leq \|\mu\|_{PM} + (k-1)R(\mu) + \epsilon$ as desired. □

<u>End of proof of Theorem 4</u>: We assume that $\alpha(E) < \infty$. Let $\mu \in M(E)$, $\mu \neq 0$, and $\epsilon > 0$, $k \geq 2$ be given. Choose $n_1, \ldots, n_k$ as in Lemma 6, so that for $\vec{a} \in \{-1, +1\}^k$,

$$\|\nu(\vec{a})\|_{PM} \leq \|\mu\|_{PM} + (k-1)\, R(\mu) + \epsilon.$$

For each $\vec{a}$, $\|\nu(\vec{a})\|_M = \int \left|\Sigma a_j\, e^{in_jt}\right| d|\mu|(t)$, and

$$\int \|\nu(\vec{a})\|_M \, d\mu_0(\vec{a}) = \int \left( \int \left| \sum a_j e^{in_jt} \right| d\mu_0(\vec{a}) \right) d|\mu|(t).$$

Applying Lemma 5, for fixed t, to $\vec{c} = (e^{in_jt})_{j=1,\ldots,k}$ we get $\int |\Sigma a_j e^{in_jt}|\, d\mu_0(\vec{a}) \geq \frac{\sqrt{3}}{3} \|\vec{c}\| = \sqrt{\frac{k}{3}}$, hence $\int \|\nu(\vec{a})\|_M \, d\mu_0(\vec{a}) \geq \sqrt{\frac{k}{3}}\, \|\mu\|_M$. So we can pick $\vec{a}_0 \in \{-1, +1\}^k$ with $\nu = \nu(\vec{a}_0)$ satisfying $\|\nu\|_M \geq \sqrt{\frac{k}{3}}\, \|\mu\|_M$.

One has then, as $\nu$ is supported by E too, $\sqrt{\frac{k}{3}}\, \|\mu\|_M \leq \|\nu\|_M \leq \alpha(E) \cdot \|\nu\|_{PM} \leq \alpha(E) \cdot (\|\mu\|_{PM} + (k-1)\, R(\mu) + \epsilon)$ for all $\epsilon > 0$, hence as $\|\mu\|_{PM} \leq \|\mu\|_M$ we get finally

$$\frac{R(\mu)}{\|\mu\|_M} \geq \frac{\sqrt{\frac{k}{3}} - \alpha(E)}{(k-1) \cdot \alpha(E)}$$

which is $> 0$ for k big enough. □

Corollary 7. The set $\{E \in K(T) : E \text{ is Helson}\}$ is a $\underset{\sim}{\Sigma}^0_3$ (i.e. $G_{\delta\sigma}$) subset of K(T).

Proof. By the preceding result we have

$$E \text{ is Helson} \Leftrightarrow \exists \epsilon > 0\ \forall \mu \in M(E) \left[\tfrac{1}{2} < \|\mu\|_M \leq 1 \Rightarrow \forall m\ \exists |n| > m\ |\hat{\mu}(n)| > \epsilon\right]$$

and this easily implies that the class of Helson sets is $\underset{\sim}{\Sigma}^0_3$ in K(T). □

We state now the main result of this section.

__Theorem__ __8__ (Körner [1]). There exists a Helson set of multiplicity (in fact one with $\alpha(E) = 1$).

We will give a proof of this theorem (following the treatment in Graham-McGehee [1]) which is due to Kaufman, and uses the shrinking method. It gives moreover the following local version.

__Theorem__ __9__ (Kaufman [3]). Any closed set of multiplicity contains a Helson set of multiplicity (again with Helson contant 1).

The first reduction consists in replacing a given M-set E by a subset of it which is nowhere dense in $\mathbb{T}$ and is still an M-set. This can be easily done for example by applying 1.3 to E and $E_n = \{x_n\}$, where $\{x_n\}$ is dense in $\mathbb{T}$. Then one obtains some closed $E_1 \subseteq E - \{x_n : n \in N\}$ which is an M-set and clearly nowhere dense.

__So we assume from now on that__ E __is nowhere dense__. Then E (in the topology it inherits from $\mathbb{T}$) has a clopen basis, so the continuous functions on E with finite range are dense in C(E). Let $\mu \in M(E)$. Then for each $\epsilon > 0$ we can find such an f, say $f = \sum_{j=1}^{k} f_j \chi_{V_j}$, with $V_j$ disjoint clopen in E, and $|f_j| \leq 1$, such that $|\int f d\mu| \geq \|\mu\|_M - \epsilon$. Now $\sum_j |\mu(V_j)| \geq \Sigma |f_j| |\mu(V_j)| \geq |\Sigma f_j \mu(V_j)| \geq \|\mu\|_M - \epsilon$, hence if we let $w_j = \frac{|\mu(V_j)|}{\mu(V_j)}$ (with $w_j = 1$, if $\mu(V_j) = 0$), we get that $h = \Sigma w_j \chi_{V_j}$ is continuous on E, takes finitely many values $w_j$ of absolute value 1, and $\int h d\mu \geq \|\mu\|_M - \epsilon$. And without losing more than $\epsilon$, we can assume that $w_j = e^{ir_j}$ with

$r_j \in [0, 2\pi] \cap 2\pi Q$, and $V_j = E \cap I_j = E \cap \bar{I}_j$ for a finite family $I_j$ of rational (relative to $2\pi$) intervals with $\bar{I}_j \cap \bar{I}_{j'} = \emptyset$ for $j \neq j'$. Thus we can find a continuous function $g : \mathbb{T} \to [0, 2\pi]$ such that $g$ takes only finitely many values in a nbhd of $E$ and $\int e^{-ig(x)} d\mu \geq \|\mu\|_M - 2\epsilon$. As the number of functions $g$ constructed this way is countable we have finally the following

__Lemma__ __10__. Let $E \in K(\mathbb{T})$ be nowhere dense. There is a countable set $\{g_n\}_{n \in \mathbb{N}}$ of continuous functions, $g_n : \mathbb{T} \to [0, 2\pi]$ such that each $g_n$ takes only finitely many values in a nbhd of $E$, and such that for each $\mu \in M(E)$,

$$\|\mu\|_M = \sup_n \left|\int e^{-ig_n} d\mu\right|.$$

The idea is now to shrink, for each $g_n$ as above, the set $E$ to a subset $E'$ on which $e^{-ig_n(x)}$ is very close to an average $\frac{1}{N} \cdot \sum_{j=1}^{N} e^{-in_j x}$ of exponentials. The main step is given by the following key lemma.

__Lemma__ __11__. Let $S \in PF \cap PM(E)$, and $g$ be a real valued continuous function on $\mathbb{T}$ which takes only finitely many values in some nbhd of $E$. Fix $\epsilon > 0$. There exists a pseudofunction $S'$ with $\|S - S'\|_{PM} \leq \epsilon$, $\operatorname{supp}(S') \subseteq \operatorname{supp}(S)$, and integers $n_1, \ldots, n_N$ such that on $E' = \operatorname{supp}(S')$

$$\left|e^{-ig(x)} - \frac{1}{N} \sum_{j=1}^{N} e^{-in_j x}\right| \leq \epsilon.$$

Assuming this lemma we can give easily the

__Proof__ __of__ __Theorem__ __9__. As we said before, we may assume that the given M-set $E$ is nowhere dense. Let $S \in PF \cap PM(E)$, $S \neq 0$, and let $\{g_n\}$ be given by Lemma 10. We define a sequence $\{S_k\}$ of pseudofunctions with $S_0 = S$, and for $k \geq 1$ $\|S_k - S_{k-1}\|_{PM} \leq \epsilon \cdot 2^{-k}$, $\operatorname{supp}(S_k) = F_k \subseteq \operatorname{supp}(S_{k-1})$, and $\left| e^{-ig_{k-1}(x)} - \frac{1}{N(k)} \cdot \sum_{j=1}^{N(k)} e^{-in_j^k x} \right| \leq \epsilon \cdot 2^{-k}$ on $F_k$, using Lemma 11. Let $S' = \lim_k S_k$, which exists and is in PF. Then $F = \operatorname{supp}(S')$ is contained in $\bigcap_k F_k \subseteq E$, and $\|S' - S\|_{PM} \leq \epsilon$, so $S' \neq 0$ if $\epsilon$ is small enough. Thus F is an M-set. If now $\mu \in M(F)$ and $\eta > 0$ is given, one can find k with $\|\mu\|_M \leq |\int e^{-ig_k} d\mu| + \eta$, since $\mu \in M(E)$ as well. But as $F \subseteq F_k$, we have

$$\left|\frac{1}{N(k)} \sum_{j=1}^{N(k)} \hat{\mu}(n_j^k)\right| \geq \left|\int e^{-ig_k} d\mu\right| - \epsilon \cdot 2^{-k} \cdot \|\mu\|_M$$

and as we may assume $\epsilon \cdot 2^{-k} \cdot \|\mu\|_M \leq \eta$ for k big enough, we get for some $n_j^k$ $|\hat{\mu}(n_j^k)| \geq \|\mu\|_M - 2\eta$ and a fortiori $\|\mu\|_{PM} \geq \|\mu\|_M - 2\eta$. As $\eta$ was arbitrary, this shows that F is Helson with $\alpha(F) = 1$, as desired. □

It remains to prove Lemma 11. Below, if $x \in \mathbb{T}$ i.e. $x = e^{i\theta}$, we let

$$|x| = \operatorname{dist}(\theta, 2\pi\mathbb{Z})$$

<u>Lemma</u> <u>12</u>. Let $N \geq 1$, and $0 < \epsilon < 1$ be given. Consider in the space $\mathbb{T}^N$ the nbhd of $\vec{0}$,

$$V_{N,\epsilon} = \{\vec{x} : \operatorname{card}\{i : |x_i| < \epsilon\} \geq N(1 - \epsilon)\}.$$

Let $\lambda$ be the Lebesgue measure on $\mathbb{T}$, and for $0 \leq t \leq 1$ let $\sigma_t = t\lambda + (1 - t)\delta_0$ ($\delta_0$ the Dirac measure supported by 0) and $\mu_t^N = \sigma_t \times \ldots \times \sigma_t$ on $\mathbb{T}^N$. There exists a number $c(\epsilon) < 1$ such that for $0 \leq t \leq \epsilon$, $\mu_t^N(\mathbb{T}^N - V_{N,\epsilon}) \leq c(\epsilon)^N$. In particular $\mu_t^N(V_{N,\epsilon}) \to 1$ uniformly for $0 \leq t \leq \epsilon$ as $N \to \infty$.

<u>Proof</u>. Let $V_\epsilon = \{x \in \mathbb{T} : |x| < \epsilon\}$, and set $p = \sigma_t(V_\epsilon) = (1 - t) + t \cdot \frac{\epsilon}{\pi}$, $q = 1 - p$. Then $a = \mu_t^N(\mathbb{T}^N - V_{N,\epsilon}) = \sum_{s > \epsilon \cdot N} \sum_{\substack{S \subseteq [1,N] \\ \text{card } S = s}} \mu_t^N(\{\vec{x} : S = \{j : |x_j| \geq \epsilon\}),$ hence $a = \sum_{s > \epsilon \cdot N} \binom{N}{s} p^{N-s} q^s$. Now for $0 \leq p < 1$, the function $p^{1-\epsilon} \cdot (1 - p)^\epsilon$ is strictly increasing on $[0, 1 - \epsilon]$, and strictly decreasing on $[1 - \epsilon, 1]$, with maximum $(1 - \epsilon)^{1-\epsilon} \cdot \epsilon^\epsilon = p_1^{1-\epsilon} \cdot q_1^\epsilon$, where $p_1 = 1 - q_1 = 1 - \epsilon$. For $0 \leq t \leq \epsilon$, $p = p(t)$ ranges over $[1 - \epsilon + \frac{\epsilon^2}{\pi}, 1]$, hence $c(t) = \dfrac{p^{1-\epsilon} q^\epsilon}{p_1^{1-\epsilon}\, q_1^\epsilon} \leq c(\epsilon) < 1$. Now $a = \sum_{s > \epsilon \cdot N} \binom{N}{s} (p^{1-\epsilon} q^\epsilon)^N \cdot p^{\epsilon N - s} \cdot q^{s - \epsilon N}$, and as $s > \epsilon N$, $p_1 < p$ and $q_1 > q$, we have

$$a \leq \sum_{s > \epsilon \cdot N} \binom{N}{s} \left[p^{1-\epsilon} q^\epsilon\right]^N \cdot p_1^{\epsilon N - s} \cdot q_1^{s - \epsilon N}$$

$$\leq c(\epsilon)^N \sum_{s > \epsilon \cdot N} \binom{N}{s} p_1^{N-s} q_1^s \leq c(\epsilon)^N$$

and the lemma is proved. □

For $F$ a continuous function on $\mathbb{T}^N$, define for $\vec{m} \in \mathbb{Z}^N$ $\hat{F}(\vec{m}) = \int e^{-i\vec{m}\cdot\vec{x}}\, d\lambda_N(\vec{x})$, where $\vec{m} \cdot \vec{x} = \sum m_j x_j$ and $\lambda_N$ is Lebesgue measure on $\mathbb{T}^N$ (i.e. $\lambda_N = \lambda \times \ldots \times \lambda$ (N times)). Let $F \in A(\mathbb{T}^N)$ if $\hat{F}$ is in $\ell^1(\mathbb{Z}^N)$, in which case the series $\sum \hat{F}(\vec{m}) e^{i\vec{m}\cdot\vec{x}}$ absolutely converges to $F$. Also for $\mu \in M(\mathbb{T}^N)$ (= the space of measures on $\mathbb{T}^N$) let $\hat{\mu}(m) = \int e^{-i\vec{m}\cdot\vec{x}} d\mu(x)$.

The crucial technical lemma and the reason for going to higher dimensional

$A(\mathbf{T}^N)$, since it fails for $A(\mathbf{T})$, is now

<u>Lemma</u> <u>13</u>. Let $0 < \epsilon < 1$. There exists $N(\epsilon) \in \mathbb{N}$ so that for $N > N(\epsilon)$ there is a function $F = F_{N,\epsilon}$ in $A(\mathbf{T}^N)$ such that

(i) $\mathrm{supp}(F) \subseteq V_{N,\epsilon}$,

(ii) $\hat{F}(\vec{0}) = 1$, and $\forall \vec{m} \in \mathbf{Z}^N - \{\vec{0}\}\ (|\hat{F}(\vec{m})| < \epsilon)$.

<u>Proof</u>. It is enough to get F supported by $V_{N,\epsilon}$ with $|\hat{F}(\vec{m})| < \epsilon \cdot |\hat{F}(\vec{0})|$ for $\vec{m} \neq \vec{0}$ (for we can then normalize F). Suppose, towards a contradiction, that this fails. Therefore for infinitely many N's, every $F \in A(\mathbf{T}^N)$ supported by $V_{N,\epsilon}$ satisfies

$$|\hat{F}(\vec{0})| \leq \epsilon^{-1} \cdot \sup_{\vec{m} \neq \vec{0}} |\hat{F}(\vec{m})|.$$

Fix such an N. This gives a linear subspace of $c_0(\mathbf{Z}^N - \{\vec{0}\})$ on which $F \mapsto \hat{F}(0)$ is continuous, with norm $\leq \epsilon^{-1}$. By Hahn-Banach, it can be extended to a continuous linear functional with the same norm on $c_0(\mathbf{Z}^N - \{\vec{0}\})$. So there is some $\{g_{\vec{m}}\}_{\vec{m} \neq 0}$ in $\ell^1(\mathbf{Z}^N - \{\vec{0}\})$ with $\sum_{\vec{m} \neq 0} |g_{\vec{m}}| \leq \epsilon^{-1}$ and $\hat{F}(\vec{0}) = \sum_{\vec{m} \neq 0} g_{\vec{m}} \hat{F}(\vec{m})$ for each $F \in A(\mathbf{T}^N)$ supported by $V_{N,\epsilon}$. Now for each $\vec{x} \in V_{N,\epsilon}$, $\delta_{\vec{x}}$ (the Dirac measure on $\vec{x}$) is a w*-limit of such F's (in the space $M(\mathbf{T}^n) = C(\mathbf{T}^n)^*$ of measures on $\mathbf{T}^n$), hence $1 = \hat{\delta_{\vec{x}}}(\vec{0}) = \sum_{\vec{m} \neq 0} g_{\vec{m}} \cdot e^{-i\vec{m} \cdot \vec{x}}$ for $\vec{x} \in V_{N,\epsilon}$. Integrating with respect to $\mu_t^N$, $0 \leq t \leq \epsilon$, gives $\left|\mu_t^N(V_{N,\epsilon}) - \sum_{\vec{m} \neq 0} g_{\vec{m}} \cdot \widehat{\mu_t^N}(\vec{m})\right| \leq \sum_{\vec{m} \neq 0} |g_{\vec{m}}| \cdot \mu_t^N(\mathbf{T}^N - V_{N,\epsilon})$, and by Lemma 12, $\left|\mu_t^N(V_{N,\epsilon}) - \sum_{\vec{m} \neq 0} g_{\vec{m}} \cdot \widehat{\mu_t^N}(m)\right| \leq \epsilon^{-1} \cdot c(\epsilon)^N$. Now $\widehat{\mu_t^N}(\vec{m}) = \prod_{j=1}^{N} \int e^{-im_jx_j} d\sigma_t(x_j) = \prod_{j=1}^{N} \widehat{\sigma}_t(m_j)$, and

$$\widehat{\sigma_t}(m_j) = \begin{cases} 1, & \text{if } m_j = 0, \\ 1 - t, & \text{if } m_j \neq 0 \end{cases}$$

so that letting $k(\vec{m}) = \text{card}\{j : m_j \neq 0\}$ (thus $k(\vec{m}) \neq 0$ for $\vec{m} \neq 0$), we have

$$\begin{aligned} \sum_{\vec{m} \neq 0} g_{\vec{m}} \cdot \widehat{\mu_t^N}(\vec{m}) &= \sum_{\vec{m} \neq 0} g_{\vec{m}} \cdot (1 - t)^{k(\vec{m})} \\ &= \sum_{k=1}^{N} \Big( \sum_{k(\vec{m})=k} g_{\vec{m}} \cdot (1 - t)^k \Big) \\ &= \sum_{k=1}^{N} c(N, k) \cdot (1 - t)^k \end{aligned}$$

with $\sum_{k=1}^{N} |c(N, k)| \leq \epsilon^{-1}$.

Consider now, for $|z| \leq 1$, the polynomials $f_N(z) = \sum_{k=1}^{N} c(N,k)z^k$, where N varies among the counterexamples to the lemma. They are uniformly bounded by $\epsilon^{-1}$, hence form a normal family and so a subsequence of them converges to some function f, analytic in $|z| < 1$. As $f_N(0) = 0$ for all N, $f(0) = 0$. But for $0 \leq t \leq \epsilon$, $f_N(1 - t)$ converges uniformly to $\lim \mu_t^N(V_{N,\epsilon}) = 1$ by Lemma 12. This contradicts the analyticity of f and proves the lemma. □

We finally give the

Proof of Lemma 11. Let $S \in PF$, $g : \mathbb{T} \to \mathbb{R}$ continuous, taking only finitely many values in a nbhd V of $E = \text{supp}(S)$, where $V = \cup_{j=1}^{p} I_j$, $I_j$ open intervals with $\overline{I_j} \cap \overline{I_{j'}} = \emptyset$ if $j \neq j'$, and $\epsilon > 0$. First we claim that there is a $K \in \mathbb{N}$, and for each $n \in \mathbb{Z}$ a function $\varphi_n \in A$ such that $\varphi_n = e^{-ing}$ on V and $\|\varphi_n\|_A \leq K$. This is because we can choose $\varphi^j \in A$ with $\varphi^j = \chi_{I_j}$ on V, hence if $g = g_j =$ constant on

$I_j$, we have $e^{-ing} = \sum_j e^{-ing_j} \cdot \varphi^j$ on V, and $\varphi_n = \sum_j e^{-ing_j} \cdot \varphi^j$ satisfies $\|\varphi_n\|_A \le \sum_j \|\varphi^j\|_A = K$, independently of n. Note that we can take $\varphi_0 = 1$.

Let now $N = N(\epsilon) + 1$ and $F = F_{N,\epsilon}$ be given by Lemma 13. Fix some $\vec{n} = (n_j)_{1 \le j \le N}$ in $\mathbf{Z}^N$. Consider

$$H(x) = F(n_1x - g(x), \ldots, n_Nx - g(x))$$

and

$$\tilde{H}(x) = \sum_{\vec{m} \in \mathbf{Z}^N} \hat{F}(\vec{m}) \cdot \varphi_{\Sigma m_j}(x) \cdot e^{i(\vec{m}\cdot\vec{n})x}.$$

Since F is the sum of its Fourier series, one easily checks that for $x \in V$, $H(x) = \tilde{H}(x)$. Notice now that $\tilde{H} \in A$. This is because

$$\sum_{\vec{m} \in \mathbf{Z}^N} \|\hat{F}(m) \cdot \varphi_{\Sigma m_j}(x) \cdot e^{i(\vec{m}\cdot\vec{n})x}\|_A \le K \cdot \sum_{\vec{m} \in \mathbf{Z}^N} |\hat{F}(m)| < \infty.$$

Consider then $S' = \tilde{H} \cdot S$. Clearly $S'$ is in PF and has support contained in E. Also if $x \in \mathrm{supp}(S')$, $x \in \mathrm{supp}(\tilde{H}) \cap E$, hence $x \in \mathrm{supp}(\tilde{H}) \cap V = \mathrm{supp}(H) \cap V$. So $(n_jx - g(x))_{1 \le j \le N}$ is in $\mathrm{supp}(F) \subseteq V_{N,\epsilon}$. This implies that for at least $(1 - \epsilon)N$ values of j, $|n_j x - g(x)| \le \epsilon$, and hence

$$\left|e^{-ig(x)} - \frac{1}{N}\sum_j e^{-in_jx}\right| \le 2 \cdot \frac{N\epsilon}{N} + \epsilon \cdot \frac{N(1-\epsilon)}{N} < 3\epsilon$$

on $\mathrm{supp}(S')$. This was independent of the choice of $\vec{n}$. It remains to choose $\vec{n}$ so that $\|S - S'\|_{PM}$ is small. Note that for all $\vec{n}$, since $\hat{F}(0) = 1$ and $\varphi_0 = 1$, $S' - S = \tilde{H} \cdot S - S = (\sum_{\vec{m} \ne 0} \hat{F}(\vec{m}) \cdot \varphi_{\Sigma m_j}(x) \cdot e^{i(\vec{m}\cdot\vec{n})x}) \cdot S$. First choose a finite set

$L = \{\vec{m}_1, \ldots, \vec{m}_\ell\} \subseteq Z^N - \{0\}$ such that $\sum_{\substack{\vec{m} \neq 0 \\ \vec{m} \notin L}} |\hat{F}(\vec{m})| < \epsilon$. For $k = 1, \ldots, \ell$, let $S_k = \varphi_{\sum_j (\vec{m}_k)_j} \cdot S$. Each $S_k$ is in PF, and $\|S_k\|_{PM} \leq K \cdot \|S\|_{PM}$. Consider, for $k = 1, \ldots, \ell$, the set $Z_k(\vec{n}) = \{p \in \mathbf{Z} : |S_k(p - \vec{m}_k \cdot \vec{n})| > 1/\ell\}$. Each of these sets is finite, and we claim thaf for some choice of $\vec{n}$ they are disjoint: it is enough to choose $\vec{n}$ outside the union of the (finitely many) hyperplanes $(\vec{m}_{k_1} - \vec{m}_{k_2}) \cdot \vec{x} = a_1 - a_2$, with $k_1 \neq k_2$ and $a_1, a_2$ varying in $\left\{a : |S_{k_1}(a)| > \frac{1}{\ell}\right\}$ and $\left\{a : |S_{k_2}(a)| > \frac{1}{\ell}\right\}$ respectively. We fix now $\vec{n}$ with this property. We get:

$$S' - S = \Big[\sum_{\vec{m} \neq 0} \hat{F}(\vec{m}) \cdot \varphi_{\Sigma m_j}(x) \cdot e^{i(\vec{m} \cdot \vec{n})x}\Big] \cdot S$$

$$= \Big[\sum_{\substack{\vec{m} \neq 0 \\ \vec{m} \notin L}} \hat{F}(\vec{m}) \cdot \varphi_{\Sigma m_j}(x) \cdot e^{i(\vec{m} \cdot \vec{n})x}\Big] \cdot S \tag{1}$$

$$+ \sum_{k=1}^{\ell} \hat{F}(\vec{m}_k) \cdot (e^{i(\vec{m}_k \cdot \vec{n})x} \cdot S_k) \tag{2}$$

Expression (1) is easily bounded in the PM-norm by $K \cdot \|S\|_{PM} \cdot \sum_{\substack{\vec{m} \neq 0 \\ \vec{m} \notin L}} |\hat{F}(m)| \leq \epsilon \cdot K \cdot \|S\|_{PM}$ by the choice of L. Now $\sum_{k=1}^{\ell} \hat{F}(\vec{m}_k) \cdot (e^{i(\vec{m}_k \cdot \vec{n})x} \cdot S_k)(p) = \sum_{k=1}^{\ell} \hat{F}(\vec{m}_k) \cdot S_k(p - \vec{m}_k \cdot \vec{n})$, hence $\left|\sum_{k=1}^{\ell} \hat{F}(\vec{m}_k) \cdot S_k(p - \vec{m}_k \cdot \vec{n})\right| \leq \epsilon \cdot \sum_{k=1}^{\ell} \left|S_k(p - \vec{m}_k \cdot \vec{n})\right|$ (as $|\hat{F}(\vec{m})| < \epsilon$ for $\vec{m} \neq 0$). Now by our choice of $\vec{n}$, there is for each p at most one value k for which $|S_k(p - \vec{m}_k \cdot \vec{n})| > \frac{1}{\ell}$, and for that value it is in any case bounded by $K \cdot \|S\|_{PM}$. This gives that expression (2) is bounded in the PM-norm by $\epsilon \cdot (K\|S\|_{PM} + 1)$, and finally $\|S' - S\|_{PM} \leq$

$\epsilon \cdot (2K\|S\|_{PM} + 1)$. As $\epsilon$ was arbitrary, we are done. □

<u>Remark</u>. If $\varphi \in A$ is such that $\hat{\varphi}(0) = 1$, $|\hat{\varphi}(m)| \leq \epsilon$ for $m \neq 0$, then by applying Lemma 1.6 to $\varphi - 1$ we obtain $\|\varphi_n \cdot S - S\|_{PM} \to \|\widehat{(\varphi - 1)}\|_{PM} \cdot \|S\|_{PM} \leq \epsilon \cdot \|S\|_{PM}$. (We made use of this in the proof of 1.7). Thus for large enough n, $\|\varphi_n \cdot S - S\|_{PM} \leq 2\epsilon \cdot \|S\|_{PM}$. The last part of the proof of Lemma 11 is a multidimensional version of this fact: We started with $F \in A(T^N)$ with $\hat{F}(\vec{0}) = 1$, $|\hat{F}(\vec{m})| \leq \epsilon$ for $\vec{m} \neq 0$, and found $\vec{n} \in Z^N$ such that if $F_{\vec{n}}(x) = F(n_1x, \ldots, n_Nx)$ then $\|F_{\vec{n}} \cdot S - S\|_{PM}$ is bounded by a fixed multiple of $\epsilon$. Actually things were a bit more complicated because instead of $F_{\vec{n}}(x)$ we had to look at $F(n_1x - g(x), \ldots, n_Nx - g(x))$, but this was a minor annoyance.

Now that by Theorems 4 and 9 we know that for each $E \in M$, there is a subset of E in $U_1'$ but not in U, we can go back to the discussion in VI.3, and get by VI.3.6 the following

<u>Corollary 14</u>. The class $U_1$ is not a $\sigma$-ideal.

<u>Corollary 15</u> (Malliavin [1]). Every closed set of multiplicity contains a set which is not of synthesis.

<u>Corollary 16</u> (Piatetski-Shapiro [2]). $U_1 \subsetneqq U_0$.

Thus we have $U \subsetneqq U_1 \subsetneqq U_0$. (Actually Piatetski-Shapiro shows even that $(U_1)_\sigma = (U_1')_\sigma \subsetneqq U_0$—see Graham-McGehee [1], p. 104). Although we know that $U_1' \subsetneqq U$, this does not rule out the possibility that $U_1' \cap U$ is contained in some

Borel subset of U. However this is not the case, in view of the following result (which is a special case of 4.2 which was proved earlier than this general theorem).

Corollary 17 (Kechris-Louveau). The Piatetski-Shapiro rank $[E]_{PS}$ is unbounded on $U'_1 \cap U$, in fact on $(U'_1 \cap U) \cap K(E)$ for every $E \notin U$.

Proof. Let $E \notin U$, and find $E' \subseteq E$ with $E' \in U'_1 - U$ given by Theorem 9. Clearly $U \cap K(E') \subseteq U'_1$, since $U'_1$ is hereditary. And by 2.2, the Piatetski-Shapiro rank is unbounded on $U \cap K(E')$. □

Another consequence of Theorem 9 deals with the relationship between the ranks $rk_1(E)$ and $[E]^1_{PS}$. We proved Piatetski-Shapiro's Theorem VI.2.3(ii) by proving that for every E $rk_1(E) \leq [E]^1_{PS}$, using in turn the inclusion

$$I^{\alpha}(E) \subseteq I(E^{(\alpha)}_{U'_1}).$$

We show now that these inclusions and inequalities cannot be reversed, in a strong sense.

Proposition 18. There exists a closed set $E \subseteq \mathbf{T}$ with $rk_1(E) = 2$ but $E \notin U_1$, and also for any countable $\alpha$, $I(E^{(1)}_{U'_1}) \not\subseteq I^{\alpha}(E)$.

Proof. Let $E_0$ be a Helson set of multiplicity, that we may assume is of pure multiplicity by restricting it to the support of a non-zero pseudofunction. Let $E = E_0^h$ be the Herz transform of $E_0$ (see VI.3). We claim first that $E^{(1)}_{U'_1} = E_0$. As any $x \in E - E_0$ is isolated in E, $E^{(1)}_{U'_1} \subseteq E_0$. But if V is open and $V \cap E_0 \neq \emptyset$,

$V \cap E_0$ contains $I \cap E_0 \neq \emptyset$ for some open dyadic interval I, and by the definition of the Herz transform, $(E_0 \cap \bar{I})^h = E_0^h \cap \bar{I}$. But $E_0 \cap \bar{I}$ is in M and $(E_0 \cap \bar{I})^h$ is of synthesis, hence is not in $U_1$. So $E_0 \subseteq E^{(1)}_{U'_1}$, and thus $E_0 = E^{(1)}_{U'_1}$. Now $E_0$ is Helson, hence in $U'_1$, and therefore $rk_1(E) = 2$. Moreover since E is of synthesis but not in U, $E \notin U_1$. Finally if for some $\alpha$ $I(E^{(1)}_{U'_1}) = I(E_0) \subseteq I^{\alpha}(E)$, one would get $I^1(E_0) = A \subseteq I^{\alpha+1}(E)$, i.e. $E \in U_1$, a contradiction. □

## §4. The solution to the Borel Basis Problem

We have already seen in §1 that U is a calibrated $\sigma$-ideal, and in §2 that for each $E \notin U$, $U \cap K(E)$ is $\underset{\sim}{\Pi}^1_1$-complete. So in order to apply the results in VI.1, it is enough, using also the theorem of Körner and Kaufman, to prove the following

<u>Lemma 1</u> (Debs-Saint Raymond [1]). Let E be a closed subset of $\mathbb{T}$ which is in $U'_1$. Then there exists a $G_\delta$ set G dense in E such that $K(G) \subseteq U'$, i.e. all closed subsets of G are in $U'$.

<u>Proof</u>. By the hypothesis there is a sequence $f_n \in I(E)$ w*-converging to 1. Let $\{x_p\}_{p\in\mathbb{N}}$ be a dense countable set in E. Fix $n \in \mathbb{N}$, and define by induction on p a sequence $f_{n,p} \in I(E)$ and open sets $V_{n,p}$ with $x_p \in V_{n,p}$ such that

(i) $f_{n,p}$ is 0 on $\bigcup_{q\leq p} V_{n,q}$,

(ii) $\|f_{n,p} - f_n\|_A \leq 2^{-n}$,

(iii) $\|f_{n,p} - f_{n,p-1}\|_A \leq 2^{-p}$.

Let $f_{n,0} = f_n$. Suppose $f_{n,p}$ has been defined satisfying (i), (ii), (iii). As $f_{n,p}\,(x_{p+1})$

$= 0$, we know that we can find for any $\epsilon$ a function $\varphi \in A$ vanishing in a neighborhood of $x_{p+1}$, say $V^{\epsilon}_{n,p+1}$, such that $\|f_{n,p} - \varphi f_{n,p}\|_A < \epsilon$ (II.2.2). By choosing $\epsilon$ small enough, $f_{n,p+1} = \varphi \cdot f_{n,p}$ and $V_{n,p+1} = V^{\epsilon}_{n,p+1}$ will clearly satisfy (i), (ii) and (iii).

Let now $V_n = \bigcup_p V_{n,p}$, and $h_n = \lim_p f_{n,p}$. Clearly $h_n = 0$ on $V_n$, and $\|h_n - f_n\|_A \leq 2^{-n}$, so that $h_n \xrightarrow{w^*} 1$. Let $G = \bigcap_n (E \cap V_n)$. Then G is a $G_\delta$ subset of E which is dense in E, as $x_p \in G$ for all p. Moreover if $F \in K(G)$, then for all n, $h_n \in J(F)$ as $h_n$ is 0 on $V_n \supseteq F$. So $F \in U'$ and we are done. □

It is an interesting open problem to find out if the set G in Lemma 1 is in $\mathcal{U}$ or in $\mathcal{M}$.

<u>Theorem</u> <u>2</u> (Debs-Saint Raymond [1]). Let E be a closed set of multiplicity. Then there is a $G_\delta$ set $G \in U^{int}$, $G \subseteq E$, which is not contained in the union of countably many closed sets of uniqueness. As a consequence, for every M-set E, $U \cap K(E)$ has no Borel basis.

In particular, (within any M-set), there are U-sets which are not countable unions of sets in $\bigcup_n H^{(n)}$, U′, etc.

<u>Proof</u>. By Kaufman's Theorem 3.9, any given $E \in M$ contains a closed subset E′ in $M \cap U'_1$, and we may of course assume $E' \in M^p \cap U'_1$. Let G be the $G_\delta$ dense subset of E′ given by Lemma 1. Then G is in $U^{int}$ (in fact in $U'^{int}$). But as $E' \in M^p$, every closed subset of E′ in U is nowhere dense in E′, hence G cannot be covered by a sequence of sets in U (by the Baire Category Theorem).

We can apply now Theorem VI.1.11. By 1.3, 2.2 and the fact just proved $U \cap K(E)$ has no Borel basis. □

<u>Remark</u>. One can give also a version of the preceding proof of the non-basis theorem which is based directly on VI.1.8 instead of VI.1.11: Given $E \in M^p \cap U'_1$ and a sequence $f_n \in I(E)$ $w^*$-converging to 1 apply VI.1.8 to E, $I = U \cap K(E)$, $D = E$ and $J_n$ defined by

$$F \in J_n \Leftrightarrow \exists f \in J(F) \cap I(E) \, (\|f - f_n\|_A < \tfrac{1}{n}).$$

Clearly $J_n$ is nonempty open, hereditary. If $F \in J_n$ and $x \in E$, let $f \in J(F) \cap I(E)$ be such that $\|f - f_n\|_A < \frac{1}{n}$. Since $f(x) = 0$, let (by II.2.2) $\varphi \in A$ vanish in a nbhd of x and be such that $\|f - f\varphi\|_A < \epsilon$, where $\|f - f_n\|_A + \epsilon < \frac{1}{n}$. Then $f\varphi$ witnesses that $F \cup \{x\} \in J_n$. Since $U \cap K(V)$ is true $\underset{\sim}{\Pi}^1_1$ for all nonempty open V in E and every $F \in \cap_n J_n$ is clearly in U′, using the calibration of U we conclude that $U \cap K(E)$ has no basis.

It is an open problem whether one can find a $G_\delta$ (or Borel) set of uniqueness G which cannot be covered by countable many closed sets of uniqueness.

The preceding result is clearly relevant to the Characterization Problem: Not only there is no definably simple characterization of the U-sets, but one cannot hope to find a simply describable family of U-sets (like the $\cup_n H^{(n)}$, U′ etc. sets) so that every U-set can be decomposed into a countable union of sets in the family. Apparently it was unknown before even whether every U-set is the union of countably many sets in $\cup_n H^{(n)}$—see Lyons [4]—a problem raised in Piatetski-Shapiro

[1]. In fact in [4] Lyons "catalogs" the "known" (then) U-sets and notices that it is open whether all U-sets are among them. One can check (see also Chapter X.1) that all these previously "known" U-sets are countable unions of U′-sets, so that the preceding theorem implies the existence of many "new" U-sets.

In fact by tracing back the various proofs one can easily get constructions of such U-sets: Suppose $B \subseteq U$ is some Borel family of U-sets and we want to construct a set $F \in U - B_\sigma$, within some given $E \in M$. First find $\alpha < \omega_1$ with $B \subseteq U^{(\alpha)}$, by boundedness. For most specific B, like $\cup_n H^{(n)}$, this $\alpha$ would be actually 1. Then apply Kaufman's method of 3.9 to construct $\emptyset \neq E' \subseteq E$ in $U'_1 \cap M^p$, and after that the Debs-Saint Raymond construction to get G and the open sets $V_n$ of Theorem 1. Next use the construction of 2.2 to produce, for each open V with $E' \cap V \neq \emptyset$, a set $E_V \in K(E' \cap V) \cap (U - U^{(\alpha)})$. (If $\alpha = 1$ this would come easily from Kaufman's construction in 2.1). Finally apply the Kechris-Louveau-Woodin construction of Lemma VI.1.7, using the $E_V$'s, $J_n = K(V_n)$ and G to produce a set F which by calibration (1.3) is in U, but is not in $(U^{(\alpha)})_\sigma$, thus not in $B_\sigma$.

It is perhaps worth pointing out that as opposed to most constructions of closed sets in harmonic analysis, the above construction builds closed sets as appropriate unions of $G_\delta$ and $F_\sigma$ sets. This approach appears to be novel in this area.

<u>Remark</u>. As we noticed after VI.1.7, the construction there does not rely on definability properties of a basis B, but on the fact that $(B_\sigma - B) \cap K(V) \neq \emptyset$ for all nonempty open in E sets V. This can be used to prove the following fact: Let

B be a hereditary basis for U. Then there exists a hereditary basis $B^*$ of the $\sigma$-ideal $(U_1')_\sigma$ such that $B = U \cap B^*$. In other words, the basis theorem of Piatetski-Shapiro, that $U_1' \cap U$ is a basis for U is in some sense optimal, since the only possible improvement would be by getting a better basis for $(U_1')_\sigma$, not for U.

To prove the above claim, define $B^*$ as

$$B^* = \{E \in (U_1')_\sigma\colon\ K(E) \cap U = K(E) \cap B\}.$$

Clearly $B \subseteq B^*$, and $B^* \cap U \subseteq B$, thus $B = B^* \cap U$. So it is enough to show that $U_1' \subseteq B^*_\sigma$. If not, one gets an $E \in U_1'$, $E \neq \emptyset$ which is not locally in $B^*_\sigma$, hence in particular is in $M^p$ and moreover for each open V with $E \cap V \neq \emptyset$, $K(E \cap V) \cap (U - B) \neq \emptyset$. The preceding outline of constructions gives then, as we can apply Lemma VI.1.7, a closed set in $U - B_\sigma$ (inside E), contradicting the hypothesis that B was a basis for U.

Remark. Let $E^{(\alpha)} = E^{(\alpha)}_{U'}$ be the derived sets corresponding to the ideal U′ and the associated Cantor-Bendixson rank which we will denote by rk(E). Recall that we have (VI.2.3(i))

$$rk_1(E) \leq [E]^1_{PS}$$

but $rk_1(E) \geq [E]^1_{PS}$ fails. On the contrary we have

$$(*) \quad rk(E) \geq [E]_{PS}$$

but $rk(E) \leq [E]_{PS}$ fails. (Otherwise U′ would be a basis for U). To prove (*), note that it is enough to verify that

$$J(E^{(\alpha)}) \subseteq J^{\alpha}(E).$$

Again, by an easy induction, it is enough to prove the case $\alpha = 1$, i.e.

$$J(E^{(1)}) \subseteq J^{1}(E).$$

So let $f \in J(E^{(1)})$, i.e. $f = 0$ on $V$, a nbhd of $E^{(1)}$. If $x \in E - E^{(1)}$, let $V_x$ be a nbhd of $x$ for which $\overline{E \cap V_x} \in U'$. Since $E \subseteq V \cup \bigcup_{x \in E - E^{(1)}} V_x$, find $x_1, \ldots, x_n \in E - E^{(1)}$ so that $E \subseteq V \cup V_{x_1} \cup \ldots \cup V_{x_n}$. Then, since $\overline{V_{x_i} \cap E} \in U'$, find $f_j^{(x_i)}$ such that $f_j^{(x_i)} = 0$ on a nbhd $V^j_{x_i}$ of $\overline{V_{x_i} \cap E}$ and $f_j^{(x_i)} \xrightarrow{w^*} 1$ as $j \to \infty$. Then, as in V.4.7, we can find $g_j \xrightarrow{w^*} f$, where each $g_j$ is of the form $f \cdot f_{j_1}^{(x_1)} \ldots f_{j_n}^{(x_n)}$, thus $g_j$ vanishes on $V \cup V^{j_1}_{x_1} \cup \ldots \cup V^{j_n}_{x_n} \supseteq E$, so that $f \in J^1(E)$.

We conclude with an open problem: Given a $\underset{\sim}{\Pi}^1_1$ $\sigma$-ideal I in K(E), where E is compact, metrizable, we say that I has a <u>Borel</u> <u>co-basis</u> if there is Borel $B \subseteq K(E) - I$ such that for every $F \notin I$ there is $K \in B$ with $K \subseteq F$. If I has a Borel basis, then it has a Borel co-basis, namely $K(E) - I^{loc}$ (by VI.1.3). The converse is however false. So this leads to the following question: Does U have a Borel co-basis? A negative answer would rule out a further potential type of characterization for U.

# Chapter VIII. Extended Uniqueness Sets

In this chapter we study mainly the $\sigma$-ideal of closed sets of extended uniqueness $U_0$. In §1 we review the basic properties of Rajchman measures and $U_0$-sets, and introduce a simple (in particular Borel) class of $U_0$-sets, denoted by $U_0'$, which is analogous to the class $U'$ of U-sets of rank 1. In §2 the result of Kechris-Louveau is proved that $U_0'$ is a basis for $U_0$. Thus $U_0$ has a Borel basis and in conjunction with the Debs-Saint Raymond Theorem that U has no Borel basis, this establishes a structural difference between the concepts of uniqueness and extended uniqueness sets. We study also in this section the classification of $U_0$-sets in a transfinite hierarchy according to the associated with $U_0'$ Cantor-Bendixson derivation. In §3 the basis theorem of §2 is combined with earlier results, particularly in VI.1 and VII.2, to give the solution to the Category Problem, due to Debs-Saint Raymond, and several other applications. We also develop in this section an alternative direct method of proof of this result of Debs-Saint Raymond, which we then apply in §4 to the study of the class $U_1$ and its relation with U. This brings forward some connections between the Union and Interior Problems and the non-synthesis phenomena.

## §1. The class $U_0'$

In Chapters II and IV, we defined the class $U_0$ of closed sets of extended uniqueness as consisting of all closed $E \subseteq \mathbb{T}$ for which no non-zero measure $\mu$ with support contained in E is a pseudofunction.

Let M(E) be the space of measures supported by a closed set E. Recall that for $\mu \in M(\mathbb{T})$, we say that $\mu$ is a Rajchman measure if $\hat{\mu}(n) \to 0$, as $|n| \to \infty$, i.e. $\hat{\mu} \in PF$. So E is a closed set of extended uniqueness if M(E) contains no non-zero Rajchman measures.

We will review now some facts about Rajchman measures and sets of extended uniqueness.

If $\mu \in M(\mathbb{T})$, the total variation $|\mu|$ of $\mu$ is the positive measure on $\mathbb{T}$ defined by

$$|\mu|(P) = \sup\left\{\sum |\mu(P_n)| : \{P_n\} \text{ a Borel partition of } P\right\}$$

for Borel sets $P \subseteq \mathbb{T}$. A Borel set $P \subseteq \mathbb{T}$ is $\mu$-null if $|\mu|(P) = 0$, or equivalently $\mu(Q) = 0$ for all Borel $Q \subseteq P$. The class of closed $\mu$-null sets will be denoted by Null($\mu$).

A measure $\nu$ on $\mathbb{T}$ is absolutely continuous with respect to another measure $\mu$, in symbols $\nu \prec\prec \mu$, if every $\mu$-null Borel set is $\nu$-null, or equivalently by the Radon-Nikodym Theorem

$$d\nu = f d|\mu|, \text{ for some } f \in L^1(|\mu|).$$

The following is the same as II.5.4.

Proposition 1. Let $\mu \in M(\mathbb{T})$ be a Rajchman measure and let $\nu \ll \mu$. Then $\nu$ is a Rajchman measure.

Also II.5.6 states (among other things),

Proposition 2. Let $E \in K(\mathbb{T})$. Then the following are equivalent definitions of $E$ being in $U_0$:

(i) E supports no non-0 Rajchman measure,

(ii) E supports no non-0 positive Rajchman measure,

(iii) For all Rajchman measures $\mu$, $E \in \mathrm{Null}(\mu)$,

(iv) For all positive Rajchman measures $\mu$, $E \in \mathrm{Null}(\mu)$.

Recall also that we defined (in II.5) the class $\mathcal{U}_0$ as consisting of those sets $A \subseteq \mathbb{T}$ for which $\mu(A) = 0$ for all positive Rajchman measures. Then $U_0 = \mathcal{U}_0 \cap K(\mathbb{T})$. We have the following easy facts: The class $U_0$ is a $\sigma$-ideal of closed sets. The class $\mathcal{U}_0$ is a $\sigma$-ideal. Also for universally measurable sets (in particularly Borel, $\Sigma^1_1$, etc.) P, $P \in U_0^{int} \Leftrightarrow P \in \mathcal{U}_0$. Finally the $\sigma$-ideal $U_0$ is calibrated (according to the definition in VI.1).

We also define (as for U) the classes

$M_0 = K(\mathbb{T}) - U_0$ = class of closed sets of restricted multiplicity,

$M_0^p = \{E \in K(\mathbb{T}) : \text{For every open } V, E \cap V \neq \emptyset \Rightarrow \overline{E \cap V} \notin U_0\}$

$= $ class of closed sets of <u>pure restricted multiplicity,</u>

$U_0^{loc} = K(\mathbb{T}) - M_0^p$

$= $ class of closed sets of <u>local extended uniqueness</u>.

<u>Proposition 3</u>. Let $E \in K(\mathbb{T})$. Then $E \in M_0^p \Leftrightarrow E$ is the support of a (positive) Rajchman measure.

<u>Proof</u>. The empty set $=$ supp(0) is in $M_0^p$ by definition. If $\mu$ is a non-zero Rajchman measure and $E = \mathrm{supp}(\mu)$, let $V$ be open with $V \cap E \neq \emptyset$. Let $\varphi \in C^\infty(\mathbb{T})$ be such that $\mathrm{supp}(\varphi) \subseteq V$ and $\varphi \cdot \hat{\mu} \neq 0$. Then $\varphi \cdot \hat{\mu} = \hat{\nu}$, where $d\nu = \varphi d\mu$, so $\nu$ is a non-zero Rajchman measure with support contained in $V \cap E$, thus $\overline{V \cap E} \in M_0$ and $E \in M_0^p$. Conversely let $E \in M_0^p$ and let $\{V_n\}$ be an open basis. For each $n$ with $E \cap V_n \neq \emptyset$, let $\mu_n$ be a probability Rajchman measure with support contained in $\overline{E \cap V_n}$. Let $\mu = \sum_n 2^{-n} \mu_n$. Then $\mu$ is a positive Rajchman measure with $\mathrm{supp}(\mu) = E$. □

By VII.1.7 every $M_0$-set contains a continuum of pairwise disjoint $M_0$-sets. Thus the $U_0$-thin sets are all in $U_0$.

From the Solovay and Kaufman results of IV.2, we know that $U_0$ is $\underset{\sim}{\Pi}_1^1$-complete, and in fact no $\underset{\sim}{\Sigma}_1^1$ set separates $H_\sigma$ (and a fortiori $U$) from $M_0$. Also by VII.3.16, the inclusion $U \subseteq U_0$ is proper.

The fact (see Proposition 2) that we can restrict our attention to positive measures in studying $U_0$ will be important in the sequel. Let us note some simple properties of positive measures: If $\mu \in M(\mathbb{T})$ then clearly $\|\mu\|_{PM} = \|\hat{\mu}\|_{PM} \leq$

$\|\mu\|_M$, but in general $\|\mu\|_{PM}$ can be much smaller than $\|\mu\|_M$. Also the weak*-topology of $M(\mathbb{T}) = C(\mathbb{T})^*$ is quite different from the one that $M(\mathbb{T})$ inherits as a subset of $PM = A^*$ and its weak*-topology (when $M(\mathbb{T})$ is embedded in PM via $\mu \mapsto \hat{\mu}$). For instance, $\mu_n \xrightarrow{w^*} \mu$ in $M(\mathbb{T})$ means ($\hat{\mu}_n(k) \to \hat{\mu}(k)$ and $\sup\|\mu_n\|_M < \infty$), while $\mu_n \xrightarrow{w^*} \mu$ in PM means ($\hat{\mu}_n(k) \to \hat{\mu}(k)$ and $\sup\|\mu_n\|_{PM} < \infty$). On the other hand, if

$$M^+(\mathbb{T})$$

denotes the class of positive measures on $\mathbb{T}$, then for $\mu \in M^+(\mathbb{T})$

$$\|\mu\|_M = \mu(\mathbb{T}) = \hat{\mu}(0) = \|\mu\|_{PM}$$

and also the weak*-topologies above coincide on $M^+(\mathbb{T}) \cap B_a(PM)$, for each $a > 0$. Moreover $M^+(\mathbb{T})$, viewed as a subset of PM, is a w*-closed set in PM. This is because for $S \in PM$,

$$S \in M^+(\mathbb{T}) \Leftrightarrow \forall f \in A(|\langle f, S\rangle| \le \|f\|_C \cdot |S(0)|) \;\&\; \forall f \in A(f \ge 0 \Rightarrow \langle f, S\rangle \ge 0)$$

where $\|f\|_C = \|f\|_{C(\mathbb{T})} = \sup_{x\in\mathbb{T}} |f(x)|$. (The first condition above puts S in $C(\mathbb{T})^* = M(\mathbb{T})$, since A is dense in $C(\mathbb{T})$, and the second makes S positive).

We define now—in analogy with the quantities $\eta(E)$ and $\eta_1(E)$ corresponding to U and $U_1$—for each $E \neq \emptyset$, $E \in K(\mathbb{T})$,

$$\eta_0(E) = \inf\left\{\frac{R(\mu)}{\|\mu\|_{PM}} : \mu \in M^+(E),\ \mu \neq 0\right\}$$

where

$$M^+(E) = M(E) \cap M^+(\mathbb{T})$$

is the class of positive measures supported by E. (Recall that $R(\mu) = R(\hat{\mu}) = \overline{\lim}\, |\hat{\mu}(n)|$). We have clearly that $\eta(E) \leq \eta_1(E) \leq \eta_0(E)$. So if we define (again by analogy with $U'$ and $U'_1$),

$$U'_0 = \{E \in K(\mathbb{T}) : \eta_0(E) > 0\} \cup \{\emptyset\}$$

we have

$$U'_0 \subseteq U_0$$

and

$$U' \subseteq U'_1 \subseteq U'_0.$$

(We have seen in Chapter VII that $U' \subsetneqq U'_1$, and it can be proved that $U'_1 \subsetneqq U'_0$ as well. This follows from the theorem of Piatetski-Shapiro that $(U'_1)_\sigma \subsetneqq U_0$—see the remarks after VII.3.16—and the result in the next section that $(U'_0)_\sigma = U_0$). The first to define $\eta_0$, $U'_0$ was Lyons, in connection with his work which we discuss in IX.2—see his [1], [4], where by the way $\eta_0$ is denoted $s^+$ and $U'_0$ by $\{s^+ > 0\}$ for Borel sets in general).

We will show first that $U'_0$ is an ideal in $K(\mathbb{T})$. This is a corollary of the next two results of Lyons and it was first observed by Host and Parreau.

<u>Lemma</u> <u>4</u> (Lyons [1]). Let $\mu$ be a positive measure on $\mathbb{T}$ and $\nu \in M(\mathbb{T})$, $\|\nu\|_M \leq 1$ be such that $|\nu| \leq \mu$, i.e. $|\nu|(E) \leq \mu(E)$, for all Borel E. Let $n_1,\ldots,n_k \in \mathbb{Z}$, and define $w_j \in \mathbb{C}$, $|w_j| = 1$ for $j = 1,\ldots,k$ by $|\hat{\nu}(n_j)| = w_j\hat{\nu}(n_j)$ ($w_j = 1$, if $\hat{\nu}(n_j) = 0$). Then

$$\Big(\frac{1}{k}\sum_{j=1}^{k}|\hat{\nu}(n_j)|\Big)^2 \leq \frac{\hat{\mu}(0)}{k} + \frac{2}{k^2}\sum_{1\leq j<j'\leq k} \operatorname{Re}\, w_{j'}\bar{w}_j\hat{\mu}(n_{j'} - n_j).$$

Proof. We have

$$\Big(\frac{1}{k}\sum_j |\hat{\nu}(n_j)|\Big)^2 = \frac{1}{k^2}\Big(\sum_j w_j \int e^{-in_jx}\,d\nu\Big)^2$$

$$\leq \frac{1}{k^2}\Big(\int \Big|\sum_j w_j e^{-in_jx}\Big|\,d|\nu|\Big)^2$$

$$\leq \frac{1}{k^2}\int \sum_j w_j e^{-in_jx}\cdot\overline{\sum_{j'} w_{j'}e^{-in_{j'}x}}\,d\mu$$

$$= \frac{\hat{\mu}(0)}{k} + \frac{2}{k^2}\sum_{j<j'} \operatorname{Re}\; w_{j'}\bar{w}_j\hat{\mu}(n_{j'} - n_j). \qquad \square$$

Corollary 5 (Lyons [1]). Let $\nu \in M(\mathbb{T})$, $\|\nu\|_M \leq 1$, $\mu \in M^+(\mathbb{T})$ and $|\nu| \leq \mu$. Then

$$R(\nu) \leq (R(\mu))^{1/2}.$$

Proof. Suppose that for infinitely many $0 < n_1 < n_2 < \dots < n_k < \dots$ we have $|\hat{\nu}(n_j)| \geq t$. (A similar argument applies if $0 > n_1 > \dots > n_k > \dots$). Using the preceding lemma for $n_1, \dots, n_k$, it follows that

$$\frac{2}{k^2}\sum_{1\leq j<j'\leq k} |\hat{\mu}(n_{j'} - n_j)| \geq t^2 - \frac{\hat{\mu}(0)}{k}$$

hence

$$R(\mu) \geq \overline{\lim_k}\,\frac{2}{k^2}\sum_{1\leq j<j'\leq k} |\hat{\mu}(n_{j'} - n_j)| \geq t^2. \qquad \square$$

Proposition 6. The class $U'_0$ is a $\underset{\sim}{\Sigma}^0_3$ (i.e. $G_{\delta\sigma}$) ideal of closed sets.

Proof. Clearly $U'_0$ is hereditary. Also

$$E \in U'_0 \Leftrightarrow \exists \epsilon > 0 \ \forall \mu \in M^+(E) \ [||\mu||_M = 1 \Rightarrow \forall n \in \mathbb{N} \ \exists m(|m| > n \text{ and } |\hat{\mu}(m)| > \epsilon)]$$

and an immediate computation gives that $U'_0 \in \underset{\sim}{\Sigma}^0_3$. So let $E_1$ and $E_2$ be two $U'_0$-sets, towards proving that $E_1 \cup E_2 \in U'_0$. Let $\mu \in M^+(E_1 \cup E_2)$ be a probability measure. Then $\mu_1 = \mu \restriction E_1$ (i.e. $\mu_1(A) = \mu(E_1 \cap A)$) and $\mu_2 = \mu \restriction E_2$ satisfy $\mu_i \leq \mu$, $||\mu_i||_M \leq 1$ and $\mu_i \in M^+(E_i)$. So we get by Corollary 5 that

$$R(\mu)^{1/2} \geq R(\mu_1) \geq \eta_0(E_1) \cdot \mu(E_1)$$

and

$$R(\mu)^{1/2} \geq R(\mu_2) \geq \eta_0(E_2) \cdot \mu(E_2)$$

therefore

$$\eta_0(E_1 \cup E_2) \geq \left(\frac{\min(\eta_0(E_1), \eta_0(E_2))}{2}\right)^2 > 0. \qquad \square$$

It is immediate to check that $U'_0$ is closed under translations. And as for $U'$ and $U'_1$ we have

Proposition 7. The class $U'_0$ is closed under dilations and contractions.

Proof. As in V.4.9 (for $U'$) and in VI.2.8 (for $U'_1$), it is enough to show that for $E, F \subseteq (0, 2\pi)$ with $F = tE$, $t > 0$, $\eta_0(E) = 0 \Rightarrow \eta_0(F) = 0$. And using the same technique as in V.4.9, this reduces to showing that if $\mu \in M^+(E)$, the pseudomeasure $\hat{\mu}_t$ defined by $\hat{\mu}_t(n) = \int e^{-itnx} \, d\mu(x)$ is (the Fourier transform of) a positive measure, which is immediate. □

Similarly the class $U_0$ is closed under dilations and contractions.

The sets in $U'_0$ will play an essential role in our study of $U_0$. As we shall see in the next section they form a Borel basis for $U_0$.

## §2. The existence of a Borel basis for $U_0$ and its associated rank

We start with the main result of this section

Theorem 1 (Kechris-Louveau [1]). Every $U_0$-set is a countable union of sets in $U'_0$, i.e. $U_0 = (U'_0)_\sigma$. In particular $U_0$ has a Borel (in fact $\underset{\sim}{\Sigma}^0_3$ basis).

Proof. Let $F \in U_0$ and put $W = \cup\{V : V$ is open in $\mathbb{T}$ and $V \cap F$ can be covered by countably many sets in $U'_0\}$. If $F \subseteq W$ we are done. Otherwise $E = F - W$ is nonempty and it has the property that for every open $V$ with $E \cap V \neq \emptyset$, $\overline{E \cap V} \notin U'_0$. To get a contradiction it is enough to show that if $E \in K(\mathbb{T})$ has this property then $E \in M_0$.

Given a closed set $E \subseteq \mathbb{T}$ we will denote by

$$\mathrm{PROB}(E)$$

the space of probability measures supported by $E$ with the weak*-topology. So PROB(E) is compact, metrizable. Let also

$$\mathcal{R}^{\epsilon}(E) = \{\mu \in \text{PROB}(E) : R(\mu) < \epsilon\}.$$

Then we have

Lemma 2. Let $E \in K(\mathbb{T})$, $E \neq \emptyset$ be such that for all open $V \subseteq \mathbb{T}$ with $V \cap E \neq \emptyset$ we have $\overline{E \cap V} \notin U_0'$. Then for each $\epsilon > 0$, $\mathcal{R}^{\epsilon}(E)$ is dense in PROB(E).

Proof. A simple application of Hahn-Banach shows that the finite support probability measures are dense in PROB(E). Since $\mathcal{R}^{\epsilon}(E)$ is convex, it is enough to show that every Dirac measure $\delta_x$, $x \in E$, can be approximated by measures in $\mathcal{R}^{\epsilon}(E)$. Note now the obvious fact that if $V_n$ is a sequence of nbhds of x with $\text{diam}(V_n) \to 0$ and $\mu_n$ are probability measures supported by $\overline{V}_n$ then $\mu_n \xrightarrow{w^*} \delta_x$. By the hypothesis on E, for each $\epsilon > 0$ one can find $\mu_n \in \mathcal{R}^{\epsilon}(E)$ supported by $\overline{V}_n$, so we are done. □

We will construct now for any E satisfying the above hypothesis a sequence $\mu_1, \mu_2, \ldots \in \text{PROB}(E)$ and a sequence of integers $0 < n_1 < n_2 < \ldots$ such that

$$\|\mu_k\|_{PM}^{n_i} = \sup\{|\hat{\mu}_k(m)| : |m| \geq n_i\} < 2^{-i}, \ \forall k \geq i.$$

Then if $\mu$ is a $w^*$-limit of a subsequence of the $\mu_k$'s, $\mu \in \text{PROB}(E) \cap PF$, i.e. $E \in M_0$.

We proceed by induction: Start with any $\mu_1 \in \text{PROB}(E)$ and $R(\mu_1) < \frac{1}{2}$. Then choose $n_1$ with $\|\mu_1\|_{PM}^{n_1} < \frac{1}{2}$.

Assume now $\mu_1, \ldots, \mu_i$ and $n_1, \ldots, n_i$ have been constructed. We claim that for each $m \geq n_i$ there is $m' > m$ and $\mu \in \mathrm{PROB}(E)$ with

(a) $\|\mu\|_{PM}^{n_j, n_{j+1}} = \sup\{|\hat{\mu}(m)| : n_j \leq |m| \leq n_{j+1}\} < 2^{-j}$, $j = 1, \ldots, i-1$;

(b) $\|\mu\|_{PM}^{n_i, m} < \eta + \delta < 2^{-i}$, where $\eta = \|\mu_i\|_{PM}^{n_i}$;

(c) $\|\mu\|_{PM}^{m'} < 2^{-i-1}$.

To see this notice that the $\mu$'s satisfying (a), (b) form an open set in PROB(E) which is non-empty, since $\mu_i$ belongs to it. So by the density of $\mathcal{R}^{\epsilon}(E)$ with $\epsilon = 2^{-i-1}$, there is $\mu \in \mathcal{R}^{\epsilon}(E)$ satisfying (a), (b). Choose then $m'$ to satisfy (c).

Iterating this, find $m_1 = n_i < m_2 < \ldots < m_N < \ldots$ and $\mu^{(1)}, \mu^{(2)}, \ldots, \mu^{(N)}, \ldots$ in PROB(E) such that (a) holds for each one of them and $\|\mu^{(k)}\|_{PM}^{m_{k+1}} < 2^{-i-1}$, $\|\mu^{(k)}\|_{PM}^{n_i, m_k} < \eta + \delta$. Choose then N large enough so that

$$\frac{1 + (\eta + \delta)(N - 1)}{N} < 2^{-i}$$

and put

$$\mu_{i+1} = \frac{1}{N}(\mu^{(1)} + \ldots \mu^{(N)}),$$
$$n_{i+1} = m_{N+1}.$$

Then $\|\mu_{i+1}\|_{PM}^{n_j, n_{j+1}} < 2^{-j}$, $j = 1, \ldots, i-1$. Also $\|\mu_{i+1}\|_{PM}^{n_{i+1}} < 2^{-i-1}$. Finally, if $n_i \leq |t| \leq n_{i+1} = m_{N+1}$, then for all but at most one $1 \leq k \leq N$ we have

$$|\widehat{\mu^{(k)}}(t)| < \eta + \delta$$

thus

$$|\widehat{\mu_{i+1}}(t)| \leq \frac{1}{N}(1 + (N - 1)(\eta + \delta)) < 2^{-i}$$

i.e. $\|\mu_{i+1}\|_{PM}^{n_i,n_{i+1}} < 2^{-i}$, so $\|\mu_{i+1}\|_{PM}^{n_j} < 2^{-j}$, if $j \leq i + 1$ and we are done. □

Here are some immediate applications of the preceding basis theorem to the computation of the complexity of certain classes of closed sets.

Corollary 3. The class $M_0^p$ of closed sets of pure restricted multiplicity (i.e., by 1.3 the class of supports of Rajchman measures) is a $\underset{\sim}{\Pi}_3^0$ (i.e. $F_{\sigma\delta}$) set in $K(\mathbb{T})$ (and thus $U_0^{loc} = K(\mathbb{T}) - M_0^p$ is $\underset{\sim}{\Sigma}_3^0$).

Proof. For $E \in K(\mathbb{T})$ we have

$E \in M_0^p \Leftrightarrow$ for every rational (relative to $2\pi$) open interval I

$(I \cap E \neq \emptyset \Rightarrow \overline{E \cap I} \notin U_0)$

$\Leftrightarrow$ (by the basis theorem)

For every rational (relative to $2\pi$) open interval I

$(I \cap E \neq \emptyset \Rightarrow \overline{E \cap I} \notin U_0')$

$\Leftrightarrow$ For every rational (relative to $2\pi$) open interval I

$(I \cap E \neq \emptyset$ & E does not contain the endpoints of

$I \Rightarrow \overline{E \cap I} \notin U_0')$.

Since $U_0'$ is $\underset{\sim}{\Sigma}_3^0$ and the map $E \mapsto \overline{E \cap I}$ is continuous on the open set $\{E \in K(\mathbb{T}) :$

E does not contain the endpoints of I}, this shows that $M_0^p$ is $\underset{\sim}{\Pi}_3^0$. □

Recall now that in III.3 we have associated with $\vec{\xi} = \{\xi_n\}_{n\in\mathbb{N}-\{0\}}$ in $(0, \frac{1}{2})^{\mathbb{N}-\{0\}}$ the symmetric perfect set $E_{\vec{\xi}} = E_{\xi_1,\xi_2,\ldots}$.

<u>Theorem</u> <u>4</u> (Kechris-Louveau). (i) Let $E_{\vec{\xi}}$ be a symmetric perfect set ($\vec{\xi} = \xi_1, \xi_2, \ldots; 0 < \xi_i < \frac{1}{2}$). Then

$$E_{\vec{\xi}} \in U_0 \Leftrightarrow E_{\vec{\xi}} \in U_0' \; (\Leftrightarrow \eta_0 \, (E_{\vec{\xi}}) > 0).$$

(ii) The set

$$\{\vec{\xi} \in (0, \tfrac{1}{2})^{\mathbb{N}-\{0\}} : E_{\vec{\xi}} \in U_0\}$$

is a $\underset{\sim}{\Sigma}_3^0$ (i.e. $G_{\delta\sigma}$) set in $(0, \frac{1}{2})^{\mathbb{N}-\{0\}}$.

<u>Proof</u>. (ii) is obvious from (i), the fact that $\vec{\xi} \mapsto E_{\vec{\xi}}$ is continuous, and the computation that $U_0'$ is $\underset{\sim}{\Sigma}_3^0$. To prove (i), notice that if $E_{\vec{\xi}} \in U_0$, then by the basis theorem and the Baire Category Theorem, there is an open interval I with $E_{\vec{\xi}} \cap I \neq \emptyset$ and $\overline{E_{\vec{\xi}} \cap I} \in U_0'$. Then for some k there is one of the $2^k$ closed intervals, say $\bar{J}$, that make up the $k^{th}$ closed set $E_k$ in the construction of $E_{\vec{\xi}}$ (see III.3), such that $E_{\vec{\xi}} \cap \bar{J} \subseteq \overline{E_{\vec{\xi}} \cap I}$, so that $E_{\vec{\xi}} \cap \bar{J} \in U_0'$ as well. But clearly $E_{\vec{\xi}}$ is the union of $2^k$ translates of $E_{\vec{\xi}} \cap \bar{J}$, so by 1.6 $E_{\vec{\xi}} \in U_0'$. □

The preceding result suggests perhaps that there could be some reasonably explicit characterization of those sequences $\vec{\xi}$ for which $E_{\vec{\xi}}$ is in $U_0$, extending the

Salem-Zygmund Theorem and the result of Y. Meyer (see Meyer [1]) that if $\Sigma\ \xi_n^2 < \infty$, then $E_{\vec{\xi}}$ is in U (thus in $U_0$). One tempting conjecture is that the natural "Lebesgue measure" on $E_{\vec{\xi}}$ (i.e. the one induced from the standard probability measure on $2^{\mathbb{N}-\{0\}}$ via the 1–1 correspondence $\{\epsilon_i\} \mapsto 2\pi \sum_{i=1}^{\infty} \epsilon_i\ \xi_1 \ldots \xi_{i-1}$ $(1 - \xi_i)$) decides whether $E_{\vec{\xi}}$ is in $U_0$ or not, i.e. $E_{\vec{\xi}} \in U_0$ iff this measure is not a Rajchman measure iff (by computations as in III.4.4)

$$\underline{\lim}_{\ n\to\infty} \sum_{i=1}^{\infty} \sin^2(\pi n\ \xi_1 \ldots \xi_{i-1}\ (1 - \xi_i)) < \infty.$$

Of course another possibility is that some other canonical measure associated with $E_{\vec{\xi}}$ might decide this question. A related question is whether for $E_{\vec{\xi}} \in U_0$ the infimum $\eta_0(E) = \inf\{R(\mu) : \mu$ a probability measure on $E\} > 0$ is attained.

The problem whether $E_{\vec{\xi}} \in U$ iff $E_{\vec{\xi}} \in U_0$ is also open, to our knowledge. (This holds for $\vec{\xi}$ constant by the Salem-Zygmund Theorem). Finally recall (IV.2.9) that it is not even known if $\{\vec{\xi} \in (0, \frac{1}{2})^{\mathbb{N}-\{0\}} : E_{\vec{\xi}} \in U\}$ is Borel or not. (Since U has no Borel basis the method in the proof of Theorem 4(ii) does not work).

By Proposition 1.6, $U_0'$ is a $\underset{\sim}{\Sigma}_3^0$ (thus Borel) ideal in $K(\mathbb{T})$. We can associate with it the usual Cantor-Bendixson derivative: For each $E \in K(\mathbb{T})$, we let $E_0^{(1)} = E_{U_0'}^{(1)}$ be the set $\{x \in E$: for every open V with $x \in V$, $\overline{E \cap V} \notin U_0'\}$, and define by induction on $\alpha$, $E_0^{(0)} = E$, $E_0^{(\alpha+1)} = (E_0^{(\alpha)})_0^{(1)}$, and $E_0^{(\lambda)} = \bigcap_{\alpha<\lambda} E_0^{(\alpha)}$ for limit $\lambda$. We define then the Cantor-Bendixson rank $rk_0(E)$ associated with that derivation, as in VI.1, by

$$\mathrm{rk}_0(E) = \begin{cases} \text{least } \alpha \ (E_0^{(\alpha)} = \emptyset), \text{ if such an } \alpha \text{ exists,} \\ \omega_1 \text{ otherwise,} \end{cases}$$

and we know by Theorem VI.1.4 that $\mathrm{rk}_0$ is a $\underset{\sim}{\Pi}_1^1$-rank on the $\sigma$-ideal $(U_0')_\sigma = U_0$ generated by $U_0'$.

We introduce now another ranking on the $U_0$-sets. For each $E \in K(\mathbb{T})$, let $M^+(E)$ be the set of positive measures supported by E, viewed as a subset of PM (via $\mu \to \hat{\mu}$). Them $M^+(E)$ is w*-closed in PM. As in V.3, we consider the R-rank on $M^+(E)$: For $Z \subseteq PM$, let $Z_{(1)} = \{S \in Z : \exists \{S_n\} \ (S_n \in Z \ \& \ S_n \xrightarrow{w^*} S \ \& \ R(S_n) \to 0\}$, and define inductively $M_\alpha^+(E)$ by

$$M_0^+(E) = M^+(E),$$

$$M_{\alpha+1}^+(E) = (M_\alpha^+(E))_{(1)}$$

$$M_\lambda^+(E) = \cap_{\alpha<\lambda} M_\alpha^+(E), \text{ for limit } \lambda.$$

Then each $M_\alpha^+(E)$ is w*-closed in PM. For this notice that if $Z \subseteq M^+(\mathbb{T})$ then $Z_{(1)}$ is w*-closed in PM : If $\mu \notin Z_{(1)}$, $\mu \in M^+(\mathbb{T})$ and $a > \|\mu\|_{PM} = \hat{\mu}(0)$, then since $B_a(PM) = \{S \in PM : \|S\|_{PM} \leq a\}$ is metrizable in the weak*-topology, there is a w*-open set $V \subseteq PM$ such that $\mu \in V$ and for some $\epsilon > 0$ every $\mu \in Z \cap V \cap B_a(PM)$ has $R(\mu) \geq \epsilon$. Then $\{\nu \in M^+(\mathbb{T}) : \hat{\nu}(0) < a \ \& \ \nu \in V\}$ is open in $M^+(\mathbb{T})$ with the weak*-topology, contains $\mu$ and is disjoint from $Z_{(1)}$. Thus $Z_{(1)}$ is closed in $M^+(\mathbb{T})$ with the weak*-topology, and since $M^+(\mathbb{T})$ is itself closed in the weak*-topology of PM it follows that $Z_{(1)}$ is as well.

So we have a decreasing transfinite sequence $M_\alpha^+(E)$ of weak*-closed subsets of PM, which must therefore stabilize at some countable ordinal. (Since

$M^+_\alpha(E) \cap B_1(PM)$ is a decreasing sequence of closed subsets of the compact, metrizable space $B_1(PM)$ with the weak*-topology, $M^+_\alpha(E) \cap B_1(PM)$ stabilizes at some countable ordinal and thus so does $M^+_\alpha(E)$, since $M^+_\alpha(E)$ is closed under multiplication by positive numbers). We define then the $U_0$-rank $[E]_0$ by

$$[E]_0 = \begin{cases} \text{least } \alpha(M^+_\alpha(E) = \{0\}), \text{ if there is such an } \alpha, \\ \omega_1 \text{ otherwise.} \end{cases}$$

We show now that $[E]_0$ and $rk_0(E)$ coincide.

<u>Theorem 5</u> (Kechris-Louveau). Let $E \in K(\mathbb{T})$. For each $\alpha$, $M^+_\alpha(E) = M^+(E_0^{(\alpha)})$. In particular $[E]_0 = rk_0(E)$ is a $\underset{\sim}{\Pi}^1_1$-rank on $U_0$ and $U_0' = \{E \in U_0 : [E]_0 \leq 1\}$.

<u>Proof</u>. That $[E]_0 = rk_0(E)$ follows immediately from the equality $M^+_\alpha(E) = M^+(E_0^{(\alpha)})$. And since $U_0'$ is an ideal it follows that $U_0' = \{E \in U_0 : rk_0(E) \leq 1\} = \{E \in U_0 : [E]_0 \leq 1\}$. Moreover it is enough to prove the above equality for $\alpha = 1$, for then by induction

$$\begin{aligned} M^+_{\alpha+1}(E) = \left[M^+_\alpha(E)\right]_{(1)} = \left[M^+(E_0^{(\alpha)})\right]_{(1)} = \\ = M^+_1(E_0^{(\alpha)}) = M^+\left[(E_0^{(\alpha)})_0^{(1)}\right] \\ = M^+\left[E_0^{(\alpha+1)}\right] \end{aligned}$$

and for limit $\lambda$,

$$M^+_\lambda(E) = \bigcap_{\alpha<\lambda} M^+_\alpha(E) = \bigcap_{\alpha<\lambda} M^+\left[E_0^{(\alpha)}\right]$$

$$= M^+\Big(\bigcap_{\alpha<\lambda} E_0^{(\alpha)}\Big)$$

$$= M^+\big(E_0^{(\lambda)}\big).$$

So it remains to show that $M^+(E)_{(1)} = M^+\big(E_0^{(1)}\big)$. Let $\mu \in M^+(E)_{(1)}$, say with $\frac{1}{2} < \hat{\mu}(0) < 1$ without loss of generality, and let $\mu_n \in M^+(E)$ be such that $\mu_n \xrightarrow{w^*} \mu$ and $R(\mu_n) \to 0$. We want to prove that $\mathrm{supp}(\mu) \subseteq E_0^{(1)}$. Suppose not, and let V open be such that $1 > \mu(V) > a > 0$ and $\overline{E \cap V} \in U_0'$. As $\mu_n \xrightarrow{w^*} \mu$, $1 > \mu_n(V) > a$ for n big enough, so $\nu_n = \mu_n \upharpoonright V$ satisfies $1 > \hat{\nu}_n(0) > a$ for n big enough. Also by 1.5 $R(\nu_n) \leq R(\mu_n))^{1/2}$, hence $R(\nu_n) \to 0$. But this contradicts that $\overline{E \cap V} \in U_0'$ (since $\mathrm{supp}(\nu_n) \subseteq E \cap V$) and we are done.

Suppose now that $\mu \in M^+(E_0^{(1)})$, in order to show that $\mu \in M^+(E)_{(1)}$. We can assume of course that $\mu \in \mathrm{PROB}(E_0^{(1)})$. And since the probabilities with finite support in $E_0^{(1)}$ are dense in the compact, metrizable space $\mathrm{PROB}(E_0^{(1)})$ it is enough to show that any $\delta_x$, $x \in E_0^{(1)}$ is in $M^+(E)_{(1)}$. But this is immediate as in the proof of Lemma 2, since for every nbhd V of x $\overline{V \cap E} \notin U_0'$, thus for each $\epsilon$ there is a probability measure $\nu$ supported by $\overline{V \cap E}$ with $R(\nu) < \epsilon$. □

We will see the main applications of the basis theorem to structural results about $U_0$ and $\mathcal{U}_0$ in the next section. Here we just notice that one can prove that $[E]_0$ is unbounded on $H_\sigma$-sets inside any $M_0$-set, thus giving an alternative "rank argument" for the following result which is implicit in Kaufman [7].

<u>Theorem 6</u> (Kaufman [7]). Let E be an $M_0$-set. Then $U_0 \cap K(E)$ is $\underset{\sim}{\Pi}_1^1$-complete.

Proof. By restricting E to the support of a non-0 positive Rajchman measure we can assume that $E \neq \emptyset$ is in $M_0^p$, i.e. $\overline{E \cap V} \notin U_0$ for all open V with $E \cap V \neq \emptyset$.

We can apply then Theorem VI.1.6 : It is enough to prove that for any open V with $E \cap V \neq \emptyset$, $E \cap V$ contains an $H_\sigma$-set F with $rk_0(F) > 1$, i.e. $F \notin U_0'$. Let $x \in E \cap V$ and choose a sequence of open sets with $x \in V_n \subseteq \bar{V}_n \subseteq V$ and $\mathrm{diam}(V_n) \to 0$. Since $\overline{E \cap V_n} \notin U_0$, there is a probability measure $\mu_n \in M^+(\overline{E \cap V_n}) \cap PF$, and we can apply Theorem VII.2.1 to this $\mu_n$ to get a probability measure $\nu_n$ with $\mathrm{supp}(\nu_n) \subseteq \mathrm{supp}(\mu_n) \subseteq \overline{E \cap V_n}$, $R(\nu_n) < \frac{1}{n}$ and $\mathrm{supp}(\nu_n)$ a finite union of H-sets. Then if $F = \{x\} \cup \bigcup_n \mathrm{supp}(\nu_n)$, $F \subseteq \overline{E \cap V}$ and $\eta_0(F) = 0$, i.e. $F \notin U_0'$. □

Again this gives the $\underset{\sim}{\Pi}_1^1$-completeness of $U_0$ by just constructing $U_0$-sets. Notice also that for $E = \mathbb{T}$ the preceding proof does not need VII.2.1 but essentially only VI.2.5, since in the proof of Lemma VI.2.6 there, one actually establishes that $\eta_0(E) \leq \epsilon$, thus in VI.2.5 one has (in each nonempty open set) $H_\sigma$-sets not in $U_0'$.

We have also the usual facts on the closure properties of

$$U_0^{(\alpha)} = \{E \in U_0 : [E]_0 \leq \alpha\}$$

Proposition 7. The class $U_0^{(\alpha)}$ is an ideal in $K(\mathbb{T})$, closed under translations, dilations and contractions.

Proof. Closure under translations, dilations and contractions follows as in VI.2.8

noticing that if E, F are closed subsets of $(0, 2\pi)$ with $E = tF$, $t > 0$ and $\mu \in M^+(E)$ then $\mu_t \in M^+(F)$. For closure under unions notice that if $E, F \in U_0^{(\alpha)}$ then by Theorem 5, $E_0^{(\alpha)} = F_0^{(\alpha)} = \emptyset$. But one can easily verify that $(E \cup F)_0^{(\beta)} = E_0^{(\beta)} \cup F_0^{(\beta)}$ for all $\beta$, thus $(E \cup F)_0^{(\alpha)} = \emptyset$, i.e. by Theorem 5 again $E \cup F \in U_0^{(\alpha)}$. □

Remark. In Debs-Saint Raymond [1] another proof of the existence of a Borel basis for $U_0$ is given, based on an idea similar to that of the Piatetski-Shapiro decomposition theorem VI.2.3(ii).

Let for each $E \in K(\mathbb{T})$

$$I_{neg}(E) = \{f \in A : \mathrm{Re}\ f \leq 0 \text{ on } E\}.$$

Then one can show that

$$E \in U_0 \Leftrightarrow \overline{I_{neg}(E)}^{w^*} = A$$

$$\Leftrightarrow 1 \in \overline{I_{neg}(E)}^{w^*}.$$

Although $I_{neg}(E)$ is not a subspace of A, it is a convex cone, i.e., it is closed under $+$ and multiplication by nonnegative reals, and many of the results in Chapter V are still true in this context. For instance V.2.2–V.2.5 go through with essentially the same proofs. So letting $I_{neg}^{\alpha}(E)$ be the $\alpha^{th}$ iterate of $I_{neg}(E)$ is the operation of taking $w^*$-sequential limits, we have

$$E \in U_0 \Leftrightarrow \exists\alpha < \omega_1(1 \in I^{\alpha}_{neg}(E))$$

and to each $E \in U_0$ we can assign the ordinal

$$[E]^*_0 = \text{least } \alpha \text{ with } 1 \in I^{\alpha}_{neg}(E).$$

Let also

$$\begin{aligned} U^*_0 &= \{E \in U_0 : [E]^*_0 \leq 1\} \\ &= \{E \in U_0 : \exists\{f_n\} (f_n \in A, \text{Re } f_n \leq 0 \text{ on } E, \text{ and } f_n \xrightarrow{w^*} 1)\}. \end{aligned}$$

Then Debs-Saint Raymond prove by induction on $[E]^*_0$ that every $E \in U_0$ is a countable union of sets in $U^*_0$ and this class is easily Borel.

As pointed out recently by R. Lyons, one can give here an argument closely resembling the Piatetski-Shapiro proof of his decomposition theorem as presented for example in Graham-McGehee [1]. The following appear in his paper [5].

One defines first for each $Y \subseteq A$,

$$\text{hull}_{neg}(Y) = \{x \in \mathbb{T} : \forall f \in Y(\text{Re } f(x) \leq 0)\}.$$

Then as in V.4.5 it can be checked that if Y is closed under +, multiplication by nonnegative real functions in A and taking of real parts, then

$$1 \in Y \Leftrightarrow \text{hull}_{neg}(Y) = \emptyset.$$

(The only difference in the argument here from that of V.4.5, is that one does not multiply $f_x$ by $\overline{f_x}$, but by a nonnegative function which is 1 around x and 0 outside a small nbhd of x).

Assume now that every $F \in U_0$ with $[F]_0^* < \alpha$ is a countable union of sets in $U_0^*$, and let $[E]_0^* = \alpha$. Then $\alpha$ is successor, say $\alpha = \beta + 1$. Let $K = \text{hull}_{neg}(I_{neg}^{\beta}(E))$. Then $K \subseteq E$ (since $I_{neg}^{\beta}(E) \supseteq I_{neg}(E)$) and $I_{neg}(K) \supseteq I_{neg}^{\beta}(E)$), so since $1 \in I_{neg}^{\beta+1}(E) \subseteq I_{neg}^{1}(K)$ clearly $K \in U_0^*$. It is then enough to check that if $L \in K(\mathbb{T})$, $L \subseteq E - K$ then $[L]_0^* < \alpha$. But $\text{hull}_{neg}(I_{neg}^{\beta}(L)) \subseteq L \cap \text{hull}_{neg}(I_{neg}^{\beta}(E)) = L \cap K = \emptyset$, therefore $1 \in I_{neg}^{\beta}(L)$ (since $Y = I_{neg}^{\beta}(L)$ has all the needed closure properties), and $[L]_0^* \le \beta < \alpha$.

Again the transfinite induction can be avoided here exactly as in the comments following the proof of VI.2.3.

Now of course it ought to be the case that $U_0^* = U_0'$ and $[E]_0^* = [E]_0$. This was indeed proved by Lyons. The main points are the following: Let Y be a convex cone in a Banach space X and let

$$Y^{neg} = \{x^* \in X^* : \forall y \in Y \ \text{Re} <y,x^*> \le 0\}.$$

(Note that if Y is a subspace, $Y^{neg} = Y^{\perp}$). Define for $x^* \in X^*$,

$$\|x^*\|_Y = \sup\{\text{Re} <y,x^*> : y \in Y \cap B_1(X)\}.$$

Then we have again

$$\|x^*\|_Y = \mathrm{dist}(x^*, Y^{neg}).$$

Moreover, if one defines for $Y \subseteq X^*$ a convex cone in the dual of a separable Banach space X, the quantity

$$s(Y) = \inf\left\{\frac{\|x\|_Y}{\|x\|} : 0 \neq x \in X\right\}$$

where now

$$\|x\|_Y = \sup\{\mathrm{Re} \langle x, y^*\rangle : y^* \in Y \cap B_1(X^*)\}$$

then Theorem V.2.8 goes through as well. (For part ii) replace in the argument there K by $K \cap - K$). Finally the remarks following V.2.9 work too, so that in particular

$$\begin{aligned} s(Y) > 0 &\Leftrightarrow \eta(Y^{neg}) > 0 \\ &\Leftrightarrow Y^{(1)} = X^* \end{aligned}$$

where as usual

$$\eta(Y^{neg}) = \inf\left\{\frac{\mathrm{dist}(y^{**},X)}{\|y^{**}\|_{**}} : y^{**} \in Y^{neg}, y^{**} \neq 0\right\}.$$

Applying this to $Y = I_{neg}(E)$, we have $Y^{neg} = M^+(E)$ and thus $U_0^* = U_0'$.

Finally as in V.3.11, it follows that if $Y \subseteq X^*$ is a convex cone and $X^*$ is separable, then $(Y^{neg})_{(\alpha)} = (Y^{(\alpha)})^{neg}$. But for such Y,

$$\bar{Y} = X^* \Leftrightarrow Y^{neg} = \{0\}$$

and so

$$\overline{I^{\alpha}_{neg}(E)} = A \Leftrightarrow (I^{\alpha}_{neg}(E))^{neg} = \{0\}$$
$$\Leftrightarrow ((I_{neg}(E))^{neg})_{(\alpha)} = \{0\}$$
$$\Leftrightarrow M^{+}_{\alpha}(E) = \{0\}.$$

Since $I^{\alpha}_{neg}(E)$ is closed under $+$, multiplication by nonnegative real functions in A and taking of real parts we have

$$I^{\alpha}_{neg}(E) = A \Leftrightarrow 1 \in I^{\alpha}_{neg}(E)$$
$$\Leftrightarrow 1 \in \overline{I^{\alpha}_{neg}(E)}$$
$$\Leftrightarrow \overline{I^{\alpha}_{neg}(E)} = A$$
$$\Leftrightarrow M^{+}_{\alpha}(E) = \{0\}$$

so that $[E]^*_0 = [E]_0$ and we are done.

It has been also shown by Lyons that $U'_0$ coincides with the following class of closed sets defined in Piatetski-Shapiro [1], and denoted there by $U^3$:

$$E \in U^3 \Leftrightarrow \exists\{h_n\}\ \exists a > 0\ (h_n \in A\ \&\ h_n \xrightarrow{w^*} 0\ \&\ \underline{\lim}\ \mathrm{Re}\ h_n(x) \geq a > 0 \text{ for all } x \in E).$$

Indeed if $E \in U^3$ with witnesses $h_n$, a, consider any $\mu \in M^+(E)$ and $S \in PF$. Then $\langle \mathrm{Re}\ h_n, \mu - S\rangle = \langle \mathrm{Re}\ h_n, \mu\rangle - \langle \mathrm{Re}\ h_n, S\rangle$ and $\underline{\lim} \langle \mathrm{Re}\ h_n, \mu\rangle = \underline{\lim} \int \mathrm{Re}\ h_n\, d\mu \geq \int \underline{\lim}\ \mathrm{Re}\ h_n(x)\, d\mu(x) \geq a \cdot \mu(E)$, while $\langle \mathrm{Re}\ h_n, S\rangle \to 0$, as $\mathrm{Re}\ h_n \xrightarrow{w^*} 0$ as well. But also if $\sup\|h_n\|_A = K < \infty$ we have $K \cdot \|\mu - S\|_{PM} \geq \langle \mathrm{Re}\ h_n, \mu - S\rangle$, thus $\|\mu - S\|_{PM} \geq \frac{a \cdot \mu(E)}{K}$, and so $R(\mu) \geq \frac{a \cdot \mu(E)}{K}$, i.e. $\eta_0(E) >$

0 and $E \in U_0'$. Conversely, if $E \in U_0'$, then by the preceding facts there is $f_n \in A$ with $f_n \xrightarrow{w^*} 1$ and $\mathrm{Re}\, f_n \leq 0$ on E. Then $1 - f_n \xrightarrow{w^*} 0$ and $\mathrm{Re}(1 - f_n) \geq 1$ on E, so $E \in U^3$.

Finally, Lyons brought to our attention a passage from Piatetski-Shapiro [1], in which the author appears to conjecture that every $U_0$-set is a countable union of sets in $U^3$. This is of course true by the above facts.

## §3. The solution to the Category Problem, and other applications

Now that we know that $U_0$ has a Borel basis, we are in a position to apply the results of section VI.1, mainly Theorem VI.1.11. This yields a complete characterization of Borel—even $\underset{\sim}{\Sigma}_1^1$—sets in $\mathcal{U}_0$.

Theorem 1 (Debs-Saint Raymond [1]). Let $P \subseteq \mathbb{T}$ be a $\underset{\sim}{\Sigma}_1^1$ set. The following are equivalent,

(i) $P \in \mathcal{U}_0$,

(ii) P is contained in a countable union of $U_0$ (therefore $U_0'$) sets.

Proof. By Theorem VI.1.11, it is enough to verify that $U_0$ is calibrated, which is immediate (see the remarks following Proposition 1.2), has a Borel basis (Theorem 2.1), and is complete $\underset{\sim}{\Pi}_1^1$ on any $M_0$-set (Theorem 2.6). □

The solution to the Category Problem is now immediate, in fact in a much stronger form (as it holds for $\mathcal{U}_0$ sets as well).

Theorem 2 (Debs-Saint Raymond [1]). Any set with the property of Baire which is in $\mathcal{U}_0$ is of the first category. In fact it is of the first category in any $M_0^p$-set E containing it, provided it still has the property of Baire in E. In particular, every set of uniqueness which has the property of Baire is of the first category.

Proof. As in the discussion (in I.6) of the Category Problem, it is enough to work with $G_\delta$ sets. Now if G is $G_\delta$, $G \in \mathcal{U}_0$, $G \subseteq E$ where E is in $M_0^p$, then by the preceding theorem $G \subseteq \cup_n F_n$, where each $F_n$ is in $U_0$, thus it is nowhere dense in E and we are done. □

The preceding (and original) proof of Theorem 1 has been derived as a consequence of the general structural and definability properties of the $\sigma$-ideal $U_0$ which have been developed until now. In view however of the basic nature of this result and its several applications (more of which we will discuss shortly) it is desirable to search for a direct proof of it. Such an alternative approach has been indeed developed by Kechris and Louveau and was motivated by the proof of the basis theorem 2.1. It leads to a very simple proof of Theorem 1 and thus of all its immediate consequences like Theorem 2 and the applications that follow. Moreover it can be used to obtain further results that we describe later in this chapter.

An alternative proof of Theorem 1 (Kechris-Louveau [1]).

We will first discuss the case where P is a $G_\delta$, which is the most useful in applications, and then explain the necessary modifications needed to prove the full result.

To start with, we notice that this can be reduced to the following equivalent statement:

(*) If $\emptyset \neq E \in M_0^p$ and $G \subseteq E$ is dense $G_\delta$ in E, then $G \in \mathcal{M}_0$.

Indeed if $P \in \mathcal{U}_0$ is $G_\delta$ let $W = \cup\{V : V$ is open in $\mathbb{T}$ and $P \cap V$ can be covered by countably many sets in $U_0\}$. Then $P \cap W$ can be covered by countably many sets in $U_0$. If $P \subseteq W$ we are done. Else let $G = P - W$. Then $G \neq \emptyset$, G is a $G_\delta$ and for any V open with $V \cap G \neq \emptyset$ we have $\overline{V \cap G} \in M_0$. If $E = \bar{G}$, then $\emptyset \neq E \in M_0^p$ and G is dense $G_\delta$ in E, so $G \in \mathcal{M}_0$ a contradiction.

In order to prove (*) we argue as follows:

Let $\emptyset \neq F$ be an arbitrary set in $M_0^p$ and let $V \subseteq F$ be dense, open in F (with its relative topology). Put

$$\mathcal{R}(V) = \{\mu \in \mathrm{PROB}(F) \cap \mathrm{PF} : \mathrm{supp}(\mu) \subseteq V\}.$$

We claim that $\mathcal{R}(V)$ is convex and dense in PROB(F) : Convexity is obvious. For density, note that the probability measures with finite support contained in V are dense in PROB(F), since V is dense in F. If now $x \in V$, then $\{x\} = \cap_n W_n$, where $W_n$ is open in F, $\bar{W}_n \subseteq V$ and $\mathrm{diam}(W_n) \to 0$. Since $F \in M_0^p$, there is $\mu_n \in \mathrm{PROB}(F) \cap \mathrm{PF}$ with $\mathrm{supp}(\mu_n) \subseteq \overline{W_n}$. Then $\mu_n \xrightarrow{w^*} \delta_x$, so $\delta_x \in \overline{\mathcal{R}(V)}^{w^*}$ and we are done.

Thus

$$\begin{aligned} \mathrm{PROB}(F) \cap \mathrm{PF} &= \overline{\mathcal{R}(V)}^{w^*} \cap \mathrm{PF} \\ &= \overline{\mathcal{R}(V)}^{w} \text{ (the weak-closure of R(V) in PF)} \\ &= \overline{\mathcal{R}(V)} \text{ (the norm-closure of R(V) in PF)} \end{aligned}$$

the last equality being true by Mazur's Theorem. So we have shown that for each $\mu \in \mathrm{PROB}(F) \cap \mathrm{PF}$ and each $\epsilon > 0$ there is $\nu \in \mathcal{R}(V)$ with $\|\mu - \nu\|_{PM} < \epsilon$.

Going back to E now, write $G = \bigcap_{n=1}^{\infty} V_n$, $V_n$ being dense open in E. Fix any $\mu_0 \in \mathrm{PROB}(E) \cap \mathrm{PF}$. Then there is $\mu_1 \in \mathcal{R}(V_1)$ with $\|\mu_0 - \mu_1\|_{PM} < 2^{-1}$. Let $\emptyset \neq W_1 \subseteq \overline{W}_1 \subseteq V_1$, $W_1$ open in E, be such that $\mathrm{supp}(\mu_1) \subseteq \overline{W}_1$. Then $E_1 = \overline{W}_1$ is also in $M_0^p$ and $V_2' = V_2 \cap E_1$ is dense open in $E_1$. So there is $\mu_2 \in \mathcal{R}(V_2')$ with $\|\mu_1 - \mu_2\|_{PM} < 2^{-2}$. Let again $\emptyset \neq W_2 \subseteq \overline{W_2} \subseteq V_2'$ be such that $W_2$ is open in $E_1$ with $\mathrm{supp}(\mu_2) \subseteq \overline{W_2}$. Put $E_2 = \overline{W_2}$ and continue as before with $V_3' = V_3 \cap E_2$, etc. Let $\mu = \lim \mu_n$ in PF. Then $0 \neq \mu \in \mathrm{PF}$ and $\mathrm{supp}(\mu) \subseteq \bigcap_n V_n = G$, i.e. $G \in \mathcal{M}_0$, and our proof is complete. □

We can mix now the preceding with a projection argument to give a proof of the full Theorem 1:

Let $P = \pi G$, where $G \subseteq \mathbb{T} \times 2^{\mathbb{N}}$ is $G_\delta$ and $\pi$ is the projection function (by IV.1.4). Assuming $P \in \mathcal{U}_0$ let $W = \bigcup \{N : N \text{ is open in } \mathbb{T} \times 2^{\mathbb{N}} \text{ and } \pi(G \cap N) \text{ can be covered by countably many sets in } U_0\}$. If $G \subseteq W$ we are done. Else $G' = G - W \neq \emptyset$, from which we will derive a contradiction. For any N open with $G' \cap N \neq \emptyset$ we have $\overline{\pi(G' \cap N)} \in M_0$. Put $E = \overline{G'}$. We say that a closed set

$F \subseteq \mathbb{T} \times 2^{\mathbb{N}}$ is "in $M_0^p$" if for any non-empty open V in F, $\overline{\pi V} \in M_0$. Thus E is "in $M_0^p$". Also if V is non-empty open in E, $\bar{V}$ is "in $M_0^p$" too.

For any closed $F \subseteq \mathbb{T} \times 2^{\mathbb{N}}$, $F \neq \emptyset$ which is "in $M_0^p$" and $V \subseteq F$ dense, open in F let

$$\mathcal{R}'(V) = \{\mu \in PROB(\pi F) \cap PF : \exists Q(Q \subseteq V, Q \text{ closed}, supp(\mu) \subseteq \pi Q)\}.$$

Then $\mathcal{R}'(V)$ is convex and dense in $PROB(\pi F)$.

So as in the proof of (*) before, if $G' = \bigcap_n V_n$, where $V_n \subseteq E$ is dense, open in E we can construct $\mu_n \in PROB(\pi E) \cap PF$, $\|\mu_n - \mu_{n+1}\|_{PM} \leq 2^{-n-1}$ and $E_n \subseteq V_n$ closed and decreasing so that $\mu_n$ is supported by $\pi E_n$. Then $\mu_n \to \mu$ in PF, so $\mu \in PF$, $\mu \neq 0$ and $supp(\mu) \subseteq \bigcap_n \pi E_n = \pi(\bigcap_n E_n) \subseteq \pi(\bigcap_n V_n) = \pi G' \subseteq P$, i.e. $P \in \mathcal{M}_0$, a contradiction. □

Let us discuss now some immediate applications of Theorem 2.

Recall first that a <u>capacity</u> on $\mathbb{T}$ is a function $\gamma$ : power($\mathbb{T}$) $\to [0, \infty)$ which is increasing ($A \subseteq B \Rightarrow \gamma(A) \leq \gamma(B)$), with $\gamma(\emptyset) = 0$, and satisfies

(i) $\gamma(\bigcup_n A_n) = \sup_n \gamma(A_n)$, if $\{A_n\}$ is <u>any increasing</u> sequence of subsets of $\mathbb{T}$,

(ii) $\gamma(\bigcap_n K_n) = \inf_n \gamma(K_n)$, if $\{K_n\}$ is a <u>decreasing</u> sequence of <u>closed</u> (= compact) subsets of $\mathbb{T}$.

A capacity $\gamma$ is <u>continuous</u> if for all closed E and $x \in \mathbb{T}$, $\gamma(E \cup \{x\}) = \gamma(E)$.

Let $h : [0, \infty) \to [0, \infty)$ be a non-decreasing, continuous at 0 function, with $h(0) = 0$ and $h(t) > 0$ for $t > 0$. The <u>Hausdorff measure</u> of a Borel set E associated with h is defined by

$$\mu_h(E) = \sup_{\epsilon>0} \inf\Big\{\sum_j h(|I_j|) : I_j \text{ open intervals with } E \subseteq \bigcup_j I_j \text{ and } |I_j| \leq \epsilon\Big\}.$$

(Note that one could have $\mu_h(E) = \infty$ for some E, h).

The following generalization of Menshov's Theorem (that there are $M_0$-sets of Lebesgue measure 0) was proved by Ivashev-Musatov [1] (for $E = \mathbb{T}$) and Kaufman [2] for Hausdorff measures. The extension to capacities and a totally different proof based on the basis theorem for $U_0$ and VI.1.7 were first noted by Kechris-Louveau. We give below an argument based directly on Theorem 2.

<u>Theorem 3</u>. Let $\gamma$ be a continuous capacity or a Hausdorff measure on $\mathbb{T}$, and let E be an $M_0$-set. Then there exists an $M_0$-set $F \subseteq E$ with $\gamma(F) = 0$.

(Actually, at least for h continuous, the case of Hausdorff measures $\mu_h$ is included in the case of continuous capacities, since then the closed sets E of $\mu_h$-measure 0 are exactly those for which $\gamma_h(E) = 0$ for some continuous capacity $\gamma_h$—see Dellacherie [1]).

<u>Proof</u>. We use a similar idea as in the construction of comeager sets of (Lebesgue) measure 0. (A set is <u>comeager</u> if its complement is of the first category).

We can assume that $\emptyset \neq E \in M_0^p$. Let first $\gamma$ be a continuous capacity, and

$\{x_m\}$ a dense subset of E. For each n we will construct a dense open set $V_n$ in (the relative topology of) E with $\gamma(V_n) \leq \frac{1}{n}$. If $G = \cap_n V_n$ then $\gamma(G) = 0$ and by Theorem 2 $G \in \mathcal{M}_0$, so there is closed $F \subseteq G$ with $F \in M_0$ and $\gamma(F) = 0$. To construct $V_n$ we find inductively on m, using the continuity of $\gamma$, open intervals $I_n^m$ with $x_m \in I_n^m$ such that $\gamma\left(\overline{I_n^0} \cup \ldots \cup \overline{I_n^m}\right) < \frac{1}{n}$ for each m and then we let $V_n = \cup_m I_n^m \cap E$.

For the case where $\gamma = \mu_h$ is a Hausdorff measure the argument is similar letting now $V_n = \cup_n I_n^m \cap E$, where $I_n^m$ are chosen so that $x_m \in I_n^m$ and $\sum_m h(|I_n^m|) \leq \frac{1}{n}$. Then there is a closed $F \subseteq G = \cap_n V_n$ in $M_0$. It remains to verify that $\mu_h(G) = 0$. Since $\sum_m h(|I_n^m|) \leq \frac{1}{n}$ we have $h(|I_n^m|) \xrightarrow[m]{} 0$, so $|I_n^m| \xrightarrow[m]{} 0$. Let $m_n$ be such that $|I_n^{m_n}| = \sup_m |I_n^m|$. Again as $h(|I_n^{m_n}|) \xrightarrow[n]{} 0$ it follows that $|I_n^{m_n}| \xrightarrow[n]{} 0$. As each $\{I_n^m\}_m$ is a covering of G we have $\mu_h(G) = 0$ and we are done. □

A further immediate consequence of Theorem 2 is the following result of Lyons [3], answering a question of Kahane and Salem.

A sequence $\{x_k\}_{k\in\mathbb{N}}$, $x_k \in [0, 2\pi]$, is <u>uniformly distributed</u> if for any interval $I \subseteq [0, 2\pi]$, letting $|I|$ = length of I,

$$\lim_N \frac{1}{N} \operatorname{card}(\{k \leq N : x_k \in I\}) = \frac{|I|}{2\pi}.$$

A set $P \subseteq [0, 2\pi]$ is called a $W^*$-<u>set</u> if there is some increasing sequence of positive integers $\{n_k\}_{k\in\mathbb{N}}$ such that for all $x \in P$, $\{n_k x \pmod{2\pi}\}$ is not uniformly distributed.

Thus the concept of a $W^*$-set is a weakening of that of an H-set. Kahane and Salem asked in 1964 if every $W^*$-set is in $\mathcal{U}_0$. Lyons answered this in the negative in 1983 by showing that for $q \geq 2$ an integer the set $W^*(\{q^k\}) = \{x : \{q^k x (\mathrm{mod}\ 2\pi)\}$ is not uniformly distributed$\}$ is in $\mathcal{M}_0$. The elements of this set are called <u>non-normal in base q</u>. However the set $W^*(\{q^k\})$ is comeager (this is due to Salat as pointed out in Lyons [1]), so one has now a new proof of Lyons' result as a direct application of Theorem 2. Here are the details

<u>Theorem 4</u> (Lyons [3]). The set $W^*(\{q^k\})$ of non-normal in base $q \geq 2$ numbers is in $\mathcal{M}_0$.

<u>Proof</u>. Let $P = \{x \in [0, 2\pi] : \overline{\lim}_N (|\frac{1}{N} \sum_{k=1}^{N} e^{iq^k x}|) = 1$. First we check that P is comeager: The complement of P is included in the union of the closed sets.

$$F_{K,r} = \{x \in [0, 2\pi] : \forall N \geq K \left|\frac{1}{N} \sum_{k=1}^{N} e^{iq^k x}\right| \leq r\}$$

for $K \in \mathbb{N}$, and rational $r < 1$. And if $F_{K,r}$ contained an interval I, it would contain a point x of the form $\frac{m}{q^{k_0}} \cdot 2\pi$ for some $k_0 \in \mathbb{N}$, and thus for $k \geq k_0$, $e^{iq^k x} = 1$, a contradiction. By Theorem 2, $P \notin \mathcal{U}_0$. It remains to show that $P \subseteq W^*(\{q^k\})$. But this follows immediately from Weyl's criterion: A sequence $\{x_k\}$ is uniformly distributed iff for all non-zero $m \in \mathbf{Z}$,

$$\lim_N \left(\frac{1}{N} \sum_{k=1}^{N} e^{imx_k}\right) = 0.$$

(To see this, notice that given $\{x_k\}$ if $\mu_N = \frac{1}{N} \sum_{k=1}^{N} \delta_{x_k}$, then the definition of uniform distribution says that $\mu_N(I) \to \lambda(I)$, where $\lambda$ is (normalized) Lebesgue measure on $[0, 2\pi]$ or equivalently $\mu_N \xrightarrow{w^*} \lambda$, which is exactly the condition in

Weyl's criterion). □

Finally let us note the following application of Theorem 1, pointed out in Debs-Saint Raymond [1]. After II.5.6 we remarked that conditions (i)–(vi) of II.5.6 are equivalent for all Borel E as well. We sketched the argument quoting a result from Zygmund [1] for the direction (vi) ⇒ (i). Using Theorem 1 one can give a different proof of (vi) ⇒ (i) as follows: Let E be Borel (or even $\underset{\sim}{\Sigma}^1_1$) and in $\mathcal{U}_0$, and assume towards a contradiction that there is a non-0 measure $\mu$ with $\sum\hat{\mu}(n)e^{int} = 0$ off E. By Theorem 1 let $E \subseteq \bigcup_n E_n$, where $E_n \in U_0$. Then $\sum\hat{\mu}(n)e^{int} = 0$ off $\bigcup_n E_n$, so that the argument in the proof of Bary's Theorem I.5.1 applies mutatis mutandis to give a contradiction.

The ideas involved in our alternate proof of Theorem 1 can be also used to establish a largeness property of the class of Rajchman measures.

Definition 5. Let $E \in K(\mathbb{T})$. A set $X \subseteq PROB(E)$ is called almost comeager if for every open $G \subseteq PROB(E)$, $G \neq \emptyset$ and every sequence $G_n \subseteq PROB(E)$ of open, dense in G and convex sets we have $X \cap (\bigcap_n G_n) \neq \emptyset$.

Notice that the class of Rajchman probability measures on $\mathbb{T}$ is meager (i.e. of the first category) in $PROB(\mathbb{T})$. This is because this class is contained in $\bigcup_n F_n$, where $F_n = \{\mu \in PROB(\mathbb{T}) : \forall\, |k| \geq n\ (|\hat{\mu}(k)| \leq \frac{1}{2})\}$. However we have

Theorem 6 (Kechris-Louveau [1]). Let $\emptyset \neq E \in M_0^p$. Then the class of Rajchman measures in PROB(E) is almost comeager in PROB(E).

Proof. Fix $\emptyset \neq G \subseteq \mathrm{PROB}(E)$, G convex, open. Let also $V \subseteq G$ be convex, open and dense in G and therefore in $Q = \overline{G}^{w^*}$. By Lemma 2.2, $R_V$ = the class of Rajchman measures in V is dense in Q and of course convex. Now $R_Q$ = the class of Rajchman measures in Q is a norm-closed subset of PF and $\overline{R_V}^{w^*} \cap PF = Q \cap PF = R_Q$. Since $\overline{R_V}^{w^*} \cap PF = \overline{R_V}^{w} = \overline{R_V}$, it follows that $R_V$ is norm-dense in $R_Q$ and, since $R_V = V \cap R_Q$, it is clearly open in $R_Q$ with its strong topology.

To summarize: For every non-$\emptyset$ open, convex $G \subseteq \mathrm{PROB}(E)$ and every open, convex $V \subseteq G$ which is dense in G, $R_V$ is open and dense in $R_Q$ in the norm-topology. So if $G \neq \emptyset$ is open in PROB(E) and $G_n$ open, dense in G and convex, then assuming without loss of generality that G itself is convex (since it contains certainly a convex, non$-\emptyset$ open subset), we have by the Baire Category Theorem in $R_Q$, which being closed in PF is Polish, $\cap_n R_{G_n} = (\cap_n G_n) \cap R_Q \neq \emptyset$ and we are done. □

Note that the preceding theorem also easily implies statement (*) : If $G = \cap_n V_n$ with $V_n$ open, dense in $E \in M_0^p$ $(E \neq \emptyset)$, then $V_n^* = \{\mu \in \mathrm{PROB}(E) : \mu(V_n) > \frac{1}{2}\}$ is convex, open and dense in PROB(E), so $\cap_n V_n^*$ contains a Rajchman measure $\mu$. Since $\mu(G) \geq \frac{1}{2}$, $G \in \mathcal{M}_0$.

Remarks (i). The first proof that every $G_\delta$ set (and thus every set having the property of Baire) in $\mathcal{U}_0$ is of the first category, does not require the full force of Theorem 1 and can be gotten by a direct application of VI.1.7 as follows (a similar argument works within any given $M_0^p$-set):

Let $G \in G_\delta$ be in $\mathcal{U}_0$ but of the second category, towards a contradiction.

Then G is dense in some open interval I. Let $E = \overline{I}$, $B = K(E) \cap U_0'$, $D = G \cap E$ and $J_n = K(V_n)$, where $G \cap E = \bigcap_n V_n$, $V_n$ open in E. Apply then Lemma VI.1.7, noticing that (*) there needs only the proof of VI.2.5. This produces $K \notin (U_0')_\sigma = U_0$ and $K_n \in U_0$ with $F \in K(\mathbb{T})$, $F \subseteq K - \bigcup_n K_n \Rightarrow F \in \bigcap_n J_n$, i.e. $K - \bigcup_n K_n \subseteq G$ thus $K - \bigcup_n K_n \in \mathcal{U}_0$. So $K \in \mathcal{U}_0$, a contradiction. Note that no definability arguments are used here.

Of course this kind of argument is quite general and shows the following:

If E is compact metrizable, $B \subseteq K(E)$ is hereditary, $I = B_\sigma$ is calibrated and for each $V \neq \emptyset$ open in E, there is $F \subseteq V$ with $F \in I - B$, then every set $P \in I^{int}$ with the property of Baire is of the first category.

(ii). Because of Theorem 1, the $\sigma$-ideal $\mathcal{U}_0$ has many similarities with the $\sigma$-ideal of countable sets. Thus one could develop a descriptive set theoretic study of $\mathcal{U}_0$ in a fashion analogous to that of the case of countable sets (see for example Kechris [1]). Among other things it can be shown for instance that Theorem 1 can be extended to sets P which are $\underset{\sim}{\Pi}^1_1$, but the proof uses strong axioms of set theory. It can be also shown that this is not provable in ZFC. We do not know however if it is equiconsistent with ZFC.

Notice also that (say among Borel sets) both the countable sets and the sets in $\mathcal{U}_0$ are characterized as the common null sets of a class of measures, the positive continuous measures for the countable case, and the positive Rajchman measures for the $\mathcal{U}_0$ case. Perhaps there is a set of properties of a class of measures, shared by these two examples, which allows a similar theory to be developed for the $\sigma$-ideal of their common null sets.

## §4. The class $U_1$ revisited

The method used in our alternative proof of Theorem 3.1 is actually applicable in certain instances in the context of pseudomeasures and allows us to prove the following extension of the Piatetski-Shapiro decomposition theorem for U (VI.2.3(ii)).

Theorem 1 (Kechris-Louveau [1]). Let $P \subseteq \mathbb{T}$ be $\undertilde{\Sigma}^1_1$. If $P \in U^{int}$, then there are $E_n \in U'_1$ with $P \subseteq \cup_n E_n$.

Proof. We will give the proof for $P \in G_\delta$. Exactly as in our alternative proof of Theorem 3.1 a projection argument handles the general case.

Since every $U_1$-set is a countable union of $U'_1$-sets it is enough to show that every $P \in U^{int}$, $P \in G_\delta$ can be covered by countably many $U_1$-sets. As in §3 this is equivalent to showing the following:

(**) If $\varnothing \neq E \in K(\mathbb{T})$ is in $M_1^p$ and $G \subseteq E$ is dense $G_\delta$ in E, then $G \notin U^{int}$.

(Recall that $E \in M_1^p \Leftrightarrow \forall$ open $V$ $(V \cap E \neq \varnothing \Rightarrow \overline{V \cap E} \in M_1)$).

The proof is similar to that of (*) in §3. We will need the following standard fact.

Lemma 2. Let $x \in \mathbb{T}$ and $V_n$ a sequence of nbhds of x with $\text{diam}(V_n) \to 0$. Let $S_n \in PM(\overline{V}_n)$ be such that $S_n(0) = 1$ and $\sup\|S_n\|_{PM} < \infty$. Then $S_n \xrightarrow{w^*} \delta_x$.

Proof. Fix $f \in A$ and $\epsilon > 0$. If $\tilde{f}(t) = f(x) - f(t)$ then $\tilde{f} \in A$ and $\tilde{f}(x) = 0$, so by II.2.2, $\tau_{x,h} \cdot \tilde{f} \to 0$ in the A-norm as $h \to 0$. It follows that for all large enough n, there is $g \in A$ with $\|g\|_A < \epsilon$ and $g = \tilde{f}$ in a nbhd of $\overline{V_n}$. Since $\text{supp}(S_n) \subseteq \overline{V_n}$, it follows that for all large enough n,

$$\begin{aligned} |\langle\tilde{f}, S_n\rangle| &= |\langle g, S_n\rangle| \\ &\leq \|g\|_A \cdot \|S_n\|_{PM} \\ &\leq K\epsilon \end{aligned}$$

for some constant K, independent of n. Thus for all large enough n,

$$\begin{aligned} |\langle f, \delta_x\rangle - \langle f, S_n\rangle| &= |f(x) - \langle f, S_n\rangle| \\ &= |\langle\tilde{f}, S_n\rangle|, \text{ since } S_n(0) = 1 \\ &\leq K\epsilon \end{aligned}$$

and the proof is complete. □

Let now $\emptyset \neq F \in M_1^p$, $V \subseteq F$ dense open in F. Put

$$\underset{\sim}{V} = \{S \in PF : \exists K \in K(\mathbb{T}) \ (K \subseteq V \ \& \ S \in N(K))\}.$$

Then $\underset{\sim}{V}$ is a subspace of $N(F) \cap PF$ and we claim that $\underset{\sim}{V}$ is $w^*$-dense in N(F). For that it is enough to show that $\overline{V}^{w^*}$ contains all the measures with finite support in

F. Since $\underset{\sim}{V}$ is a subspace it is enough to consider only $\delta_x$, $x \in F$. We can clearly assume that $x \in V$, since $V$ is dense in $F$. Let $x \in W_n$, where $W_n$ is open in $F$, $W_n \subseteq \overline{W_n} \subseteq V$ and $\mathrm{diam}(W_n) \longrightarrow 0$. Since $F \in M_1^p$, choose $S_n \in N(\overline{W_n}) \cap PF$ with $\|S_n\|_{PM} < 2$, $S_n(0) = 1$. Then $S_n \xrightarrow{w^*} \delta_x$ and since $S_n \in \underset{\sim}{V}$ we are done.

In particular $N(F) \cap PF = \overline{\underset{\sim}{V}}^{w^*} \cap PF = \overline{\underset{\sim}{V}}^{w} = \overline{\underset{\sim}{V}}$ and we can proceed exactly as in the proof of (*) in §3. □

Remark. D. Colella [1] has also independently shown that if $E \in M_1^p$ and $E_n$ is an increasing sequence of closed subsets of $E$ with $E_n \in M_1^p$ and $\cup_n E_n$ dense in $E$, then for any $S \in PF \cap N(E)$ there are $S_n \in PF \cap N(E_n)$ with $\|S_n - S\|_{PM} \to 0$.

We present now some applications.

Let $\Sigma S(n)e^{inx}$ be a non-0 trigonometric series, which converges to 0 almost everywhere. (In particular $S \in PF$). Let

$$RN_S = \{x \in \mathbb{T} : \sum_{|n| \leq N} S(n)e^{inx} = S_N(x) \text{ is unbounded}\}$$

be the so-called reduced nucleus of $S$. It is an old problem (see Bary [1]) to find out if $RN_S$ is always a $\mathcal{M}$-set.

Definition 3. A pseudomeasure $S$ satisfies synthesis if

$$f \in A,\ f = 0 \text{ on } \mathrm{supp}(S) \Rightarrow \langle f, S\rangle = 0.$$

We have now the following corollary.

Corollary 4. Let $\Sigma S(n)e^{inx}$ be a trigonometric series converging to 0 almost everywhere. If S satisfies synthesis, then the reduced nucleus $RN_S$ of S is a $\mathcal{M}$-set in fact contains a M-set, i.e. is not in $U^{int}$.

Proof. First notice that $RN_S$ is a $G_\delta$ set and it is dense in supp(S), as it follows easily from I.5.3 (see also the proof of I.5.5) and II.3.5. Finally since S is a pseudofunction satisfying synthesis we have that supp(S) $\in M_1^p$ so we are done. □

From this we obtain a further conclusion concerning the Interior and Union Problems.

Corollary 5. Let G be a set in $U^{int}$. Then there is no non-0 trigonometric series $\Sigma S(n)e^{inx}$ with S satisfying synthesis such that $\Sigma S(n)e^{inx} = 0, \forall x \notin G$. In particular if $G_n \in U^{int}$, $G_n \in \underset{\sim}{\Sigma}_3^0$ ($\equiv G_{\delta\sigma}$), then there is no such S with $\Sigma S(n)e^{inx} = 0, \forall x \notin \cup_n G_n$.

Proof. Suppose such S exists towards a contradiction. Then $RN_S \subseteq G$ and $RN_S \notin U^{int}$, so $G \notin U^{int}$. The last conclusion follows from VII.1.4. □

So if the Union Problem admits a negative solution for $\underset{\sim}{\Sigma}_3^0$ sets, this would have to incorporate a construction of non-synthesis sets. In fact, even more, if the Interior Problem admits a negative solution, then since in Corollary 4 we have only used that supp(S) $\in M_1^p$, it follows that such a negative answer would have to incorporate the existence of sets in $U_1 - U$ and thus in $U_1' - U$ (Körner's Theorem;

see VII.3.8 and VII.3.4). Perhaps the construction of Debs-Saint Raymond in VII.4.1, which shows that within sets in $U'_1$ one can always find $G_\delta$ dense subsets in $U^{int}$ is relevant here. (Recall from VII.4 that is is not known if these sets are in $\mathcal{M}$ or not).

Combining this aforementioned result of Debs-Saint Raymond with Theorem 1 we have the following characterization.

Corollary 6. Let $E \in K(\mathbb{T})$. Then the following are equivalent:

(1) $E \in M_1^p$,

(2) Every $G \subseteq E$ with the property of Baire in E which is non-meager in E is not in $U^{int}$. (Here we can restrict G to be a $G_\delta$. Also we can replace $U^{int}$ by $(U')^{int}$).

Proof. (1) ⇒ (2): Assume $E \in M_1^p$. If G is as in (2) then for some $V \subseteq E$ open in E, G contains a $G_\delta$ set G′ dense in V. Since $E' = \bar{V}$ is also in $M_1^p$ it follows that G′ and thus G is not in $U^{int}$.

(2) ⇒ (1): Let $E \notin M_1^p$, so there is $\varnothing \neq V \subseteq E$ open in E such that $\bar{V} \in U'_1$. Then by Debs-Saint Raymond (VII.4.1) there is $G \subseteq \bar{V}$ dense $G_\delta$ in $\bar{V}$ with $K(G) \subseteq U'$, so (2) fails. □

We can similarly provide an "intrinsic" characterization of the σ-ideal

$$U_1^* = (U_1)_\sigma \;(=(U'_1)_\sigma)$$

generated by the $U_1$-(or $U'_1$-) sets in terms of the $\sigma$-ideal U (or even just U′).

Corollary 7. The $\sigma$-ideal $U_1^*$ is the smallest $\sigma$-ideal I of closed sets in $\mathbb{T}$ such that every $G_\delta$ set in $U^{int}$ (or even $(U')^{int}$) can be covered by countably many sets in I.

Proof. By Theorem 1 $U_1^*$ has this covering property. If now I is any $\sigma$-ideal with this property and $E \in U_1^* - I$, towards a contradiction, there is $F \neq \emptyset$, $F \in K(E)$ which is in $U'_1$ and is I-perfect, i.e. for every non-$\emptyset$ open V in F, $\overline{V} \notin I$. Then subsets of F in I are nowhere dense in F and so all $G_\delta$ subsets of F in $U^{int}$ are meager in F. But this contradicts VII.4.1. □

Let us note here another characterization of $U_1^*$ in terms of the concept of synthesis. From the proof of VI.3.8(iii) it follows easily that for $E \in K(\mathbb{T})$

$$E \in M_1^p \Leftrightarrow \exists S \in PF(S \text{ satisfies synthesis and } E = \operatorname{supp}(S)).$$

From that one has immediately for $E \in K(\mathbb{T})$

$E \in U_1^* \Leftrightarrow$ E supports no non-0 pseudofunction which satisfies synthesis

$\Leftrightarrow$ Every trigonometric series $\Sigma S(n)e^{inx}$ for which S satisfies synthesis and $\Sigma S(n)e^{inx} = 0$, $\forall x \notin E$, is identically zero.

Thus $U_1^*$ appears as an interesting analog of U and $U_0$. Moreover $U_1^*$ has a natural extension to arbitrary subsets of $\mathbb{T}$, namely if

$$P \in \mathcal{U}_1^* \Leftrightarrow \text{Every trigonometric series } \Sigma S(n)e^{inx} \text{ for which } S \text{ satisfies synthesis and } \Sigma S(n)e^{inx} = 0,\ \forall x \notin E, \text{ is identically zero}$$

then $\mathcal{U}_1^* \cap K(\mathbb{T}) = U_1^*$. Note that by Corollary 5 $U^{int} \subseteq \mathcal{U}_1^*$.

Finally one can prove an analog of Theorem 3.3 for $M_1$-, M-sets. The proof is the same.

<u>Corollary 8</u>. Let $\gamma$ be a continuous capacity or a Hausdorff measure on $\mathbb{T}$ and let $E \notin U_1^*$. Then there is $F \subseteq E$, $F \in M$ with $\gamma(F) = 0$.

It is not known if in the conclusion of this corollary one can get $F \notin U_1^*$. However, by a recent result of Dougherty and Kechris [1] (answering a question of Kaufman), one cannot assume only that $E \in M$. In fact they show that for $E \in K(\mathbb{T})$,

$$E \notin U_1^* \Leftrightarrow \text{For all Hausdorff measures } \mu_h \text{ on } \mathbb{T} \text{ there is } F \subseteq E,\ F \in M \text{ with } \mu_h(F) = 0.$$

So one has a further intrinsic characterization of $U_1^*$ in terms of U.

There are several problems suggested by the results in this section. First from Theorem 1 it follows that every $\mathcal{U}$-set is of the first category in any $M_1^p$-set E containing it in which it has the property of Baire. Is this still true for E just an $M^p$-set? By the Debs-Saint Raymond construction an affirmative answer would solve negatively the Interior Problem. Also in Theorem 1 it is natural to ask if the conclusion goes through under the weaker assumption that $P \in (U_1^*)^{int}$, thus obtaining a full extension of the Piatetski-Shapiro theorem $(U_1 \subseteq (U_1')_\sigma)$. If so, then one would have for $\Sigma_1^1$ sets, $\mathcal{U}_1^* = (U_1^*)^{int} = (U_1^*)^{ext}$. In view of the results in this section the $\sigma$-ideal $U_1^*$ appears to be of interest. Not much is known however about it. Is it $\Pi_1^1$? Is it calibrated?

And we finish this chapter by using the methods discussed here to give a different proof of the Piatetski-Shapiro decomposition theorem for $U_1$(VI.2.3(ii)):

As usual it is enough to show that if $\emptyset \neq E$ is closed and has the property

$$\emptyset \neq V \subseteq E, \text{ V open in E} \Rightarrow \bar{V} \notin U_1'$$

then $E \in M_1$.

Let for $\epsilon > 0$, $N^\epsilon(E) = \{S \in N(E) : R(S) < \epsilon\}$. First note that $N^\epsilon(E)$ is $w^*$-dense in N(E). This follows easily as in the proof of Lemma 2.2. We need next the following:

Lemma 9. If $C \subseteq PM^\epsilon = \{S \in PM : R(S) < \epsilon\}$ is convex and $T \in \bar{C}^{w^*}$, there is a sequence $T_n \in C$, $T_n$ T with $\|T_n\|_{PM} < \|T\|_{PM} + \epsilon$.

Proof. It is enough to prove this for $T \in C^{(1)}$ (= the set of $w^*$-limits of sequences from C). Because this immediately implies that $C^{(1)}$ is closed under $w^*$-limits of sequences, so by the following strengthening of V.2.3, which can be established by the same proof, we have that $C^{(1)} = \overline{C}^{w^*}$: If X is a separable Banach space and $Y \subseteq X^*$ is convex and such that $Y \cap B_r(X^*)$ is $w^*$-closed for all $r > 0$, then Y is $w^*$-closed in $X^*$.

So let $T \in C^{(1)}$, say $S_n \xrightarrow{w^*} T$, with $S_n \in C$, $\|S_n\|_{PM} < M$. Given any $N \in \mathbb{N}$, $0 < \delta < \epsilon$ we will find $S \in C$ with $|T(m) - S(m)| < \delta$, $\forall |m| \leq N$ and $\|S\|_{PM} < \|T\|_{PM} + \epsilon$. For this note that for each $p \geq N$ there is $S \in C$, $\|S\|_{PM} < M$ and $q > p$ with $|T(m) - S(m)| < \delta$ if $|m| \leq p$ and $|S(m)| < \epsilon$ if $|m| > q$. Then we are done by an "iterating and averaging" argument as in the proof of Theorem 2.1. □

It follows that given any $S \in N(E)$, with $\|S\|_{PM} < a$, there is a sequence $T_n \in N^{\epsilon}(E)$ with $T_n \xrightarrow{w^*} S$ and $\|T_n\|_{PM} < a + \epsilon$.

Define now inductively $S_1, S_2, \ldots \in N(E)$, $0 < n_1 < n_2 < \ldots$ such that $\|S_1\|_{PM} < 1$, $\|S_k\|_{PM} < 1 + \Sigma_{i \leq k-1} 2^{-i}$ if $k \geq 2$, $\|S_k\|_{PM}^{n_i} < 2^{-i}$ if $k \geq i$ and $S_k(0) > \frac{1}{2}$. Then there is a subsequence $S_{k_i} \xrightarrow{w^*} S$. Moreover $S \in N(E)$, $S \neq 0$ and $S \in PF$, thus $E \in M_1$.

The definition of $S_1, S_2, \ldots, n_1, n_2, \ldots$ is as follows: Choose first $S_1$ so that $S_1 \in N(E)$, $R(S_1) < 1$, $\|S_1\|_{PM} < 1$, $S_1(0) > \frac{1}{2}$ and then choose $n_1$ with $\|S_1\|_{PM}^{n_1} < \frac{1}{2}$. Now for every $m \geq n_1$ there is $m' > m$ and $S \in N(E)$ with $\|S\|_{PM} < 1 + \frac{1}{2}$, $\|S\|_{PM}^{n_1,m} < \|S\|_{PM}^{n_1} + \epsilon < \frac{1}{2}$, $\|S\|_{PM}^{m'} < \frac{1}{4}$ and $S(0) > \frac{1}{2}$. By "iterating and averaging" obtain $S_2$ and then $n_2$, etc.

# Chapter IX. Characterizing Rajchman Measures

The closed sets of extended uniqueness are the common null sets of all the Rajchman measures. In this chapter we establish the "dual" theorem, due to Lyons, that the Rajchman measures are exactly the measures which annihilate all $U_0$-sets.

In the first section, we present a very general result of Mokobodzki in measure theory which implies this result of Lyons. In the second section, we give Lyons' original proof, which yields also a stronger characterization by replacing $U_0$ by a subclass of it, the so called Weyl- or W-sets.

## §1. A theorem of Mokobodzki in measure theory

In this section, we will work with positive subprobabilities on $\mathbf{T}$, i.e. positive measures satisfying $\mu(\mathbf{T})$ $(= \hat{\mu}(0)) \leq 1$. We let $M_1^+(\mathbf{T})$ be the space of such measures, which is a $w^*$-compact subset of $M(\mathbf{T})$ (and of PM). What we want to study is the set $\mathcal{R} \subseteq M_1^+(\mathbf{T})$ of measures $\mu \in M_1^+(\mathbf{T})$ which are Rajchman measures. We first look at general subsets H of $M_1^+(\mathbf{T})$. For each such H, we let

$$\mathrm{Null}(H) = \{E \in K(\mathbf{T}) : \forall \mu \in H\ (\mu(E) = 0)\}$$

and we want to find out under what conditions Null(H) characterizes H, in the sense that one has

$$H = \{\mu \in M_1^+(\mathbb{T}) : \forall\, E \in \text{Null}(H)(\mu(E) = 0)\}.$$

We need first some definitions. For convenience we will write in the sequel

$$\mu(f) \text{ instead of } \int f\, d\mu.$$

A set $H \subseteq M_1^+(\mathbb{T})$ is <u>hereditary</u> if $\mu \in H$ and $\nu \leq \mu$ imply $\nu \in H$ (here $\nu \leq \mu$ if $\nu(E) \leq \mu(E)$ for Borel E, or equivalently for every continuous—or Borel—positive function f on $\mathbb{T}$, $\nu(f) \leq \mu(f)$). A set $H \subseteq M_1^+(\mathbb{T})$ is a <u>band</u> if $\mu \in H$ and $\nu \prec\prec \mu$ imply $\nu \in H$, where $\nu \prec\prec \mu$ means $\nu$ is absolutely continuous with respect to $\mu$.

Let $\Lambda$ be a (Borel) probability measure on the compact, metrizable ( in the weak*-topology) space $M_1^+(\mathbb{T})$. The <u>barycenter</u> $b(\Lambda)$ of the measure $\Lambda$ is the unique measure in $M_1^+(\mathbb{T})$ satisfying, for any $f \in C(\mathbb{T})$,

$$b(\Lambda)(f) = \int \mu(f)\, d\Lambda(\mu).$$

The measure $\Lambda$ <u>concentrates</u> on $H \subseteq M_1^+(\mathbb{T})$ if $\Lambda(M_1^+(\mathbb{T}) - H) = 0$.

A set $H \subseteq M_1^+(\mathbb{T})$ is <u>strongly convex</u> if for every probability measure $\Lambda$ on $M_1^+(\mathbb{T})$ which concentrates on H, $b(\Lambda) \in H$.

Note that (by considering convex combinations of Dirac measures) any strongly convex set H is convex. And if $H \subseteq M_1^+(\mathbb{T})$ is closed the converse also holds, as any probability measure concentrated on H can be $w^*$-approximated by such convex combinations, using compactness.

So in particular if H is strongly convex and $L \subseteq H$ is closed, <u>the convex closed envelope</u>

$$\overline{co(L)} = \{b(\Lambda) : \Lambda \text{ a probability measure concentrated on } L\}$$

is also contained in H. This is actually a characterization of strongly convex sets:

<u>Lemma 1</u>. Suppose $H \subseteq M_1^+(\mathbb{T})$ is such that for any closed set $L \subseteq H$, $\overline{co(L)} \subseteq H$. Then H is strongly convex.

<u>Proof</u>. Let $\Lambda$ be a probability measure concentrating on H but on no closed subset of H (since otherwise there is nothing to prove), and let $\{L_n\}$ be an increasing sequence of closed sets, $L_n \subseteq H$, with $\Lambda(L_{n+1}) > \Lambda(L_n) > 0$ and $\Lambda(\bigcup_n L_n) = 1$. Let $L_0^* = L_0$, $L_{n+1}^* = L_{n+1} - L_n$. The measure $\frac{1}{\Lambda(L_n^*)} \cdot \Lambda \restriction L_n^* = \Lambda_n$ is a probability measure, concentrated on $L_n$. And one can write $\Lambda = \Sigma\, c_n \Lambda_n$, with $c_n = \Lambda(L_n^*)$, so that $0 < c_n < 1$, $\Sigma\, c_n = 1$. Let then $d_n \geq 1$ be a sequence of numbers with $d_n \to \infty$ as $n \to \infty$, and $\sum_{n>0} c_n d_n = 1$. Define then $a_n = c_n d_n$ and let for $n > 0$ $\Lambda_n' = (1 - \frac{1}{d_n})\Lambda_0 + \frac{1}{d_n} \Lambda_n$. Then $\Lambda = \sum_{n>0} a_n \Lambda_n'$. By the hypothesis, $b(\Lambda_n) \in \overline{co(L_n)} \subseteq H$, hence for each $n > 0$, $b(\Lambda_n') = (1 - \frac{1}{d_n})\, b(\Lambda_0) + \frac{1}{d_n}\, b(\Lambda_n) \in H$. But the sequence $b(\Lambda_n')$ converges, as $n \to \infty$, to $b(\Lambda_0) \in H$. So the closed set $L = \{b(\Lambda_0)\} \cup \{b(\Lambda_n') : n > 0\}$ is contained in H, and so is $\overline{co(L)}$. But $\Lambda = \Sigma_{n>0}\, a_n \Lambda_n'$ has the same barycenter as $\Lambda' = \Sigma_{n>0}\, a_n\, \delta_{b(\Lambda_n')}$ which concentrates on L, hence $b(\Lambda) \in \overline{co(L)} \subseteq H$. □

The result we will use for characterizing Rajchman measures is an easy corollary of a result of Mokobodzki in measure theory. The fact that Mokobodzki's result can be used this way was noticed by Louveau.

Theorem 2. Let $H \subseteq M_1^+(\mathbb{T})$ be a $\underset{\sim}{\Sigma}_1^1$ set. The following are equivalent:

(i) $H = \{\mu \in M_1^+(\mathbb{T}) : \forall E \in \text{Null}(H)\ (\mu(E) = 0)\}$,

(ii) H is a strongly convex, norm-closed band in $M_1^+(\mathbb{T})$.

Proof. One direction of Theorem 2 is very simple: If (i) holds, H is clearly a band, as $\text{Null}(H) \subseteq \text{Null}(\mu) \subseteq \text{Null}(\nu)$ implies $\text{Null}(H) \subseteq \text{Null}(\nu)$. Also if $\mu_n \in H$ norm-converges to $\mu$, i.e., $\|\mu_n - \mu\|_M \to 0$, $\mu(F) = 0$ for all $F \in \text{Null}(H)$, hence $\mu \in H$. And finally if $\Lambda$ is a probability measure concentrating on H and $E \in \text{Null}(H)$, one can choose a decreasing sequence of continuous functions $f_n$ with values in [0, 1] converging to $\chi_E$. Then for $\mu \in M_1^+(\mathbb{T})$, $\mu(E) = \lim_n \mu(f_n)$, hence by the Lebesgue convergence theorem

$$\begin{aligned} b(\Lambda)(E) &= \lim_n b(\Lambda)(f_n) = \lim_n \int \mu(f_n)\, d\Lambda(\mu) \\ &= \lim_n \int_H \mu(f_n) d\Lambda(\mu) \\ &= \int_H \mu(E) d\Lambda(\mu) = 0 \end{aligned}$$

hence $b(\Lambda) \in H$ and we are done. Note that in this direction we did not use the fact that H is $\underset{\sim}{\Sigma}_1^1$.

For the other direction, we will use a more general result of Mokobodzki. For each set $H \subseteq M_1^+(\mathbb{T})$, define its <u>envelope</u> env(H) by

$$\mathrm{env}(H) = \{\mu \in M_1^+(\mathbb{T}) : \forall f \in \mathrm{Bor}^+(\mathbb{T})(\mu(f) \leq \sup_{\nu \in H} \nu(f))\}$$

where $\mathrm{Bor}^+(\mathbb{T})$ is the set of bounded positive Borel functions on $\mathbb{T}$.

<u>Theorem 3</u> (Mokobodzki, see Dellacherie-Meyer [1]). Let H be a $\underset{\sim}{\Sigma}_1^1$ subset of $M_1^+(\mathbb{T})$. The envelop env(H) is the smallest set containing H which is hereditary, strongly convex and norm-closed in $M_1^+(\mathbb{T})$.

Granting Mokobodzki's Theorem, we can finish the proof of Theorem 2: We assume now that H is a strongly convex, norm-closed band, and we fix $\mu \in M_1^+(\mathbb{T})$ with $\mathrm{Null}(H) \subseteq \mathrm{Null}(\mu)$. We want to prove that $\mu \in H$. By the hypotheses on H and Mokobodzki's theorem, $H = \mathrm{env}(H)$ so it is enough to show that $\mu \in \mathrm{env}(H)$, i.e. for every function $f \in \mathrm{Bor}^+(\mathbb{T})$,

$$\mu(f) \leq \sup_{\nu \in H} \nu(f) \overset{\text{def}}{=} \gamma(f).$$

We claim that $\gamma(f) = \sup\{t : \exists \nu \in H(\nu(\{f > t\}) > 0)\} \overset{\text{def}}{=} \gamma'(f)$. To see this, note that if $t < \gamma(f)$, hence $\nu(f) > t$ for some $\nu \in H$, then $\nu(\{f > t\}) > 0$ for that $\nu$, hence $\gamma'(f) \geq t$. So $\gamma'(f) \geq \gamma(f)$. Conversely if for some $\nu \in H$, $\nu(\{f > t\}) > 0$, then $\nu' = \dfrac{1}{\nu(\{f > t\})} \cdot \nu \upharpoonright \{f > t\}$ is absolutely continuous with respect to $\nu$, hence $\nu' \in H$ and $\nu'(f) \geq t$. So $\gamma(f) \geq \gamma'(f)$. We finally show that $\mu(f) \leq \gamma'(f)$. Suppose it fails. Then for some t, $\mu(f) > t > \gamma'(f)$. Then $\mu(\{f > t\}) > 0$, hence for some closed $E \subseteq \{f > t\}$, $\mu(E) > 0$. But by the definition of $\gamma'$ any closed $E \subseteq \{f > t\}$

is in Null(H), and we get that Null(H) $\not\subseteq$ Null($\mu$), a contradiction finishing the proof.□

We turn now to the proof of Mokobodzki's Theorem. The main argument in the proof is a capacitability argument.

Recall that if E is a compact, metrizable space, a capacity $\gamma$ on E is a function $\gamma$ : Power(E) $\rightarrow$ [0, $\infty$) which is increasing, with $\gamma(\emptyset) = 0$, and satisfies

(i) $\gamma(\bigcup_n A_n) = \sup_n \gamma(A_n)$ for all increasing sequences $\{A_n\}$ of subsets of E

(ii) $\gamma(\bigcap_n E_n) = \inf_n \gamma(E_n)$ for all decreasing sequences $\{E_n\}$ in K(E).

The main result about capacities is the Choquet Capacitability Theorem. In order to prove it, we need first a lemma providing a nice representation of $\underset{\sim}{\Sigma}^1_1$ sets in compact, metrizable spaces.

Lemma 4. Let E be compact, metrizable and P $\subseteq$ E a $\underset{\sim}{\Sigma}^1_1$ set. There exists a sequence $\{F_s\}_{s \in \mathrm{Seq}\,\mathbb{N}}$ of closed subsets of E, indexed by the finite sequences of natural numbers, such that

(i) $t \supseteq s \Rightarrow F_t \subseteq F_s$,

(ii) $[\ell h(s) = \ell h(t) = n \;\&\; \forall j < n(s(j) \leq t(j))] \Rightarrow F_s \subseteq F_t$,

(iii) $P = \{x \in E : \exists \epsilon \in \mathbb{N}^{\mathbb{N}} \; \forall n \; (x \in F_{\epsilon \restriction n})\}$.

(Such a sequence $\{F_s\}$ is sometimes called a privileged Suslin scheme for P).

Proof. Let $F \subseteq E \times \mathbb{N}^{\mathbb{N}}$ be a closed set with P = proj(F). For each s $\in$ Seq $\mathbb{N}$, let $K_s$ be the closure of $\{x \in E : \exists\, \epsilon \in \mathbb{N}^{\mathbb{N}}\, [\epsilon \supseteq s \text{ and } (x, \epsilon) \in F]\}$. Finally let $F_s = \bigcup_{t \leq s} K_t$, where $t \leq s$ means $\ell h(t) = \ell h(s)$ and $\forall j < \ell h(s)\, (t(j) \leq s(j))$. As the union above is finite, each $F_s$ is closed in E and it is immediate to check that the $F_s$ satisfy (i) and (ii). Moreover, if $x \in P$, then $(x, \epsilon) \in F$ for some $\epsilon$, hence for all n, $x \in K_{\epsilon \upharpoonright n}$ and a fortiori $x \in F_{\epsilon \upharpoonright n}$. Suppose now $\epsilon \in \mathbb{N}^{\mathbb{N}}$ is such that $\forall n(x \in F_{\epsilon \upharpoonright n})$. For each $n \in \mathbb{N}$, let $B_n$ be the ball $\{y \in E : \text{dist}(x, y) \leq 2^{-n}\}$. By definition of the $F_s$, one can find for each n an $\epsilon_n \in \mathbb{N}^{\mathbb{N}}$ and a point $x_n \in B_n$ such that (i) $(x_n, \epsilon_n) \in F$, and (ii) $\epsilon_n \upharpoonright n \leq \epsilon \upharpoonright n$. The sequence $\{x_n\}$ converges to x. Moreover one easily checks that a subsequence of the $\epsilon_n$'s converges to some $\epsilon' \in \mathbb{N}^{\mathbb{N}}$. But as F is closed, this implies $(x, \epsilon') \in F$, hence $x \in P$ and (iii) is proved. □

Theorem 5 (Choquet Capacitability Theorem). Let E be compact metrizable, $\gamma$ a capacity on E, and P a $\Sigma^1_1$ subset of E. Then

$$\gamma(P) = \sup\{\gamma(F) : F \in K(E),\, F \subseteq P\}$$

Proof. As $\gamma$ is increasing, the left hand side is clearly $\geq$ than the right hand side. So it is enough to show that if $t \in [0, \infty)$ is such that $\gamma(P) > t$, one can find a closed set $F \subseteq P$ with $\gamma(F) \geq t$.

Let $\{F_s\}_{s \in \text{Seq}\,\mathbb{N}}$ be a privileged Suslin scheme for P, as given by Lemma 4, and let for $s \in \text{Seq}\,\mathbb{N}$

$$P_s = \{x \in E : \exists \epsilon \in \mathbb{N}^{\mathbb{N}}\, (\epsilon \supseteq s \text{ and } \forall n(x \in F_{\epsilon \upharpoonright n}))\}.$$

By the definition, $P_{\emptyset} = P$, and $P_s \subseteq F_s$. Moreover by property (ii) of the scheme $\{F_s\}$, $P_s$ is the increasing union of the sets $P_{s\hat{}(n)}$, $n \in \mathbb{N}$. Using these facts and the first property of capacities, one can easily construct by induction a sequence $\epsilon \in \mathbb{N}^{\mathbb{N}}$ such that for all n $\gamma(P_{\epsilon\restriction n}) > t$ : The case $n = 0$ is the hypothesis $\gamma(P) > t$. And if $\epsilon \restriction n$ is such that $\gamma(P_{\epsilon\restriction n}) > t$, then $\gamma(\bigcup_p P_{\epsilon\restriction n\hat{}(p)}) > t$, hence for some p $\gamma(P_{\epsilon\restriction n\hat{}(p)}) > t$. So we can set $\epsilon(n) = p$.

Now $\gamma$ is increasing, and $P_{\epsilon\restriction n} \subseteq F_{\epsilon\restriction n}$, hence $\gamma(F_{\epsilon\restriction n}) > t$. Let $F_\epsilon = \bigcap_n F_{\epsilon\restriction n}$. By the first property of the scheme, the $F_{\epsilon\restriction n}$'s are decreasing in n, and closed. So using the second property of capacities, $\gamma(F_\epsilon) \geq t$. Finally, clearly $F = F_\epsilon$ is closed and $F \subseteq P$. □

We will apply the capacitability theorem to the space $E = M_1^+(\mathbb{T})$. Fix a measure $\mu \in M_1^+(\mathbb{T})$. If f is a Borel function on $\mathbb{T}$ with $0 \leq f \leq 1$, define for each $H \subseteq M_1^+(\mathbb{T})$,

$$C_f(H) = (\sup_{\nu \in H} \nu(f)) + \mu(1 - f)$$

and let

$$C(H) = \inf\{C_f(H) : f \text{ Borel}, 0 \leq f \leq 1\}.$$

(Although we do not put $\mu$ as an index, these functions depend on $\mu \in M_1^+(\mathbb{T})$).

Note that if $H = \{\nu\}$ is a singleton, we have

$$C(H) = C(\{\nu\}) = \inf\{\nu(f) + \mu(1 - f) : f \in C(\mathbb{T}), 0 \le f \le 1\}$$
$$= (\mu \wedge \nu)(1)$$

where $\mu \wedge \nu$, the infimum of $\mu$ and $\nu$, is the largest measure in $M_1^+(\mathbb{T})$ which is smaller than both $\mu$ and $\nu$. Thus $\mu \wedge \nu = \frac{1}{2}(\mu + \nu - |\mu - \nu|)$, and also for Borel B,

$$(\mu \wedge \nu)(B) = \inf\{\mu(C) + \nu(B - C) : C \subseteq B, \text{ Borel}\}.$$

We show now that a similar result holds if H is convex compact.

Lemma 6. Let H be convex closed in $M_1^+(\mathbb{T})$. Then

$$C(H) = \inf\{C_f(H) : f \in C(\mathbb{T}), 0 \le f \le 1\} \tag{1}$$
$$= \sup\{(\mu \wedge \nu)(1) : \nu \in H\}. \tag{2}$$

Proof. Let C′(H) be the right hand side of (1), C″(H) the right-hand side of (2). Clearly as the infimum is taken over more functions, C(H) ≤ C′(H). Also

$$\sup_{\nu \in H} \inf_{\substack{f \in \mathrm{Bor}^+(\mathbb{T}) \\ 0 \le f \le 1}} (\nu(f) + \mu(1 - f)) \le \inf_{\substack{f \in \mathrm{Bor}^+(\mathbb{T}) \\ 0 \le f \le 1}} ((\sup_{\nu \in H} \nu(f)) + \mu(1 - f))$$

hence C″(H) ≤ C(H). So it remains to show that C′(H) ≤ C″(H), and this uses an easy application of the Hahn-Banach Theorem, known as the minimax lemma: If E is compact metrizable, $\mathcal{A}$ is a convex family of continuous real functions on E and $a \in \mathbb{R}$ is such that $\forall f \in \mathcal{A}\ \exists x \in E(f(x) \ge a)$, then there is a probability measure $\Lambda$ on E such that $\forall f \in \mathcal{A}(\Lambda(f) \ge a)$. To see this, we may assume that $a = 0$.

Consider in $C_{\mathbb{R}}(E)$ (= the space of continuous real functions on E) the open convex set $\mathcal{B} = \{f : f < 0\}$. Then $\mathcal{A} \cap \mathcal{B} = \emptyset$, hence by Hahn-Banach one can find a real measure $\Lambda \in M(E)$ and $c \in \mathbb{R}$ such that $\Lambda(f) < c \leq \Lambda(g)$ for all $f \in \mathcal{B}$ and $g \in \mathcal{A}$. As $\mathcal{B}$ is closed under positive multiplication, one easily gets that $\Lambda \in M^+(E)$ and $c \geq 0$, so that $\Lambda \neq 0$ and $\Lambda(g) \geq c \geq 0$ for all $g \in \mathcal{A}$ as required. Moreover one can normalize $\Lambda$ to get a probability measure with the same properties.

Applying this to E = H and the family

$$\mathcal{A} = \{\nu \mapsto \nu(f) + \mu(1 - f) : f \in C(\mathbb{T}), 0 \leq f \leq 1\}$$

which is clearly a convex family of continuous real functions on H, one gets that if $C'(H) > t$ there is a probability measure $\Lambda$ on H with

$$\int \nu(f)\, d\Lambda(\nu) + \mu(1 - f) \geq t \tag{3}$$

for all $f \in C(\mathbb{T})$, $0 \leq f \leq 1$. Now the barycenter $b(\Lambda)$ is in H, as H is convex closed, and by (3) $(b(\Lambda)\wedge\mu)(1) \geq t$. This shows $C''(H) \geq C'(H)$ and finishes the proof. □

We use Lemma 6 to show:

<u>Lemma 7</u>. The function C is a capacity on $M_1^+(\mathbb{T})$.

<u>Proof</u>. It is immediate that C is increasing, and $C(\emptyset) = 0$.

(1) Suppose $\{H_n\}$ is a decreasing sequence of closed sets, and let $H = \bigcap_n H_n$. Clearly $C(H) \leq \inf_n C(H_n)$. Let $H'_n = \overline{co(H_n)}$ and $H' = \overline{co(H)}$. As the function $\Lambda \mapsto b(\Lambda)$ is continuous, $H' = \bigcap_n H'_n$. We claim that $C(H) = C(H')$, and similarly $C(H_n) = C(H'_n)$. Clearly $C(H) \leq C(H')$. But if $\nu \in H'$, $\nu = b(\Lambda)$ for some probability measure $\Lambda$ concentrating on H, hence for any continuous $f \in C(\mathbf{T})$, $0 \leq f \leq 1$, $b(\Lambda)(f) = \int_H \nu(f) d\Lambda(\nu)$ is $\leq \sup_{\nu \in H} \nu(f)$, hence by Lemma 6,

$$C(H') \leq C(H).$$

So it remains to show that if for all n, $C(H'_n) > t$, then $C(H') \geq t$. By Lemma 6 again, there are $\nu_n \in H'_n$ with $(\nu_n \wedge \mu)(1) > t$, and by taking a subsequence if necessary, we may assume $\nu_n \xrightarrow{w^*} \nu \in M_1^+(\mathbf{T})$. But as the $H'_n$ are decreasing, $\nu \in \bigcap_n H'_n = H'$, and $(\nu \wedge \mu)(1) \geq t$, hence $C(H') \geq t$ by Lemma 6, and we are done.

(2) Suppose now $\{H_n\}$ is an increasing family of subsets of $M_1^+(\mathbf{T})$, towards proving that $C(\bigcup_n H_n) = \sup_n C(H_n)$. Clearly $C(\bigcup_n H_n) \geq \sup_n C(H_n)$. So assume for all n, $C(H_n) < t$, and let $f_n \in Bor^+(\mathbf{T})$, $0 \leq f_n \leq 1$, be such that $C_{f_n}(H_n) < t$. The sequence $\{f_n\}_{n \in \mathbb{N}}$ is bounded in $L^2(\mu)$, so by choosing a subsequence $\{n_k\}$, we can insure that $f_{n_k} \to g$ weakly in $L^2(\mu)$ and, by Mazur's Theorem, that some suitable convex combination $\{f'_k\}$ of the $f_{n_k}$ norm-converges to g in $L^2(\mu)$. (Recall here that $L^2(\mu)$ is its own dual). We may assume that $f'_k$ is a convex combination of $\{f_{n_p} : p \geq k\}$ and $\|f'_k - f'_{k+1}\|_{L^2(\mu)} \leq 4^{-k}$. It follows then that $\mu(\{x : |f'_k(x) - f'_{k+1}(x)| > 2^{-k}\}) \leq 4^{-k}$, and $\mu(\{x : \text{for some } k > k_0, |f'_k(x) - f'_{k+1}(x)| > 2^{-k}\}) \leq \frac{1}{3} 4^{-k_0}$. So $f'_k(x)$ converges $\mu$ — a.e. on $\mathbf{T}$. Let $f = \underline{\lim}\, f'_k$. The function f is Borel, and $0 \leq f \leq 1$. We claim that $C_f(\cup_n H_n) \leq t$, hence $C(\cup_n H_n) \leq t$ and the proof is finished. To prove the claim, notice that $f(x) = \lim f'_k(x)$ $\mu$ — a.e., so (since $0 \leq f'_k \leq 1$)

$\mu(1 - f) = \lim_k \mu(1 - f'_k)$. And for $\nu \in \bigcup_n H_n$, $\nu$ is in $H_n$ for all big enough n as the $H_n$ are increasing, and

$$\nu(f) = \nu\left(\sup_k \inf_m f'_{k+m}\right)$$

$$= \sup_k \nu\left(\inf_m f'_{k+m}\right)$$

$$\leq \sup_k \inf_m \nu\left(f'_{k+m}\right) = \underline{\lim}\ \nu(f'_k).$$

But for $m \geq n$ the $f'_{k+m}$ are convex combinations of the $f_p$ for $p \geq n$, and by the hypothesis for n big enough, $\nu(f_p) + \mu(1 - f_p) < t$. So this is true for the $f'_{k+m}$ too, and finally $\nu(f) + \mu(1 - f) \leq \underline{\lim}\ (\nu(f'_k) + \mu(1 - f'_k)) \leq t$ and we are done. □

Proof of Theorem 3. Let $H^*$ be the smallest hereditary, norm-closed and strongly convex set containing H. We want to show that $\mathrm{env}(H) = \{\mu \in M_1^+(T) : \forall f \in \mathrm{Bor}^+(T)\ (\mu(f) \leq \sup_{\nu\in H} \nu(f))\}$ is equal to $H^*$. Clearly env(H) is hereditary, and norm-closed, for if $f \in \mathrm{Bor}^+(T)$ and $\mu_n \to \mu$ in norm, $\mu_n(f) \to \mu(f)$. Moreover if $\Lambda$ is a probability measure concentrating on env(H), then $b(\Lambda)(f) = \int_{\mathrm{env}(H)} \mu(f)\, d\Lambda(\mu) \leq (\sup_{\nu\in H} \nu(f))$, for all $f \in \mathrm{Bor}^+(T)$, hence $b(\Lambda) \in \mathrm{env}(H)$. This shows $H^* \subseteq \mathrm{env}(H)$. To prove the converse, let $\mu \in \mathrm{env}(H)$, and consider the associated function C, which is a capacity by Lemma 7. For any $f \in \mathrm{Bor}^+(T)$ with $0 \leq f \leq 1$, $\mu(f) \leq \sup_{\nu\in H} \nu(f)$, hence $C_f(H) = (\sup_{\nu\in H} \nu(f)) + \mu(1 - f) \geq \mu(f) + \mu(1 - f) = \mu(1)$. So $C(H) \geq \mu(1)$. On the other hand, for $f = 0$, $C_f(H) = \mu(1)$, hence $C(H) = \mu(1)$.

We apply now the capacitability theorem: $C(H) = \sup\{C(L) : L \subseteq H,$ L compact$\}$. So for each n we can find a compact set $L_n \subseteq H$ with $C(L_n) >$

$\mu(1) - 2^{-n}$. Let $L'_n = \overline{co(L_n)}$. As $L_n \subseteq H$, $L'_n \subseteq H^*$, and $C(L'_n) > \mu(1) - 2^{-n}$. By Lemma 6, there is a measure $\nu_n \in L'_n$ with $(\nu_n \wedge \mu)(1) \geq \mu(1) - 2^{-n}$. As $H^*$ is hereditary, $\nu'_n = \nu_n \wedge \mu \leq \nu_n$ is in $H^*$. And finally as $\nu'_n \leq \mu$, $\|\mu - \nu'_n\|_M = (\mu - \nu'_n)(1) \leq 2^{-n}$, so $\nu'_n$ norm-converges to $\mu$. Hence $\mu \in H^*$ and we are done. $\square$

We can apply now the preceding results to the example we had in mind.

<u>Theorem</u> <u>8</u> (Lyons [2]). The Rajchman measures are characterized by their common null sets, i.e. $\mu \in M(\mathbb{T})$ is Rajchman iff $U_0 \subseteq \mathrm{Null}(\mu)$.

<u>Proof</u>. By definition of $U_0$, $U_0 \subseteq \mathrm{Null}(\mu)$ for every Rajchman measure. For the converse, we may restrict our attention to $M_1^+(\mathbb{T})$, for this implies the result for $\mu \in M^+(\mathbb{T})$ by scalar multiplication, hence for $|\mu|$ for any $\mu$ with $U_0 \subseteq \mathrm{Null}(\mu)$. But $\mu \prec\prec |\mu|$, hence by VIII.1.1 this implies the result for $\mu$.

So let

$$\mathcal{R} = \{\mu \in M_1^+(\mathbb{T}) : \mu \text{ is Rajchman}\}.$$

By its very definition, one easily checks that $\mathcal{R}$ is a $\underset{\sim}{\Pi}_3^0$ (i.e. $F_{\sigma\delta}$) subset of $M_1^+(\mathbb{T})$, hence by Theorem 2 it is enough to show that $\mathcal{R}$ is a strongly convex, norm-closed band in $M_1^+(\mathbb{T})$. It is a band by Proposition VIII.1.1, and is clearly norm-closed, for if $\mu_n \in \mathcal{R} \to \mu$ in (the $M(\mathbb{T})$ —) norm, $\mu_n \to \mu$ in PM, and $\mu_n \in PF$, hence $\mu \in PF$. Finally let $\Lambda$ be a probability measure concentrating on $\mathcal{R}$. Then

$$\widehat{b(\Lambda)}(n) = b(\Lambda)(e^{-int}) = \int_{\mathcal{R}} \mu(e^{-int})d\Lambda(\mu) = \int_{\mathcal{R}} \hat{\mu}(n)d\Lambda(\mu).$$

The sequence of continuous functions $\mu \mapsto \hat{\mu}(n)$ on $\mathcal{R}$ converges pointwise to 0 as $|n| \to \infty$, and is bounded, hence by the Lebesgue convergence theorem $\int_{\mathcal{R}} \hat{\mu}(n) d\Lambda(\mu) \to 0$, as $|n| \to \infty$, i.e. $b(\Lambda)$ is a Rajchman measure, and $\mathcal{R}$ is strongly convex. □

## §2. W-sets and Lyons' characterization of Rajchman measures

In the preceding section, we proved Lyons' result that Rajchman measures are characterized by their common null sets by establishing a more general purely measure-theoretic fact. Lyons' original, more direct, proof of his result gives more information about Rajchman measures.

Let us say (with Lyons) that a subfamily B of $U_0$ is large in $U_0$ if it characterizes the Rajchman measures, i.e. $\mu \in M(\mathbf{T})$ is Rajchman iff $B \subseteq \mathrm{Null}(\mu)$. (So Lyons' Theorem 1.8 could be restated as "$U_0$ is large in $U_0$"). Using Theorem 1.8, any basis B for $U_0$ must be large in $U_0$, as measures are $\sigma$-additive. So one gets at once from the basis theorem VIII.2.1

**Proposition 1.** The class $U_0'$ is large in $U_0$.

We will show now that a smaller class is still large in $U_0$. (For negative results, e.g. that $\bigcup_n H^{(n)}$ or the family of Helson sets are not large in $U_0$, we refer the reader to Lyons [4]. It is not known whether U is large in $U_0$).

**Definition 2.** A set $P \subseteq \mathbf{T}$ is a Weyl set (or W-set) if there is some increasing sequence $(n_k)_{k \in \mathbb{N}}$ of natural numbers such that for each $x \in P$ the sequence

$\frac{1}{N}\sum_{k=1}^{N}\delta_{n_k x (\text{mod } 2\pi)}$ w*-converges in $M_1^+(\mathbb{T})$ to some (probability) measure $\nu_x$ different from Lebesgue measure $\lambda$.

Remark: This notion is clearly related to the notion of W*-set we defined in VIII.3 (see in particular Theorem VIII.3.4). We have that $\frac{1}{N}\sum_{k=1}^{N}\delta_{n_k x(\text{mod } 2\pi)}$ w*-converges to Lebesgue measure just in case the sequence $\{n_k x(\text{mod } 2\pi)\}$ is uniformly distributed. Hence W*-sets are those sets P for which for some sequence $\{n_k\}$ and any x in P $\frac{1}{N}\sum_{1}^{N}\delta_{n_k x(\text{mod } 2\pi)}$ does not converge to $\lambda$, whereas W-sets are those for which it converges to something else. By VIII.3.4 there are closed W*-sets in $M_0$. However by the next proposition all Borel W-sets are in $\mathcal{U}_0$.

Proposition 3. (i) A set $P \subseteq \mathbb{T}$ is a W-set iff for some increasing sequence $\{n_k\}_{k\in\mathbb{N}}$:

(a) for each $x \in P$ the polynomials $P_m^N(x) = \frac{1}{N}\sum_{k=1}^{N} e^{-imn_k x}$ converge as $N \to \infty$, and

(b) for each $x \in P$, there is an $m \neq 0$ such that $\lim_N P_m^N(x) \neq 0$.

(ii) If $P \subseteq \mathbb{T}$ is a Borel W-set, $P \in \mathcal{U}_0$.

Proof. (i) is a mere restatement of the definition: If

$$\nu_x^N = \frac{1}{N}\sum_{k=1}^{N}\delta_{n_k x(\text{mod } 2\pi)}$$

w*-converges to $\nu_x$ as $N \to \infty$, then $P_m^N(x) = \widehat{\nu_x^N}(m)$ converges to $\widehat{\nu_x}(m)$, so (a) holds. And as $\nu_x \neq \lambda$, but $\widehat{\nu_x}(0) = 1 = \hat{\lambda}(0)$, one must have some $m \neq 0$ with $\widehat{\nu_x}(m) \neq 0$, i.e. (b) holds too. Conversely if $P_m^N(x) = \widehat{\nu_x^N}(m)$ does converge, $\nu_x^N$ w*-

converges to some $\nu_x$ as $N \to \infty$, and (b) insures that $\nu_x \neq \lambda$.

(ii). Let P be a Borel W-set, as witnessed by $\{n_k\}_{k\in\mathbb{N}}$. Decompose P into countably many Borel sets P′ on which, for some $a > 0$ and some $m \in \mathbb{Z} - \{0\}$, one of $\pm$ Re $\widehat{\nu_x}(m)$ or $\pm$ Im $\widehat{\nu_x}(m)$ is $\geq a$. Then if $\mu$ is in $M_1^+(\mathbb{T})$ and $\mu(P) > 0$, $\mu(P') > 0$ for one of the sets P′, corresponding to say a, m and + Re. Let $\nu = \mu \restriction P'$. One gets then $\int \mathrm{Re}\, \widehat{\nu_x}(m)\, d\nu(x) \geq a \cdot \mu(P') > 0$, hence

$$\overline{\lim_N}\, |\hat{\nu}(N)| \geq \lim_N \left|\int P_m^N(x)\, d\nu(x)\right| = \left|\int \widehat{\nu_x}(m) d\nu(x)\right| \geq a \cdot \mu(P') > 0$$

so that $\nu$ and therefore $\mu$ (since $\nu \prec\prec \mu$) is not a Rajchman measure. This shows $P \in \mathcal{U}_0$. □

<u>Remark</u>: The proof above gives that if W is large in $U_0$, so is $W \cap U_0'$, without using the fact that $U_0'$ is a basis for $U_0$. This is because it clearly shows that if $E \in W$ then $E = \cup_n B_n$, where for each n $B_n$ is Borel and there is $a_n > 0$ so that $\eta_0(F) \geq a_n$ for each closed $F \subseteq B_n$ (in particular $K(B_n) \subseteq U_0'$).

We have finally

<u>Theorem</u> <u>4</u> (Lyons [2]). The family of W-sets is large in $U_0$, i.e. $\mu \in M(\mathbb{T})$ is a Rajchman measure iff all closed W-sets are $\mu$-null. (Thus the family $W \cap U_0'$ is large too).

We will establish Theorem 4 directly, without using Theorem 1.8, so that this result will be a byproduct of Theorem 4.

Lyons' proof of Theorem 4 is based on an "extraction of subsequences" theorem of Komlós [1] (rediscovered independently by Lyons) which deals with Cesàro convergence in $L^2(\mu)$ and $\mu$- a.e., and is more precise than the extractions we used in the proof of 1.7 (that the Mokobodzki function C is a capacity). Note that although the ideas in the two proofs seem very close, it is unclear whether one can get W-sets (or other kinds of $U_0$-sets) from Mokobodzki's method.

Theorem 5 (Komlós [1]). Let $\mu$ be a measure in $M_1^+(T)$, and $\{f_n\}$ a bounded sequence of functions in $L^2(\mu)$. Then there exists a subsequence $\{g_n\}$ of the sequence $\{f_n\}$ such that for every subsequence $\{g_{n_k}\}_{k\in\mathbb{N}}$ the Cesàro means $\frac{1}{N}\sum_{k=1}^{N} g_{n_k}$ converge $\mu$ — a.e. on $\mathbb{T}$.

Proof. We give here the proof from Lyons [2]. First by going to a subsequence, we may assume that $f_n$ weakly converges in $L^2(\mu)$ to some g. We claim first that there is a subsequence $\{g_n\}$ of $\{f_n\}$ such that for all its subsequences $\{g_{n_k}\}_{k\in\mathbb{N}}$, $\|\frac{1}{N}\sum_{k=1}^{N} g_{n_k} - g\|^2_{L^2(\mu)} = O(N^{-1})$, with a constant independent of the subsequence. To see this, we may assume $g = 0$. Let C be such that $\|f_n\|_{L^2(\mu)} \le C$ for all n, and choose inductively the subsequence $\{g_n\}$ of $f_n$ such that (denoting by $<g, f>$ the inner product in $L^2(\mu)$) $|<g_k, g_{n+1}>| < (n+1)^{-3}$, $\forall k \le n$ — which is possible as $f_n \to 0$ weakly in $L^2(\mu)$, thus, given g, $|<g, f_m>| \to 0$ as $m \to \infty$. Then for any subsequence $\{g_{n_k}\}$, one gets

$$\|\frac{1}{N}\sum_{k=1}^{N} g_{n_k}\|^2_{L^2(\mu)} \le \frac{C^2}{N} + \frac{2}{N^2}\sum_{1\le k<j\le N} |<g_{n_k}, g_{n_j}>|$$

$$\le \frac{C^2}{N} + \frac{2}{N^2}\sum_{1\le k<j\le N} j^{-3}$$

$$\le \frac{C^2}{N} + \frac{2}{N^2}\left(\sum_{j=1}^{N} j^{-2}\right) = 0\left(\frac{1}{N}\right).$$

We show now that the sequence $\{g_n\}$ works: Let $\{g_{n_k}\}$ be any subsequence of $\{g_n\}$. Then

$$\int \sum_{j=1}^{\infty} \left| \frac{1}{j^{4/3}} \sum_{k=1}^{j^{4/3}} g_{n_k} \right|^2 d\mu \le \sum_{j=1}^{\infty} \left\| \frac{1}{j^{4/3}} \sum_{k=1}^{j^{4/3}} g_{n_k} \right\|^2_{L^2(\mu)}$$

$$\le C' \cdot \sum_{j=1}^{\infty} j^{-4/3} < \infty$$

for some constant $C'$, by the choice of $\{g_n\}$. So the integrand is finite $\mu$ - a.e., and $j^{-4/3} \left( \sum_{k=1}^{j^{4/3}} g_{n_k} \right) \to 0$ $\mu$ - a.e. (*). Let

$$h_k(x) = \begin{cases} g_{n_k}(x), \text{ if } |g_{n_k}(x)| \le k^{2/3}, \\ 0, \text{ otherwise.} \end{cases}$$

Then for $j^{4/3} \le N < (j+1)^{4/3}$ one gets

$$\left| \frac{1}{N} \sum_{k=1}^{N} g_{n_k} \right| \le \frac{j^{4/3}}{N} \left| \frac{1}{j^{4/3}} \sum_{k=1}^{j^{4/3}} g_{n_k} \right| \tag{1}$$

$$+ \frac{1}{N} \sum_{j^{4/3} < k \le N} |h_k| \tag{2}$$

$$+ \frac{1}{N} \sum_{j^{4/3} < k \le N} |g_{n_k} - h_k|. \tag{3}$$

By (*), the first term tends to 0 $\mu$ — a.e., and the second term is bounded by $\frac{(j+1)^{4/3} - j^{4/3}}{j^{4/3}} (j+1)^{8/9}$, hence goes to 0 as $N \to \infty$. Finally

$$\sum_{k=1}^{\infty} \mu(\{x : g_{n_k}(x) \ne h_k(x)\}) = \sum_{k=1}^{\infty} \mu(\{x : |g_{n_k}(x)| > k^{2/3}\})$$

is bounded by $C^2 \cdot \sum_{k=1}^{\infty} k^{-4/3} < \infty$. Let $E_k = \{x : g_{n_k}(x) \ne h_k(x)\}$. Then $\{x : x$ is in infinitely many $E_k$'s$\} = \bigcap_n \bigcup_{k=n}^{\infty} E_k$ has $\mu$-measure 0, since $\mu\left(\bigcup_{k=n}^{\infty} E_k\right) \to 0$ as $n \to \infty$. This means that for $\mu$- almost all x, $h_k(x) = g_{n_k}(x)$ for k big enough, hence the third term also tends to 0 $\mu$ - a.e. □

<u>Proof of Theorem 4</u>. We want to show that if $\mu$ is not a Rajchman measure, $|\mu|(E) > 0$ for some closed set $E \in W$. It clearly suffices to prove it for $\mu \in M_1^+(\mathbb{T})$, and with some Borel W-set P (for then a closed subset E of P will satisfy $\mu(E) > 0$ and $E \in W$).

So let $\mu \in M_1^+(\mathbb{T})$, $\mu \notin \mathcal{R}_0$, so that $R(\mu) \neq 0$, and choose $p_k \in \mathbb{N}$ increasing such that $\lim \hat{\mu}(p_k)$ exists, and $\lim |\hat{\mu}(p_k)| = R(\mu)$. Let $\alpha \in \mathbb{C}$, $|\alpha| = 1$ be such that $\lim \hat{\mu}(p_k) = \alpha R(\mu)$. Applying Theorem 5 successively to the sequences $\{e^{-imp_k}\}_{k \in \mathbb{N}}$ in $L^2(\mu)$, for all $m \in \mathbb{Z}$, and taking a diagonal subsequence gives a subsequence $\{n_k\}$ of $\{p_k\}$ such that for all m, $P_m^N(x) = \frac{1}{N} \sum_{k=1}^{N} e^{-imn_k x}$ converges $\mu$ — a.e. Consider then the Borel set P defined by

$$P = \{x : \forall m(P_m^N(x) \text{ converges}) \;\&\; \lim_N P_1^N(x) \neq 0\}.$$

Clearly P is a W-set, as witnessed by $\{n_k\}$. Moreover $\mu(P) > 0$, since otherwise $\lim_N P_1^N(x) = 0$ $\mu$ — a.e., thus $\lim_N \int P_1^N(x)\, d\mu(x) = 0$. But

$$\lim_N \int P_1^N(x) d\mu(x) = \lim_N \left(\frac{1}{N} \sum_{k=1}^{N} \hat{\mu}(n_k)\right) = \alpha R(\mu) \neq 0. \qquad \square$$

# Chapter X. Sets of Resolution and Synthesis

In this chapter, we study first an important subclass of the class of closed sets of uniqueness, the so called sets of resolution, i.e. those sets which are hereditarily sets of synthesis. We discuss then definability questions related to sets of synthesis themselves, which also throw some light on the differences between the ideals J(E) and I(E) introduced earlier.

Our treatment in this chapter will be less self-contained and we will make crucial use of some results in the literature without giving the proofs here. The results, for the kind of questions that we are looking at, are still at a preliminary stage with several basic problems still awaiting solution. So this chapter should be viewed best as a short introduction to this subject.

## §1. Sets of resolution

R. Lyons reviews in his paper [4] the various examples or classes of U-sets known at the time of the writing of this paper, and observes that they fall in one of the following three classes or are countable unions of sets in these classes:

(i) The $H^{(n)}$-sets, $n = 1, 2, \ldots$

(ii) A class of sets dubbed "Meyer sets" by Lyons, which include the ultrathin symmetric sets $E_{\xi_1, \xi_2 \ldots}$ with $\Sigma\ \xi_n^2 < \infty$, and some earlier examples of McGehee [1].

(iii) The sets of resolution, where we have the following

Definition 1. A closed set $E \subseteq \mathbf{T}$ is called a set of resolution if every closed subset of E is a set of synthesis. Denote by RE the class of sets of resolution.

According to Malliavin's Theorem VII.3.15 every set of resolution is a U-set.

We have seen in III.1.4 that all $H^{(n)}$-sets are actually in $U'$, and Lyons verifies in [4] that the Meyer sets are also in $U'$. Finally all sets of resolution are in $U'_\sigma$ because they are in $(U'_1)_\sigma$ by VI.2.3(ii) and for sets of synthesis E, $E \in U'_1 \Leftrightarrow E \in U'$.

It follows that all the "known" until recently U-sets are actually countable unions of $U'$-sets. Thus from the Debs-Saint Raymond result VII.4.2, which implies the existence of U-sets not in $U'_\sigma$, we have now many new examples of U-sets.

In this section, we concentrate on the class of sets of resolution. Clearly every countable closed set is in RE, by VI.3.3. In order to see some examples of perfect sets in RE, let us introduce another interesting notion.

Definition 2. A closed set $E \subseteq \mathbf{T}$ is called a set without true pseudomeasure if every pseudomeasure supported by E is actually a measure, i.e., $M(E) = PM(E)$. We denote by WTP the class of sets without true pseudomeasure. (This class is also denoted as SVP from the French initials).

Recall that E is a set of synthesis if $I(E) = \bar{J}(E)$, or equivalently $N(E)$ $(= I(E)^{\perp}) = PM(E)$ $(= J(E)^{\perp})$. Since $M(E) \subseteq N(E) \subseteq PM(E)$ for all E, it follows that every WTP-set is a set of synthesis and since obviously every closed subset of a WTP-set is also in WTP, it follows that every WTP-set is in RE, i.e.

$$\text{WTP} \subseteq \text{RE}.$$

Recall also that E is a Helson set iff $M(E) = N(E)$ (VII.3.3), therefore $E \in$ WTP iff E is a Helson set of synthesis.

Perfect sets in WTP were first constructed by Kahane and Salem—see e.g. Kahane-Salem [1, p. 126]. We give below a different proof based on definability theory. First we need to calculate the complexity of WTP.

<u>Theorem</u> 3. The class WTP of closed sets without true pseudomeasure is $\Sigma_3^0$ (i.e. $G_{\delta\sigma}$) in $K(\mathbb{T})$.

<u>Proof</u>. Suppose $E \in K(\mathbb{T})$, and for some $n > 0$ $PM(E) \cap B_{1/n}(PM) \subseteq B_1(M(\mathbb{T}))$, where as usual we view $M(\mathbb{T})$ as a subset of PM via the map $\mu \mapsto \hat{\mu}$. Then clearly $PM(E) \subseteq M(\mathbb{T})$, hence $E \in$ WTP. Conversely suppose $E \in$ WTP. Then $PM(E) \subseteq \bigcup_n B_n(M(\mathbb{T}))$. Now $B_n(M(\mathbb{T}))$ is compact in the weak*-topology of $M(\mathbb{T})$, which therefore coincides with the weak*-topology inherited from PM. In particular, $B_n(M(\mathbb{T}))$ is norm-closed in PM and of course so is $PM(E)$. Applying the Baire Category Theorem and using translation gives an $n > 0$ with $PM(E) \cap B_{1/n}(PM) \subseteq B_1(M(\mathbb{T}))$. This shows the equivalence

$$(*) \qquad E \in WTP \Leftrightarrow \exists n(PM(E) \cap B_{1/n}(PM) \subseteq B_1(M(T)))$$
$$\Leftrightarrow \exists n \ \forall S \in B_1(PM)[S \in B_1(M(T)) \vee S \notin B_{1/n}(PM) \vee S \notin PM(E)].$$

Since $\{(E, S) \in B_1(PM) \times K(T) : S \in PM(E)\}$ is closed (by IV.2.3) it follows easily that (*) gives a $\underset{\sim}{\Sigma}_3^0$ definition of WTP. □

It has been shown by Varopoulos [3], extending work of Drury, that the Helson sets form an ideal. Using that it can be shown that WTP is an ideal as well (see Lindahl-Poulsen [1, p. 116]). However WTP is not a $\sigma$-ideal, in fact there are countable sets which are not Helson, i.e. not in WTP. (For example, using the so-called Rudin-Shapiro measures, see e.g. Kahane [1], one can construct sequences $x_n \to 0$ with $E = \{0\} \cup \{x_n : n \in \mathbb{N}\}$ not Helson. One can also give a definability argument; see the remarks following Theorem 12).

There is however a simple condition under which countable closed sets are in WTP. Let us introduce first the following

<u>Definition 4</u>. A set $P \subseteq \mathbb{T}$ is called <u>independent</u> if, viewing P as a subset of $[0, 2\pi]$, we have for all distinct $x_1, \ldots, x_k \in P$ and all $n_1, \ldots, n_k \in \mathbb{Z}$

$$n_1 x_1 + \ldots + n_k x_k = 0 \ (\text{mod } 2\pi) \Rightarrow n_1 = \ldots = n_k = 0.$$

An important property of finite independent sets is the following classical theorem of Kronecker (see e.g. Katznelson [1]):

<u>Theorem 5</u> (Kronecker). If E is a finite independent set, then given any $f : E \to \mathbb{C}$

with $|f(x)| = 1$, for all $x \in E$, and $\epsilon > 0$, there is $n \in \mathbb{Z}$ with

$$|e^{inx} - f(x)| < \epsilon, \quad \forall x \in E.$$

It follows immediately that if E is a finite independent set and $\mu \in M(E)$ then

$$\|\mu\|_M = \|\mu\|_{PM}.$$

This is because $\|\mu\|_M = \sup\{|\int f d\mu| : |f(x)| = 1, \forall x \in E\}$. (This is a general fact, see VII.3.10). We have then easily

Proposition 6. Every countable, independent closed set is in WTP.

Proof. It is enough to show that every such set is Helson. But every measure with countable support E is the limit in the norm of $M(\mathbb{T})$ of a sequence of measures with finite support contained in E. So we are done by our preceding remarks and VII.3.3. □

We are now ready to give the promised proof of

Theorem 7 (Kahane-Salem). There exist perfect sets without true pseudomeasure.

Proof. First notice that it is easy to construct by a direct splitting argument a perfect independent set $F \subseteq (0, 2\pi)$ (see for example Kahane-Salem [1]). Consider then $K_\omega(F) = \{E \subseteq F : E \text{ is countable}\}$. By IV.3.5 $K_\omega(F)$ is not Borel in $K(\mathbb{T})$. Since $K_\omega(F) \subseteq \text{WTP} \cap K(F)$ and WTP is Borel, it follows that

$(K(F) \cap WTP) - K_\omega(F) \neq \emptyset$, so there is an uncountable closed set in WTP and thus its perfect kernel is a perfect set in WTP. □

There is an alternative way to produce perfect WTP-sets, via the so-called Kronecker sets.

Definition 8. A closed set $E \subseteq K(\mathbf{T})$ is called a Kronecker set if for every continuous $f : E \to \mathbb{C}$ with $|f(x)| = 1$ for all $x \in E$ (i.e. $f : E \to \mathbf{T}$), and every $\epsilon > 0$, there is $n \in \mathbb{Z}$ with

$$|e^{inx} - f(x)| < \epsilon, \quad \forall x \in E$$

i.e.

$$\|e^{inx} - f\|_{C(E)} < \epsilon.$$

In other words, E is Kronecker if the exponentials are dense in $\{f \in C(E): f : E \to \mathbf{T}\}$ in the norm of $C(E)$.

Thus every finite independent set is Kronecker and it is easy to check that every Kronecker set is independent and nowhere dense so for example by VII.3.10 again, we have $\|\mu\|_M = \|\mu\|_{PM}$ for all $\mu \in PM(E)$, if E is Kronecker. Thus all Kronecker sets are Helson. We actually have the much stronger

Theorem 9 (Varopoulos [1]). Every Kronecker set is in WTP.

Proof. Let $E \subseteq \mathbf{T}$ be a Kronecker set and $S \in PM(E)$. Denote by $X_E$ the subspace of $C(\mathbf{T})$ consisting of all $f \in C(\mathbf{T})$ which take only finitely many values in a nbhd of E. Since E is nowhere dense, $X_E$ is dense in C(T). For $f \in X_E$ define $S^*(f) = \langle f^*, S\rangle$ for any $f^* \in A$ such that $f = f^*$ is a nbhd of E. This is clearly well-defined. Moreover $S^*$ is linear on $X_E$.

Claim. There is a constant K such that $|S^*(f)| \leq K \cdot \|f\|_C$, $\forall f \in X_E$.

Granting this, $S^*$ can be extended to a continuous linear functional on C(T), i.e. to an element $\mu \in M(\mathbf{T})$. Then we claim that $S = \hat{\mu}$, which completes the proof (modulo the claim). Indeed we show that for any $\varphi \in C^\infty(\mathbf{T})$, $\langle\varphi, S\rangle = \langle\varphi, \mu\rangle$. To see this it is enough to check that any such $\varphi$ can be approximated as closely as we want in the norm of A by a function in $A \cap X_E$. Since $E \subsetneq (0, 2\pi)$, given $\epsilon > 0$ cover E by finitely many open intervals $(a_1, b_1), \ldots, (a_n, b_n)$ with $0 < a_1 < b_1 < a_2 < b_2 < \cdots < a_n < b_n < 2\pi$, so that if $V = \cup_n(a_n, b_n)$, $\int_V |\varphi'(x)|\,dx < \epsilon$ and $\int_V |\varphi'(x)|^2 dx < \epsilon$. This can be done as E has measure 0. Then construct $\psi \in A \cap X_E$ as follows:

Let $\delta = \sum_{i=1}^n [\varphi(a_i) - \varphi(b_i)]$, so that $|\delta| < \epsilon$. In $[0, a_1]$, let $\varphi = \psi$. In $[a_1, b_1]$ let $\psi = \varphi(a_1)$. In $[b_1, a_2]$, let $\psi = \varphi + (\varphi(a_1) - \varphi(b_1))$. In $[a_2, b_2]$, let $\psi = \varphi(a_2) + (\varphi(a_1) - \varphi(b_1))$. In $[b_2, a_3]$, let $\psi = \varphi + (\varphi(a_2) - \varphi(b_2)) + (\varphi(a_1) - \varphi(b_1))$, etc. So in $[a_n, b_n]$, $\psi$ will be equal to $\varphi(a_n) + (\varphi(a_1) - \varphi(b_1)) + \cdots + (\varphi(a_{n-1}) - \varphi(b_{n-1}))$ and $\psi(b_n) = \varphi(b_n) + \delta$. Since we can clearly assume (by taking $\epsilon$ small enough) that $b_n + |\delta| < 2\pi$, define $\psi = \varphi - \frac{\delta}{|\delta|}(x - b_n) + \delta$ ($\frac{\delta}{|\delta|} = 1$, if $\delta = 0$) on $[b_n, b_n + |\delta|]$ and finally $\psi = \varphi$ on $[b_n + |\delta|, 2\pi]$.

Using II.2.1 we can easily check now that $\|\varphi - \psi\|_A \leq C \cdot \sqrt{\epsilon}$, where C is some constant, so we are done.

It remains to prove our claim. Given $f \in X_E$, let $\alpha_1, \ldots, \alpha_n$ be the finitely many distinct values that f takes in some nbhd V of E, let $V_j = f^{-1}[\{\alpha_j\}] \cap V$, so that $V_j$ is open and $E \subseteq \cup_j V_j$, and let $\varphi_j \in C^\infty(\mathbf{T})$, $0 \le \varphi_j \le 1$ be such that $\mathrm{supp}(\varphi_j) \subseteq V_j$ and $\varphi_j = 1$ in a nbhd of $E \cap V_j$. In particular $\mathrm{supp}(\varphi_j) \cap \mathrm{supp}(\varphi_{j'}) = \emptyset$ if $j \ne j'$. Let $S_j = \varphi_j \cdot S$. Thus $\mathrm{supp}(S_j) \subseteq V_j \cap E$ and $S = \Sigma S_j$, as $\Sigma \varphi_j = 1$ in a nbhd of E. Then

$$\begin{aligned} S^*(f) &= \langle \sum \alpha_j \varphi_j, S\rangle \\ &= \sum \alpha_j \langle 1, \varphi_j \cdot S\rangle \\ &= \sum \alpha_j\, S_j(0) \end{aligned}$$

thus

$$|S^*(f)| \le (\max |\alpha_j|) \cdot \sum_{j=1}^{n} \|S_j\|_{PM} \le \|f\|_C \cdot \sum_{j=1}^{n} \|S_j\|_{PM}.$$

We will complete the proof by showing that $\sum_{j=1}^{n} \|S_j\|_{PM} = \|S\|_{PM}$, so that we can take $K = \|S\|_{PM}$. The proof is based on the following

Lemma 10. For each $\epsilon > 0$ there is $\delta > 0$ such that for all $\theta \in \mathbf{R}$, $k \in \mathbf{Z}$, $E \in K(\mathbf{T})$

$$\|e^{i\theta} - e^{ikx}\|_{C(E)} \le \delta \Rightarrow \exists f \in A(f(x) = e^{i\theta} - e^{ikx} \text{ in a nbhd of } E \text{ and } \|f\|_A \le \epsilon).$$

Assuming the lemma we proceed as follows: Let $\epsilon > 0$ and choose $n_j \in \mathbf{Z}$, $\theta_j \in \mathbf{R}$ so that $\sum_{j=1}^{n} e^{i\theta_j} S_j(n_j) \ge \sum_{j=1}^{n} \|S_j\|_{PM} - \epsilon$. Since E is Kronecker, given any $\delta > 0$ there is $m \in \mathbf{Z}$ with $\|g - e^{-imx}\|_{C(E)} \le \delta$, where $g(x) = e^{i\theta_j} \cdot e^{-in_jx}$ on $E \cap V_j$. Thus $\|e^{i\theta_j} e^{-in_jx} - e^{-imx}\|_{C(E_j)} \le \delta$, where $E_j = \mathrm{supp}(S_j)$. By the lemma then, we can choose $\delta$ so that for some $f_j \in A$, $f_j = e^{i\theta_j} - e^{-i(m-n_j)x}$ in a nbhd of

$E_j$ and $\|f_j\|_A \le \frac{\epsilon}{n\cdot\|S_j\|_{PM}}$. So

$$
\begin{aligned}
&|e^{i\theta_j} S_j(n_j) - S_j(m)| \\
&= |\langle e^{i\theta_j}\cdot e^{-in_jx} - e^{-imx}, S_j\rangle| \\
&= |\langle e^{-in_jx} f_j, S_j\rangle| \\
&\le \|e^{-in_jx} f_j\|_A \cdot \|S_j\|_{PM} \le \tfrac{\epsilon}{n}
\end{aligned}
$$

and

$$
\begin{aligned}
\|S\|_{PM} \ge |S(m)| &= \left|\sum_{j=1}^n S_j(m)\right| \\
&\ge \sum_{j=1}^n e^{i\theta_j} S_j(n_j) - \epsilon \\
&\ge \sum_{j=1}^n \|S_j\|_{PM} - 2\epsilon
\end{aligned}
$$

and we are done.

It remains only to give the

<u>Proof of Lemma 10</u>. Fix $\epsilon > 0$ and choose $h > 0$ so that $\|\tau_h(1 - e^{ix})\|_A \le \epsilon$ by Lemma II.2.2, where $\tau_h = \tau_{0,h}$ is the usual trapezoidal function. So if $\tilde{f} = \tau_h(1 - e^{ix})$, $\|\tilde{f}\|_A \le \epsilon$ and $\tilde{f} = 1 - e^{ix}$ in $[-h, h]$ (mod $2\pi$). Take now $\delta = |1 - e^{ih/2}|$. Given $\theta \in \mathbb{R}$, $k \in \mathbb{Z}$ and $E \in K(\mathbb{T})$ with $\|e^{i\theta} - e^{ikx}\|_{C(E)} \le \delta$ put

$$f(x) = e^{i\theta}\tilde{f}(kx - \theta).$$

Then $\|f\|_A \le \epsilon$. Moreover for $x \in E$, $|1 - e^{i(kx-\theta)}| \le |1 - e^{ih/2}|$ so $kx - \theta$ (mod $2\pi$) is in $[-h/2, h/2]$ (mod $2\pi$), thus $V = \{x : kx - \theta \text{ (mod } 2\pi) \in (-h, h) \text{ (mod } 2\pi)\}$ is a nbhd of E, and for $x \in V$ we have $f(x) = e^{i\theta}\tilde{f}(kx - \theta) = e^{i\theta}(1 - e^{i(kx-\theta)}) = e^{i\theta} - e^{ikx}$, so we are done. □

Although it is not hard to construct directly perfect Kronecker sets, we will establish below their abundant existence by a Baire Category argument. It was Kaufman [1] who first used such an argument to show that the set of all continuous $f : 2^{\mathbf{N}} \to \mathbf{T}$ for which $f[2^{\mathbf{N}}]$ is perfect Kronecker is comeager in the Polish space of all such functions with the usual uniform metric. We do this below for the space $K(\mathbf{T})$ itself. Our original presentation has been considerably simplified by Debs, who noticed in fact the following.

<u>Theorem</u> <u>11</u>. The class of perfect Kronecker sets is a dense $G_\delta$ in $K(\mathbf{T})$.

<u>Proof</u>. Recall that the class of perfect sets in a dense $G_\delta$ in $K(\mathbf{T})$. So it is enough to show that the class of Kronecker sets is also a dense $G_\delta$. For the density, note that given finitely many points $x_1, \ldots, x_n \in \mathbf{T}$, one can easily find (for example by induction on n) independent points $y_1, \ldots, y_n \in \mathbf{T}$, with $y_i$ as close to $x_i$ as we want. Since finite independent sets are Kronecker we are done. To verify that the class of Kronecker sets is a $G_\delta$ notice that since every such set is nowhere dense, we can restrict in Definition 8 the f's to take only finitely many values of the form $e^{i2\pi r}$, where $r \in \mathbb{Q}$, in a nbhd of E. So for $E \in K(\mathbf{T})$,

E is Kronecker $\Leftrightarrow$ E is nowhere dense & $\forall\epsilon\ \forall I_0 \ldots I_m\ \forall r_0 \ldots \forall r_m$ [($\epsilon$ is positive in $\mathbb{Q}$ & $I_0 \ldots I_m$ are pairwise disjoint, rational relative to $2\pi$, closed intervals & $r_0 \dots r_m \in \mathbf{Q}$ &

$E \subseteq I_0 \cup \dots \cup I_m$)

$\Rightarrow \exists n \in \mathbf{Z}\ \forall k \leq m\ \forall x \in E \cap I_k\ (|e^{i2\pi r_k} - e^{inx}| < \epsilon)]$

which is clearly a $G_\delta$ and completes the proof. □

<u>Remark</u>. Similarly one can obtain further results such as: For non-∅ pairwise disjoint perfect sets $P_1, \dots, P_n \subseteq \mathbf{T}$ there is a Kronecker set K so that $K \cap P_j$ is non-∅ perfect (see Kahane [1], p. 92). Instead of working with K(**T**) here one works with the closed subspace $\{E \in K(\mathbf{T}) : E \cap P_j \neq \emptyset$ for all j$\}$ and shows as before that the Kronecker sets form a dense $G_\delta$ in this subspace. Since the sets K for which $K \cap P_j$ is perfect for all j form also a dense $G_\delta$, we are done.

<u>Remark</u>. Another class of sets studied in this context are the so-called Dirichlet sets. A set $E \in K(\mathbf{T})$ is called <u>Dirichlet</u> if there is a sequence $n_1 < n_2 < \dots$ of positive integers with $\|e^{in_jx} - 1\|_{C(E)} \to 0$ as $j \to \infty$. (We have seen this concept in the proof of VII.2.1). The motivation for this definition comes from the classical Dirichlet Theorem (which says that if $x_1, \dots, x_k \in \mathbf{R}$ and $N \in \mathbf{N}$ is given, there is $1 \leq q \leq N^k$ with $\text{dist}(qx_i, \mathbf{Z}) \leq \frac{1}{N}$, for $i = 1, \dots, k$), which immediately implies that finite sets are Dirichlet.

Clearly every Dirichlet set is an H-set, thus a U′-set. But Kahane (see e.g. Kahane [1, p. 97]) has shown that actually $\eta(E) = 1$ for all Dirichlet E. This follows for example by the following more general fact due to Lyons [4]: Let $E \in K(\mathbf{T})$ be an H-set and $n_1 < n_2 < \dots$, I an open non-∅ interval in **T** with $(n_k E) \cap I = \emptyset$ for all k. Then $\eta(E) \geq \frac{\lambda(I)}{2 - \lambda(I)}$. (Lyons [4] actually proves similar inequalities for $H^{(n)}$-sets). Since for each Dirichlet set one can find such I with $\lambda(I)$ as close to 1 as we want, the fact about Dirichlet sets follows immediately. To prove the above lower bound for $\eta(E)$ consider the triangular function $\psi_{a,h} = \psi$, where $J = (a - h, a + h)$ is an open interval with $\overline{J} \subseteq I$. Thus $\hat{\psi}(0) = 1$, $\|\psi\|_A = \frac{2\pi}{h} = \frac{2}{\lambda(J)}$ and so $\|1 - \psi\|_A = \frac{2 - \lambda(J)}{\lambda(J)}$. Let $f = 1 - \psi$, so that $f(n_k x) = 1$ in a nbhd of E and (since $\hat{f}(0) = 0$)

$$S(0) = \langle f(n_k x), S\rangle$$

$$= \sum_{\substack{n\in\mathbb{Z}\\ n\neq 0}} \hat{f}(-n)\, S(nn_k).$$

Thus (since $n_k \to \infty$)

$$|S(0)| \leq \|f\|_A \cdot \overline{\lim}\, |S(n)|$$

$$= \|f\|_A \cdot R(S).$$

By applying this to $e^{-imx} \cdot S$ we have that the same bound holds for each $|S(m)|$, i.e.

$$\|S\|_{PM} \leq \|f\|_A \cdot R(S),$$

and

$$\eta(E) \geq \frac{1}{\|f\|_A} = \frac{\lambda(J)}{2 - \lambda(J)}.$$

Since J was arbitrary with $\overline{J} \subseteq I$ we are done.

We conclude this section with some facts and questions concerning definability and structural properties of RE.

<u>Theorem</u> <u>12</u>. The class RE of sets of resolution is $\underset{\sim}{\Pi}^1_1$-complete in K(T). In fact every set P with $K_\omega(\mathbf{T}) \subseteq P \subseteq RE$ is $\underset{\sim}{\Pi}^1_1$-hard.

<u>Proof</u>. That RE is $\underset{\sim}{\Pi}^1_1$ follows immediately from the fact that the class of sets of synthesis S is $\underset{\sim}{\Pi}^1_1$, which will be proved shortly in §2 (see 2.3 and the beginning of the proof of 2.4).

The fact that any P between $K_\omega(\mathbf{T})$ and RE is $\underset{\sim}{\Pi}^1_1$-hard depends on the following result of Varopoulos [2], see e.g. Kahane [1, p. 110]:

If E is closed and uncountable then E + E (= {x + y(mod 2π) : x, y ∈ E}, viewing E as a subset of [0, 2π]) is not RE.

Define now for E ∈ K(T),

$$f(E) = E + E.$$

Then f is continuous and

$$E \in K_\omega(\mathbf{T}) \Rightarrow f(E) \in K_\omega(\mathbf{T})$$
$$E \notin K_\omega(\mathbf{T}) \Rightarrow f(E) \notin RE$$

so we are done. □

Note that Theorem 12 easily implies the existence of countable closed sets which are not Helson. This is because a countable closed set is Helson iff it is in WTP. Since by Theorem 3 WTP is Borel, we cannot have $K_\omega(\mathbb{T}) \subseteq$ WTP.

The preceding result is still a bit unsatisfactory as it does not show that the class of <u>perfect</u> RE sets is $\underset{\sim}{\Pi}^1_1$-complete or even non-Borel. This is at this stage an open problem. Another interesting problem about RE is to find a "natural" $\underset{\sim}{\Pi}^1_1$-rank on it.

Not much is known about closure properties of RE. An old problem in this subject is whether RE is an ideal or a $\sigma$-ideal. (If RE is a $\sigma$-ideal the perfect RE sets form a $\underset{\sim}{\Pi}^1_1$-complete set, since then by Varopoulos' Theorem for each perfect RE set E and $K \in K(E)$, $K \in K_\omega(E) \Leftrightarrow K + E$ is perfect RE).

Finally we have seen that RE $\subseteq$ U. Is it true that RE $\subseteq U^{(\alpha)} = \{E \in U : [E]_{PS} \leq \alpha\}$, for some $\alpha < \omega_1$? Or even RE $\subseteq U_0^{(\alpha)} = \{E \in U_0 : [E]_0 \leq \alpha\}$, for some $\alpha < \omega_1$? The answer to these questions seems to be unknown. It was pointed out however to us by Kaufman that it follows easily from work of his (see Kaufman [4]) or Körner's (see Körner [2]) that there are RE sets not in $U_0'$. Such sets can indeed be found in any open interval, so that by VI.1.6 it follows that if RE is a $\sigma$-ideal, then RE $\not\subseteq U_0^{(\alpha)}$ for each $\alpha < \omega_1$.

## §2. <u>Sets of synthesis</u>

Recall that we have denoted by S the class of sets of synthesis. Our main goal in this section is to show that S is a true $\underset{\sim}{\Pi}^1_1$ set and discuss a natural $\underset{\sim}{\Pi}^1_1$-

rank on S.

Given a closed set E we always have $M(E) \subseteq N(E) \subseteq PM(E)$ and $N(E)$ is the weak$^*$-closure of $M(E)$ in PM. (This is because if $<f, \mu> = 0$ for all $\mu \in M(E)$ then $<f, \delta_x> = 0$ for all $x \in E$, so $f \in I(E)$, i.e. $<f, S> = 0$ for all $S \in N(E)$). Thus E is a set of synthesis iff $I(E) = \overline{J(E)}$ iff $PM(E) = N(E)$ iff $\overline{M(E)}^{w^*} = PM(E)$. So if $E \in S$, then by Banach's Theorem V.2.2 (and V.2.5) there is a countable ordinal $\alpha$ such that if $M^\alpha(E) = M(E)^{(\alpha)}$ ($= \alpha^{th}$ iterate of M(E) in the operation of closure under sequential $w^*$-limits), then $M^\alpha(E) = PM(E)$. We define then the following rank on S (see Katznelson-McGehee [1]).

<u>Definition 1</u>. For each $E \in S$ let

$$[E]_S = \text{least } \alpha < \omega_1 \text{ such that } M^\alpha(E) = PM(E).$$

As usual we let $[E]_S = \omega_1$ if $E \notin S$.

One now has the following result of Katznelson-McGehee extending work of Varopoulos.

<u>Theorem 2</u> (Katznelson-McGehee [1]). The rank $[E]_S$ is unbounded on S, i.e. for every countable ordinal $\alpha$ there is $E \in S$ with $[E]_S > \alpha$.

We will not give the proof of this result here.

Let us provide also an equivalent formulation for $[E]_S$ that we will need soon.

For each closed set $E \neq \emptyset$ consider the separable Banach space

$$A/\overline{J(E)} = A^{\circ}(E).$$

(Recall that $A(E) = A/I(E)$). Thus $A^{\circ}(E)$ consists of all equivalence classes $[f]$ of the equivalence relation on A given by

$$f \sim g \Leftrightarrow f - g \in \overline{J(E)}$$

with norm

$$\begin{aligned} \|[f]\|_{A^{\circ}(E)} &= \text{dist}\,(f, \overline{J(E)}) \\ &= \inf\,\{\|f - h\|_A : h \in \overline{J(E)}\}. \end{aligned}$$

Then if to each $x^* \in (A^{\circ}(E))^*$ we assign $S(x^*) \in PM$ given by

$$\langle f, S(x^*)\rangle = \langle [f], x^*\rangle$$

the map $x^* \mapsto S(x^*)$ is an isometric isomorphism between $(A^{\circ}(E))^*$ and $(\overline{J(E)})^{\perp} = PM(E)$, so we will identify $(A^{\circ}(E))^*$ with $PM(E)$ from now on. Notice moreover that under this identification the weak*-topology of $(A^{\circ}(E))^*$ is the same as the weak*-topology that $PM(E)$ inherits from PM. It follows that $\emptyset \neq E$ is a set of synthesis iff $M(E)$ is w*-dense in $(A^{\circ}(E))^*$, and that $[E]_S$ is just the order of $M(E)$ in $(A^{\circ}(E))^*$ as defined in V.2.6.

We will also need the following simple fact

Proposition 3. There is a Borel function $d : K(\mathbf{T}) \to A^{\mathbf{N}}$ such that for each

$E \in K(\mathbf{T})$, $d(E) = \{d_n(E)\}$ is a dense subset of $\overline{J(E)}$.

Proof. By II.3.2 and Hahn-Banach, the set of rational linear combinations of triangular functions $\psi_{a,h}$ with support disjoint from E and a,h rational relative to $2\pi$ is dense in $\overline{J(E)}$. Take $d_n(E)$ to be some canonical enumeration of these functions. □

We are now ready to prove

Theorem 4 (Kechris-Solovay). The rank $[E]_S$ is a $\underset{\sim}{\Pi}^1_1$-rank on the $\underset{\sim}{\Pi}^1_1$ set S.

Proof. First let us compute that S is $\underset{\sim}{\Pi}^1_1$. For $E \in K(\mathbf{T})$ we have

$$E \in S \Leftrightarrow I(E) = \overline{J(E)}$$
$$\Leftrightarrow \forall f \in A \ \forall \epsilon > 0 \ [f \in I(E) \Rightarrow \exists n(\|f - d_n(E)\|_A < \epsilon)]$$

where $\{d_n(E)\}$ is as in Proposition 3. Since

$$f \in I(E) \Leftrightarrow \forall n \ \forall m(V_n \cap E \neq \emptyset \Rightarrow \exists x \in V_n \ (|f(x)| < \tfrac{1}{m}))$$

where $\{V_n\}$ is an open basis for $\mathbf{T}$, it follows that $\{(f, E) \in A \times K(\mathbf{T}) : f \in I(E)\}$ is $G_\delta$ in $A \times K(\mathbf{T})$, so S is $\underset{\sim}{\Pi}^1_1$.

To show that $[E]_S$ is a $\underset{\sim}{\Pi}^1_1$-rank on S, we will use the tree-rank of Chapter V. (The first proof that $[E]_S$ is a $\underset{\sim}{\Pi}^1_1$-rank was motivated by the original proof of Solovay's result V.4.3 and used again a good deal of logic).

Note first that

$$[E]_S = 0 \Leftrightarrow M(E) = PM(E)$$
$$\Leftrightarrow E \in WTP.$$

Since WTP is Borel by 1.2 it is enough to calculate that $[E]_S$ is a $\underset{\sim}{\Pi}^1_1$-rank on $S - WTP$. Now by V.3.5, with $X = A^{\circ}(E)$, $Y = M(E) \subseteq PM(E) = X^*$ and $E \notin WTP$, we have $[E]_S = \mathrm{ord}(M(E)) = rk_T(M(E))$. So it is enough to show that $rk_T(M(E))$ is a $\underset{\sim}{\Pi}^1_1$-rank on $S - WTP$.

First we will define a canonical countable dense $D_E$ in the open unit ball of $A^{\circ}(E)$ as follows: Let D be the set of all trigonometric polynomials with coefficients in $\mathbb{Q} + i\mathbb{Q}$. Let then $\tilde{D}_E = \{p \in D : |||[p]|||_{A^{\circ}(E)} < 1\}$ and let $D_E = \{[p] : p \in \tilde{D}_E\}$. Then $D_E$ is dense in the open unit ball of $A^{\circ}(E)$ and closed under rational multiplication. So we can use $D_E$ for the definition of the tree-rank as in V.3. Since moreover (for E infinite) the map $p \mapsto [p]$ is 1–1 from $\tilde{D}_E$ onto $D_E$ we can define the trees $T^{\epsilon}_{M(E)}$ (of V.3) as follows

$$\begin{aligned} T^{\epsilon}_{M(E)} = \{\emptyset\} \cup \{(p_0, \ldots, p_n) \in \mathrm{Seq}\ D : \ & \forall j \le n(p_j \in \tilde{D}_E \text{ and} \\ & |||[p_j]|||_{A^{\circ}(E)} \ge \epsilon) \ \&\ \forall j < n(|||[p_j - p_{j+1}]|||_{A^{\circ}(E)} \le 2^{-(j+3)}) \ \& \\ & \forall j \le n\ (|||[p_j]|||_{M(E)} \le 2^{-(j+1)})) \end{aligned}$$

where $||\ ||_{M(E)}$ refers to the norm as a linear functional on the space M(E) <u>with the PM-norm</u>, as in V.2, so that each $T^{\epsilon}_{M(E)}$ is a tree on Seq D. Then as in the proof of V.4.3 it is enough to check that for each $p \in D$, $\delta > 0$ the following two sets are Borel in K(T):

$$B = \{E \in K(T) : |||[p]|||_{A^{\circ}(E)} \leq \delta\}$$
$$C = \{E \in K(T) : |||[p]|||_{M(E)} \leq \delta\}.$$

For B note that $|||[p]|||_{A^{\circ}(E)} \leq \delta$ iff $\text{dist}(p, \overline{J(E)}) \leq \delta$ iff $\forall m \exists n\ (\|p - d_n(E)\|_A < \delta + \frac{1}{m})$, where $d_n(E)$ is as in Proposition 3. Finally for C we have

$$E \in C \Leftrightarrow \forall \mu \in M(E)\ (\|\mu\|_{PM} \leq 1 \Rightarrow |\langle p, \mu\rangle| \leq \delta)$$
$$\Leftrightarrow \forall N\ \forall \mu \in K_N\ (\mu \in M(E) \Rightarrow |\langle p, \mu\rangle| \leq \delta)$$

where $K_N = \{\mu \in M(T) : \|\mu\|_M \leq N \ \&\ \|\mu\|_{PM} \leq 1\}$. Then $K_N$ is compact, metrizable in the weak*-topology of M(T), and we are done. □

We now have the following immediate corollary of Theorem 4 and Theorem 2 (using V.1.7).

Corollary 5. The class S of sets of synthesis is $\underset{\sim}{\Pi}^1_1$ but not Borel in K(T).

There is a further corollary of this classification, which establishes an interesting difference between $\overline{J(E)}$ and I(E) and can be viewed as a definability version of Malliavin's Theorem on the existence of closed sets which are not of synthesis.

Recall from Proposition 3 that there is a Borel map $E \mapsto \{d_n(E)\}$ from K(T) into $A^{\mathbb{N}}$ such that $\{d_n(E)\}$ is dense in $\overline{J(E)}$. In some sense this says that $E \mapsto \overline{J(E)}$ is "Borel". We observe now that this fails for I(E).

Corollary 6. There is no Borel map $E \mapsto \{f_n(E)\}$ from $K(\mathbf{T})$ into $A^{\mathbb{N}}$ such that $\{f_n(E)\}$ is dense in $I(E)$ for all $E \in K(\mathbf{T})$.

Proof. If such a map existed, then we would have

$$E \in S \Leftrightarrow \forall n\ \forall m\ \exists k\ [\|f_n(E) - d_k(E)\|_A < \tfrac{1}{m+1}]$$

so that S would be Borel, a contradiction. □

An equivalent way to state this corollary is also the following:

Corollary 6*. The map

$$F(f, E) = \text{dist}(f, I(E))$$

from $A \times K(\mathbf{T})$ into $\mathbb{R}$ is not Borel.

Of course the map $\text{dist}(f, J(E)) = \text{dist}(f, \overline{J(E)})$ is easily Borel.

We conclude now with some open problems about sets of synthesis.

First notice that the proof of Corollary 5 is by a rank argument and although it shows that S is true $\underset{\sim}{\Pi}^1_1$, it does not show that S is actually $\underset{\sim}{\Pi}^1_1$-complete. As we pointed out in IV.1 (a little after Theorem IV.1.4), it can be shown from strong axioms of set theory that every true $\underset{\sim}{\Pi}^1_1$ set is actually $\underset{\sim}{\Pi}^1_1$-complete. But we do not know how to show in ZFC alone that S is $\underset{\sim}{\Pi}^1_1$-complete. This is certainly an awkward situation.

Next we turn to closure properties of S. It is obvious that S is not hereditary, e.g. $\mathbb{T}$ itself is in S. However it is an old problem in this area whether S is closed under finite or countable unions.

Our final question has to do with the rank $[E]_S$. One can show that all countable closed sets E have $[E]_S \leq 1$. This follows from the fact that every $S \in PM(E)$ is almost periodic (see V.5.13), and thus it is the limit in the norm of PM of a sequence of measures with finite support contained in E (see Katznelson [1], Loomis [2], Benedetto [2]). However it is an open problem whether all sets of resolution have $[E]_S \leq 1$. (Sets of synthesis with $[E]_S \leq 1$ are known as sets of <u>bounded synthesis</u>). It seems to be also unknown even if there is $\alpha < \omega_1$ with $[E]_S \leq \alpha$ for all $E \in RE$.

# List of Problems

We list below most of the open problems discussed in this book that can be stated precisely so that to admit a "yes" or "no" answer. Several other very interesting, but hard to formulate exactly, questions are included in various chapters of the book. In particular, although the Characterization Problem is not listed explicitly here, it motivates Problems 6, 11, 12, 13 below.

After each problem we include some comments and references to the sections in which the problem or related issues are discussed.

1. (Union Problem) Is the union of two or countably many Borel sets of uniqueness a set of uniqueness? [This classical problem is open even for two $G_\delta$ sets or the union of a $G_\delta$ set and a countable set; I.6, VII.1].
2. Is the union of two or countably many Borel sets of interior uniqueness a set of interior uniqueness? [This is open even for $F_{\sigma\delta}$ sets; VII.1].
3. (Interior Problem) Is every $(G_{\delta\sigma})$ set of interior uniqueness a set of uniqueness? [This classical problem is open even for $G_\delta$ sets and related to the Union Problem; VII.1].
4. Is every $G_\delta$ $\mathcal{U}$-set of the first category in any $M^p$-set containing it? [This is true if $\mathcal{U}$, $M^p$ are replaced by $\mathcal{U}_0$, $M_0^p$ or if $\mathcal{U}$, $M^p$ are replaced by $U^{int}$, $M_1^p$, but fails if $\mathcal{U}$ is replaced by $U^{int}$. It is also related to the Interior Problem; VII.4, VIII.4].
5. Is the $G_\delta$ set in the Debs-Saint Raymond construction VII.4.1 a $\mathcal{U}$-set or an $\mathcal{M}$-set? [This is related to the Interior Problem and Problem 4; VII.4, VIII.4].

6. Does U have a Borel cobasis? [VII.4].

7. Is U strongly calibrated? [This is related to Problem 2; VII.1].

8. Is $U_1$ a $\underset{\sim}{\Pi}_1^1$ set? [It is $\underset{\sim}{\Pi}_1^1$-hard; VI.2].

9. Is $U_1'$ Borel? [It is $\underset{\sim}{\Sigma}_1^1$; VI. 2].

10. Can one prove in ZFC that S is $\underset{\sim}{\Pi}_1^1$-complete? [It can be proved in ZFC that it is $\underset{\sim}{\Pi}_1^1$ but not Borel; X.2].

11. Is $\{E_{\xi_1,\xi_2,\ldots} : E_{\xi_1,\xi_2,\ldots} \in U\}$ Borel? [If U is replaced by $U_0$ the answer is positive; IV.2, VIII.2].

12. Is it always true that $E_{\xi_1,\xi_2,\ldots} \in U \Leftrightarrow E_{\xi_1,\xi_2,\ldots} \in U_0$? [This is true if $\xi_i = \xi$ for all i; III.3, VIII.2].

13. Is $\eta_0(E_{\xi_1,\xi_2,\ldots})$ attained for $E_{\xi_1,\xi_2,\ldots} \in U_0$? (Old problem) Is the question of whether $E_{\xi_1,\xi_2,\ldots} \in U_0$ determined by the Lebesgue measure on $E_{\xi_1,\xi_2,\ldots}$? [III.3, VIII.2].

14. Is the statement "Every $\underset{\sim}{\Pi}_1^1$ set in $\mathcal{U}_0$ can be covered by countable many $U_0$-sets" consistent relative to ZFC? [It can be proved from strong axioms of set theory, but not in ZFC; VIII.3, Remark (ii)].

15. Can every $\underset{\approx}{\Sigma}_1^1$ set in $(U_1^*)^{int}$ be covered by countably many $U_1$-sets? [This is true if $(U_1^*)^{int}$ is replaced by $U^{int}$; VIII.4].

16. Is the $\sigma$-ideal $U_1^* = (U_1)_\sigma$ $\underset{\sim}{\Pi}_1^1$? Is it calibrated? [This is related also to Problems 9, 15; VIII.4].

17. (Old problem) Let $\Sigma S(n)e^{inx}$ be a trigonometric series converging to 0 a.e. Is the reduced nucleus $RN_S = \{x \in \mathbf{T} : \sum_{|n|\leq N} S(n)e^{inx} = S_N(x)$ is unbounded$\}$ an $\mathcal{M}$-set? [This is true if S satisfies synthesis. This problem is related to the Union and Interior problems; VIII.4].

18. (Lyons) If $\mu$ is a probability measure such that $\mu(E) = 0$ for all $E \in U$, is $\mu$ a Rajchman measure? [The answer is positive if U is replaced by $U_0$; IX.1, IX.2].

19. (Old problem) Is S closed under finite or countable unions? [X.2].

20. (Old problem) Is RE closed under finite or countable unions? [X.1].

21. Is the Piatetski-Shapiro rank bounded on RE? Is the $U_0$-rank bounded on RE? [This is related to Problem 20; X.1].

22. Is the S-rank $[E]_S$ bounded on RE? (Old problem) Is it bounded by 1, i.e. is every RE set of bounded synthesis? [X.2].

23. Is the class of perfect RE sets $\underset{\sim}{\Pi}^1_1$-complete? [This is related to Problem 20; X.1].

# References

(The numbers after each reference indicate the pages in which it appears )

M. AJTAI and A. S. KECHRIS

[1] The set of continuous functions with everywhere convergent Fourier series, Trans. Amer. Math. Soc., to appear. 108

N. K. BARY

[1] Sur l'unicité du développement trigonometrique, Fund. Math., 9, 62–115 (1927). 3, 4, 46, 50, 302

[2] A Treatise on Trigonometric Series, Two Volumes, Macmillan, New York, 1964. 19, 41, 46, 81

J. J. BENEDETTO

[1] Harmonic Analysis on Totally Disconnected Sets, Lecture Notes in Math., Vol. 202, Springer-Verlag, 1971. 19

[2] Spectral Synthesis, Academic Press, New York, 1975. 19, 186, 348

C. CARLET and G. DEBS

[1] Un résultat sur les ensembles d'unicité du tore, Semin d'Init. à l'Analyse, G. Choquet-M. Rogalski-J. Saint Raymond, 24éme année, 1984/85, C2. 44

D. COLELLA

[1] A Bochner-Herz property in bounded synthesis, Math. Scand., to appear. 302

J. W. DAUBEN

[1] Georg Cantor, His Mathematics and Philosophy of the Infinite, Harvard Univ. Press, Cambridge, Mass., 1979. 2

MORTON DAVIS

[1] Infinite games of perfect information, Annals of Math. Studies, 52, 85–101, (1964). 120

G. DEBS and J. SAINT RAYMOND

[1] Ensembles d'unicité et d'unicité au sens large, Ann. Inst. Fourier, Grenoble, to appear. 48, 50, 208, 233, 242, 260, 261, 284, 289, 290, 297

C. DELLACHERIE

[1] Ensembles Analytiques, Capacités, Mesures de Hausdorff, Lecture Notes in Math., Vol. 295, Springer-Verlag, 1972. 294

C. DELLACHERIE and P. A. MEYER

[1] Probabilités et Potentiel, Vol. 3, Hermann, Paris, 1984. 313

J. DIESTEL

[1] Sequences and Series in Banach Spaces, Graduate Texts in Math., Vol. 92, Springer-Verlag, New York, 1984. 171

J. DIXMIER

[1] Sur un théorème de Banach, Duke Math. J., 15, 1057–1071, (1948). 157

R. DOUGHERTY and A. S. KECHRIS

[1] Where Hausdorff measure 0 implies uniqueness, preprint. 306

H. B. ENDERTON

[1] Elements of Set Theory, Academic Press, New York, 1977. 17

C. C. GRAHAM and O. C. MCGEHEE

[1] Essays in Commutative Harmonic Analysis, Grund. d. math. Wissen., Vol. 238, Springer-Verlag, New York, 1979. 19, 250, 258, 285

P. HALMOS

[1] Naive Set Theory, Van Nostrand, Princeton, 1960. 17

F. HAUSDORFF

[1] Set Theory (Translated from 3rd German Edition), Chelsea Publ. Co., New York, 1957. 19

E. HEWITT and K. STROMBERG

[1] Real and Abstract Analysis, Springer-Verlag, New York, 1975. 17

J. HOFFMANN-JORGENSEN

[1] The Theory of Analytic Spaces, Various Publ. Series, No. 10, Mathem. Inst., Aarhus, Denmark, 1970. 19

W. HUREWICZ

[1] Relative perfekte Teile von Punktmengen und Mengen(A), Fund. Math., 12, 78–109, (1928). 119, 133, 137, 201

O. S. IVASHEV-MUSATOV

[1] M-sets and Hausdorff measure, Soviet Math. Doklady, 3, 213–216, (1962). 294

ECH

Set Theory, Academic Press, New York, 1978. 17, 19

-P. KAHANE

[1] Séries de Fourier Absolument Convergentes, Springer-Verlag, Berlin-Heidelberg-New York, 1970. 19, 331, 338, 340

J.-P. KAHANE and Y. KATZNELSON

[1] Contribution à deux problèmes, concernant les founctions de la classe A, Israel J. Math., 1, 110–131, (1963). 234

J.-P. KAHANE and R. SALEM

[1] Ensembles Parfaits et Séries Trigonométriques, Hermann, Paris, 1963. 19, 62, 69, 70, 71, 81, 84, 85, 96, 103, 247, 330, 332

Y. KATZNELSON

[1] An Introduction to Harmonic Analysis, Dover, New York, 1976. 17, 19, 186, 331, 348

Y. KATZNELSON and O. C. MCGEHEE

[1] Some sets obeying harmonic synthesis, Israel J. Math., 23, 88–93, (1976). 342

R. KAUFMAN

[1] A functional method for linear sets I, Israel J. Math., 5, 185–187, (1967). 337

[2] Kronecker sets and metric properties of $M_0$-sets, Proc. Amer. Math. Soc., 36, No. 2, 519–524, (1972). 294

[3] M-sets and distributions, Asterisque 5, 225–230, (1973). 250

[4] Topics on Kronecker sets, Ann. Inst. Fourier, Grenoble, 23, 4, 65–74, (1973). 341

[5] Lipschitz spaces and Suslin sets, J. Funct. Anal., 42, 271–273, (1981). 50, 119

[6] Fourier transforms and descriptive set theory, Mathematika, 31, 336–339, (1984). 119, 123

[7] Absolutely convergent Fourier series and some classes of sets, Bull. Sc. Math., 2e série, 109, 363–372, (1985). 119, 239, 242, 282

[8] Perfect sets and sets of multiplicity, Hokkaido Math. J., 16, 51–55, (1987). 119, 242

A. S. KECHRIS

[1] The theory of countable analytical sets, Trans. Amer. Math. Soc., 202,

259–297, 1975. 299

A. S. KECHRIS and A. LOUVEAU

[1] Covering theorems for uniqueness and extended uniqueness sets, preprint. 274, 290, 297, 300

A. S. KECHRIS, A. LOUVEAU, and W. H. WOODIN

[1] The structure of $\sigma$-ideals of compact sets, Trans. Amer. Math. Soc., 301(1), 263–288, (1987). 48, 119, 132, 133, 134, 194, 203, 238 239

N. N. KHOLSHCHEVNIKOVA

[1] The sum of less-than-continuum many closed U-sets, Engl. Transl., Moscow Univ. Math. Bull. 36, No. 1, 60–64, (1981). 44, 45

J. KOMLÓS

[1] Every sequence converging to 0 weakly in $L_2$ contains an unconditional convergence subsequence, Ark. Matem., 12, 41–49, (1974). 325

T. W. KÖRNER

[1] A pseudofunction on a Helson set, I and II, Asterisque 5, 3–224 and 231–239, (1973). 250

[2] Some results on Kronecker, Dirichlet and Helson sets, Ann. Inst. Fourier, Grenoble, 20, 2, 219–324, (1970). 341

K. KURATOWSKI

[1] Topology, Two Volumes, Academic Press, New York, 1966. 17, 19, 44, 45, 104, 117, 134, 136

L.-Å. LINDAHL and F. POULSEN (Eds.)

[1] Thin Sets in Harmonic Analysis, Marcel Decker, New York, 1971. 19, 331

L. LOOMIS

[1] The spectral characterization of a class of almost periodic functions, Ann. of Math., 72, 362–368, (1960). 185, 191

[2] Abstract Harmonic Analysis, Van Nostrand, New York, 1953. 19, 186, 348

N. N. LUSIN

[1] Leçons sur les Ensembles Analytiques, 2nd Edition, Chelsea Publ. Co., New York, 1972. 19

R. LYONS

[1] A characterization of measures whose Fourier-Stieltjes transforms vanish at

infinity, Ph.D. Thesis, Univ. of Michigan, 1983. 271, 272, 296

Fourier-Stieltjes coefficients and asymptotic distribution modulo 1, Ann. of Math., 122, 155–170, (1985). 321, 324, 325

3] The measure of non-normal sets, Inv. Math., 83, 605–616, (1986). 295, 296

[4] The size of some classes of thin sets, Studia Math., to appear. 103, 262, 263, 271, 322, 328, 329, 338

[5] A new type of sets of uniqueness, preprint. 285

P. MALLIAVIN

[1] Ensembles de résolution spectrale, in Proc. Internat. Congr. of Mathematicians, Stockholm, 368–378, (1962). 258

D. A. MARTIN

[1] A purely inductive proof of Borel determinacy, in Recursion Theory, Proc. Symposia in Pure Math., Vol. 42, Amer. Math. Soc., Providence, 303–310, (1985). 120, 136

D. A. MARTIN and A. S. KECHRIS

[1] Infinite games and effective descriptive set theory, in Analytic Sets, C. A. Rogers et al., Academic Press, London, 1980. 116

S. MAZURKIEWICZ

[1] Über die Menge der differezierbaren Funktionen, Fund, Math., 27, 244–249, (1936). 108

O. C. MCGEHEE

[1] Certain isomorphisms between quotients of a group algebra, Pac. J. Math., 21, 133–152, (1967). 328

[2] A proof of a statement of Banach on the weak* topology, Mich. Math. J., 15, 135–140, (1968). 176, 184, 216

Y. MEYER

[1] Algebraic Numbers and Harmonic Analysis, North Holland, Amsterdam, 1972. 19, 103, 279

Y. N. MOSCHOVAKIS

[1] Descriptive Set Theory, North Holland, Amsterdam, 1980. 17, 19, 104, 116, 119, 120, 136, 141, 144, 145

I. I. PIATETSKI-SHAPIRO

[1] On the problem of uniqueness of expansion of a function in a trigonometric

series, (in Russian), Moscov. Gos. Univ. Uc. Zap., 155, Mat. 5, 54–72, (19
81, 82, 227, 262, 263, 288, 289

[2] Supplement to the work "On the problem...", (in Russian), ibid, 165, Mat.
79–97, (1954). 48, 69, 70, 82, 139, 174, 180, 213, 216, 217, 220, 258

C. A. ROGERS ET AL.

[1] Analytic Sets, Academic Press, London, 1980. 19

W. RUDIN

[1] Real and Complex Analysis, 3rd Edition, McGraw-Hill, New York, 1987. 17, 59

[2] Functional Analysis, McGraw-Hill, New York, 1973. 17, 151

R. SALEM

[1] Algebraic Numbers and Fourier Analysis, Heath, Boston, 1963. 19, 84, 85, 87

D. SALINGER

[1] Sur les ensembles indépendants dénombrables, C.R. Acad. Sc. Paris, Sér. A–B 272, A786–788, (1981). 39

R. M. SOLOVAY

[1] Private communication, December 1983. 50, 119, 123

[2] Private communication, February 1984. 175

K. STROMBERG

[1] An Introduction to Classical Real Analysis, Wadsworth, Belmont, CA, 1981. 17

N. TH. VAROPOULOS

[1] Sur les ensembles parfaits et les séries trigonométriques, C.R. Acad. Sci. Paris Sér. A–B 260, A3831–3834, (1965). 333

[2] Tensor algebras and harmonic analysis, Acta Math., 51–112, (1967). 340

[3] Groups of continuous functions in harmonic analysis, Acta Math., 109–152, (1970). 331

A. ZYGMUND

[1] Trigonometric Series, 2nd Ed., reprinted, Two Volumes, The Cambridge University Press, Cambridge, 1979. 19, 41, 46, 79, 297

# Symbols and Abbreviations

Continued

# Index

## A



## B



## C

## D



## E



## F



## G



## H



## I

## L



## M



## N



## O



## P

## R



## S



## S Continued



## T

## V



## W



## Y

# Errata

3 $l.6$ "Conjecture" should be "problem"

85 $l.3^-$ "$\mathrm{R}((\theta)^{(i)})^p$" should be "$\mathrm{R}(\theta^{(i)})$"

117 between $l.8^-$ and $1.7^-$
add the line "$= 0$, if $\mathrm{K}=\mathrm{L}=\emptyset$"

ge 131 $l.8^-$ replace "if $P \in \mathcal{C}, \ldots P \cup Q \in \mathcal{C}$." by "if $P_1, \ldots, P_n \in \mathcal{C}$
and $P_1 \cup \ldots \cup P_n \in \mathcal{F}$, then $P_1 \cup \ldots \cup P_n \in \mathcal{C}$."

page 165 $l.4$ "$\|x_n\|_{Y(\alpha)} \leq 2^{-(n+3)}$" should be "$\|x_n\|_{Y(\alpha)} < \epsilon \cdot 2^{-(n+3)}$"
$l.6^-$ "$\|x_n\|_{Y(\beta)} \leq 2^{-(n+3)}$" should be "$\|x_n\|_{Y(\beta)} < 2^{-(n+4)}$"

page 167 $l.7$ "$\|x_n\|_{Y(\alpha+1)} \leq 2^{-(n+3)}$" should be "$\|x_n\|_{Y(\alpha+1)} < \epsilon \cdot 2^{-(n+3)}$"
"Let ... $b \cdot \|x_n\|_{Y(\alpha+1)}$" should be "Let $a = \epsilon \cdot 2^{-(n+3)}$"
$l.6^-$ replace this line by "$\geq \|x_n\| - a$"
$l.3^-$ "$\|y_k\|_{Y(\alpha)} \leq 2^{-(n+k+3)}$" should be "$\|y_k\|_{Y(\alpha)} < 2^{-(n+k+3)} \cdot \epsilon$"

page 199 $l.1^-$ replace "$h(q) \subseteq (\bigcap_{\substack{p<q \\ p \in R}} h(p)_B^{(1)}))$." by
"$(q \neq 0 \Rightarrow h(q) \subseteq (\bigcap_{\substack{p<q \\ p \in R}} h(p)_B^{(1)})))$."

page 200 $l.6$ add "$q \neq 0$" before "$q \in R$"
$l.9$ add "$q \neq 0$" before "$q \in R$";
add "$;f(0) = 0$" after "$K_B{}^{(\alpha+1)}$"
$l.6$ and 9
replace "$\overset{\subset}{\neq}$" by "$\not\subseteq$"
$l.8^-$ replace "$q \in R$," by "$q \neq 0$, $q \in R$, $h(q) \subseteq$"
$l.7^-$ replace "so $h(q) \subseteq \ldots K_B{}^{(f(q))}$" by
"so for all $q$, $h(q) \subseteq K_B{}^{(f(q))}$"
$l.3^-$ replace "$h(q) = \bigcap_{\substack{p<q \\ p \in R}} h(p)_B^{(1)})$" by
"$(q \neq 0 \Rightarrow h(q) = \bigcap_{\substack{p<q \\ p \in R}} h(p)_B^{(1)}))$"

page 204 $l.5^-$ "$x \in X$" should be "$x \in E$"

page 225 $l.6^-$ last "$2\pi(k-1)/N$" should be "$2\pi k/N$"
$l.4^-$ "II.2.2" should be "II.2.1"

page 235 $l.2$ "$[-M, M]$" should be "$(-M, M)$"

page 239 $l.2$ change "3.2.5" to "3.3.5"

page 243 $l.1$ delete "if"

page 253 $l.8$ change to "$c_\epsilon(t) = p^{1-\epsilon}q^\epsilon / p_1^{1-\epsilon}q_1^\epsilon \leq c_\epsilon(\epsilon) = c(\epsilon) < 1$"

page 257 $l.6$ change "$a_1 - a_2$" to "$a_2 - a_1$"

page 260 $l.1^-$ change "$f_{n,0} = f_n$" to "$f_{n,-1} = f_n$"

page 265 $l.4^-$ replace "$\mathcal{I}^{loc}$" by "$\mathcal{I}^{loc} \cup \{\emptyset\}$"

page 279 $l.7$ replace by
"$\overline{\lim}_{n\to\infty} \prod_{i=1}^{\infty} \cos^2(\pi n \xi_1 \ldots \xi_{i-1}(1-\xi_i)) > 0$"

age 282 $l.9$ replace "$\mathrm{supp}(\nu_n) \subseteq E \cap V$" by "$\mathrm{supp}(\nu_n) \subseteq \overline{E \cap V}$"

age 307 $l.4^-$ "$T_n T$" should be "$T_n \to^{w^*} T$"

age 353 $l.5^-$ insert "boréliens" after "Ensembles"
$l.5^-, 4^-$ replace "to appear" by "37, 3, 217-239, 1987"